Biomechanics of
Cell Division

NATO ASI Series

Advanced Science Institutes Series

A series presenting the results of activities sponsored by the NATO Science Committee, which aims at the dissemination of advanced scientific and technological knowledge, with a view to strengthening links between scientific communities.

The series is published by an international board of publishers in conjunction with the NATO Scientific Affairs Division

A	**Life Sciences**	Plenum Publishing Corporation
B	**Physics**	New York and London
C	**Mathematical and Physical Sciences**	D. Reidel Publishing Company Dordrecht, Boston, and Lancaster
D	**Behavioral and Social Sciences**	Martinus Nijhoff Publishers
E	**Engineering and Materials Sciences**	The Hague, Boston, Dordrecht, and Lancaster
F	**Computer and Systems Sciences**	Springer-Verlag
G	**Ecological Sciences**	Berlin, Heidelberg, New York, London,
H	**Cell Biology**	Paris, and Tokyo

Recent Volumes in this Series

Series A: Life Sciences

Biomechanics of Cell Division

Edited by
Nuri Akkas

Graduate School of Natural and Applied Sciences
Ankara University
Ankara, Turkey

Plenum Press
New York and London
Published in cooperation with NATO Scientific Affairs Division

Proceedings of a NATO Advanced Research Workshop on
Biomechanics of Cell Division,
held October 12–17, 1986,
in Istanbul, Turkey

Library of Congress Cataloging in Publication Data

NATO Advanced Research Workshop on Biomechanics of Cell Division (1986:
    Istanbul, Turkey)
    Biomechanics of cell division.

    (NATO ASI series. Series A, Life sciences; vol. 132)
    "Published in cooperation with NATO Scientific Affairs Division."
    "Proceedings of a NATO Advanced Research Workshop on Biomechanics
of Cell Division, held October 12–17, 1986, in Istanbul, Turkey"—T.p. verso.
    Includes bibliographical references and index.
    1. Cell division—Congresses. 2. Biomechanics—Congresses. I. Akkas, Nuri.
II. North Atlantic Treaty Organization. Scientific Affairs Division. III. Title. IV.
Series. NATO ASI series. Series A, Life sciences; v. 132.
QH605.N27   1986   574.87′62   87-12349
ISBN 978-1-4684-1273-4   ISBN 978-1-4684-1271-0   (eBook)
DOI 10.1007/978-1-4684-1271-0

© 1987 Plenum Press, New York
Softcover reprint of the hardcover 1st edition 1987
A Division of Plenum Publishing Corporation
233 Spring Street, New York, N.Y. 10013

PREFACE

There are virtually hundreds of life scientists publishing hundreds
of papers a year on numerous aspects of the cell cycle. The following are
few of the topics covered : cell membrane organization, membrane components,
cytoskeleton and associated proteins, cell motility, actin in dividing
cells, surface modulating assemblies, microfilaments, microtubules, cleavage
furrow, fusion, etc. In all these topics, lifescientists talk about, among
others, the forces within the system, the motion within the system and the
failure of the system. The concepts of force, motion and failure are, one
way or another, all related to the structure of the cell and to the mechanics
of the cell activities. When the concepts of mechanics and structure enter
the problem then one has to talk about biomechanics; in this case,
biomechanics of cytology which we would like to call "Cytomechanics".
However, a review of the journals, books and conference proceedings related
to various aspects of cytology reveals that mechanicians have not yet entered
the field of cytology at a noticeable level. Some lifescientists have indeed
made use of the general principles of mechanics in their works; however, no
truly interdisciplinary publication has yet appeared from the collaboration
of mechanicians and lifescientists in the field of, for instance, cell
division.

The purpose of this NATO Advanced Research Workshop was to bring
together the lifescientists, who are experts on various aspects of division
in animal cells, and the mechanicians, mathematicians and physicists who
are aware of the possible application of their knowledge in this exciting
and challenging field. The latter learned about the ultrastructure of a
dividing cell, about the forces created within the cell and about the
mechanisms taking place during division as explained by the lifescientists.
The lifescientists, in turn, learned about the possible mechanistic
explanation of the division related events, such as force generation and
transmission, structural changes, material behavior and changes therein,
mechanistic factors in normal and abnormal behavior, and stability of
behavior. Because the presentations by both groups were accompanied by
informal discussion, unclear points were clarified and missing links
discovered.

The objective of the Advanced Research Workshop was to provide the forum
for an exchange of views at the frontiers of knowledge and for the formulation

of recommendations for new research directions. The expected output is, among others, to make the views of the authorities on the necessity of interdisciplinary collaboration in cell division available in a book form so that an interdisciplinary cooperation will hopefully be activated.

The members of the International Organizing Committee were Prof. R. Rappaport of Union College, New York, and Dr. J. Hyams of University College London. Their contribution in the organization of the Workshop is duly acknowledged. I am personally grateful to Prof. Rappaport for his contribution during the Workshop as the moderator of the panel discussion. The resolution made by the participants in the Workshop was written down by Profs. Rappaport and Skalak. Some relevant points of the resolution are given in the last chapter of this book.

It was a pleasing coincidence that 1986, in which the NATO Advanced Research Workshop on Biomechanics of Cell Division took place, was the fortieth anniversary of Ankara University. Thus, the Workshop was included among the celebration activities of the anniversary.

January 1987, Ankara   Nuri Akkaş

CONTENTS

LOCATION OF THE PHYSICAL MECHANISM IN THE CELL

R. Rappaport

Department of Biological Sciences
Union College
Schenectady, New York and
The Mount Desert Island Biological Laboratory
Salsbury Cove, Maine

A complete understanding of the biomechanics of cell division has been considered a noble goal for over a century. At any one time during that period, the number of investigators actively contributing new data to the field has been, by present standards, small, and progress has been slow. Cytokinesis in animal cells, the aspect of the process that will be the subject of this paper, has always been a subject of intense speculation. Many of the early cell and developmental biologists appeared very willing to publish their ideas on the subject, even though it might not be in their area of expertise. Speculation or theorization usually took the form of models that could be either physical or verbal. To be considered feasible, models had to replicate the normal events of cytokinesis and because of the apparent simplicity of the phenomenon, the same effect could be achieved by models based upon different physical mechanisms. In this circumstance, the workability of a model revealed little about the possibility that it accurately resembled the events that take place in the cell. There was also a near absence of information concerning the division-relatedness of the structures and activities and whether their roles were active or passive. In consequence, the number of possible mechanisms was immense. As late as 1928, Heilbrunn was moved to remark that, "Usually it is easier to make a new theory of cell division than to test an old one." In fact, by that time it was rather difficult to make a really new theory as the last major original theory was proposed about twenty-five years before. Most of what followed has been recycling with variations. But why that kind of statement, even though made in partial jest, should be accurate enough to cause some discomfort is presently unclear. Experimentation was then an accepted method of analysis and at that time several excellent micromanipulators were commercially available. It occasionally took a long time for the results of experimentation to affect cell division theory. Despite convincing demonstrations by Ziegler (1898) and by McClendon (1908) that cytokinesis in the absence of chromosomes is perfectly possible, essential roles for chromosomes were proposed decades later. It may be possible that this circumstance arose from simple ignorance of the literature. But it could also have resulted from a devout faith that the beauty, ingenuity and personally lovable qualities of the hypothesis or model would eventually triumph over a few ugly, inconvenient facts.

We carry a heavier burden than our predecessors. For those who wish to know the mechanism of cytokinesis that actually operates in the cell, the days of freewheeling speculation are over. Information derived from measurements and from experimentation has greatly reduced the number of possibilities. Every hypothesis has predictive properties, and the accuracy of the prediction can often be determined by forcing cells to divide under circumstances that were not specifically allowed for when the hypothesis was created. Because the cleaving echinoderm eggs that are used in experimental studies are all very similar, and because echinoderm eggs also serve as the pattern for models of typical animal cell division, the possibility that inconsistencies between the behavior of models and that of real cells can be attributed to species differences is small. The certainty that additional important information concerning the process will be forthcoming does not change the requirement that, when properly interpreted, all existing information must fit into a single, logically coherent pattern. No significant exceptions are possible.

In the study of cytokinesis, one of the current central problems is understanding the basis of the unequal distribution of active contractile components associated with the surface. Division activity requires previous interaction between the achromatic mitotic apparatus and the surface and, because we presently lack means for direct study of the local events that lead to or constitute the organizing activity that creates the division mechanism, it has been necessary instead to study the events and relations that can affect the process, both directly and indirectly.

TIMING

The causal chain of events that culminates in division follows a characteristic schedule. The cell's ability to divide when the mitotic apparatus is removed late in the mitotic cycle (Yatsu, 1912; Hiramoto, 1956) revealed not only the absence of any physical role for the mitotic apparatus, but also the existence of a pause between the time when the surface alteration elicited by the mitotic apparatus achieves irreversibility and the beginning of division mechanism function. The duration of the pause is about 4 min (Hamaguchi, 1975; Rappaport, 1981). The time when the interaction between the mitotic apparatus and the surface begins has not been determined. It has, however, been shown that an interaction period of about 1 min is sufficient to establish a furrow (Rappaport and Ebstein, 1965; Rappaport, 1965). The fact that precocious furrows form when the mitotic apparatus and the surface are located closer together than normal (Rappaport, 1975) strongly suggests that they are capable of interacting before they normally do.

Under experimental conditions, the mitotic apparatus can establish several furrows in succession (Harvey, 1935; Rappaport and Ebstein, 1965; Rappaport, 1975, 1985). This fact strongly suggests that the property of the mitotic apparatus that initiates surface contraction is preserved and functional after the time when the furrow first appears. In the sand dollar Echinarachnius parma, the mitotic apparatus has been shown to initiate as many as 13 successive furrows as it was pushed back and forth in a cylindrical cell over a 24.5 min period (Rappaport, 1985). After it has been shifted several times, its radiate structure appears diminished and the distance between the astral centers is abnormally large. Because both of these changes have been associated with reduced ability of the mitotic apparatus to establish furrows, it is not clear whether the shifted mitotic apparatus loses its ability because of an event associated with the normal cell cycle or because of the structural and geometrical changes caused by the manipulation. The cell's ability to form both precocious and belated furrows suggests that the precise time of furrowing characteristic of cleavage divisions may result from changing geometrical relations of the interactants rather than a restricted period of physiological activity.

The idea that furrow establishment is dependent upon a factor that moves toward the surface from the central region of the cell is supported by the results of experiments in which the time when the cleavage furrow appears is related to the initial distance between the mitotic apparatus and the surface. In flattened echinoderm eggs, the mitotic apparatus can easily be held in an eccentric position, with the consequence that the furrow appears first in the closer equatorial margin. There is a proportionality between the time of appearance and distance, and if it is assumed that the relation is based upon a moving front, the rate of movement can be calculated at 6-8 μm/min (Rappaport, 1973, 1982). This interpretation does not reveal just what it is that is approaching the surface. One of the oldest suggestions was that of a chemical stimulus which moved toward the periphery along the visible elements of the asters (Bütschli, 1876; Ziegler, 1903). Alternatively, it has been suggested that the capacity to establish the furrow is a property of ultrastructural and cytoskeletal elements of the mitotic apparatus, rather than a diffusable or freely moving substance. It is also possible that the mitotic apparatus could transfer to local regions of the subsurface potentially active contractile components. The consequent instability could result in self-amplifying recruitment of similar elements to the target area from other regions of the subsurface. A mechanism of this kind would require very little initial transfer to be effective. Although there are no data that preclude its existence, neither has the idea enjoyed noticeable popularity.

IDENTIFICATION OF AN IMMEDIATE DIRECT EFFECT OF THE ACTIVE PARTS OF THE MITOTIC APPARATUS

Of the several microscopically visible components of the mitotic apparatus, only the chromosomes have no active role in furrow establishment (Ziegler, 1898; McClendon, 1908; Rappaport, 1961). Both the pair of asters (Rappaport, 1961) and the spindle (Ris, 1949; Rappaport and Rappaport, 1974; Kawamura, 1977) can initiate furrowing activity. It is likely, however, that the role played by each of these components depends upon cell shape, and the overall size and proportions of the parts of the mitotic apparatus. In spherical cleaving sea urchin eggs with large asters, the presence of the spindle is not required (Hiramoto, 1971), although it appears to be able to initiate furrowing under conditions where astral involvement seems very unlikely (Rappaport and Rappaport, 1974). But in tissue cells, in which the asters are characteristically small and the spindle appears large and close to the equatorial surface, the role of the asters may be minor or non-existent (Rappaport and Rappaport, 1974).

Furrow establishment activity may occur despite chemical and physical disruption sufficient to result in microscopically visible effects on the mitotic apparatus. Wilson (1901) reported that the reduced mitotic apparatus of ether-treated sea urchin eggs could initiate furrowing in the presence of ether, although the furrows often regressed. Ethyl urethane also reduces the overall size of the mitotic apparatus. But urethane concentrations that block cleavage of spherical eggs do not prevent furrowing when the mitotic apparatus and surface are brought closer together, either by shifting the position of the mitotic apparatus (Rappaport, 1971), or reshaping the egg into a cylinder (Rappaport and Rappaport, 1984)(fig. 1). In these cases, furrows are established in the presence of concentrations of substances that prevent or reverse microtubule polymerization. Although the ability of relatively specific microtubule poisons to reduce the mitotic apparatus and interfere with

3

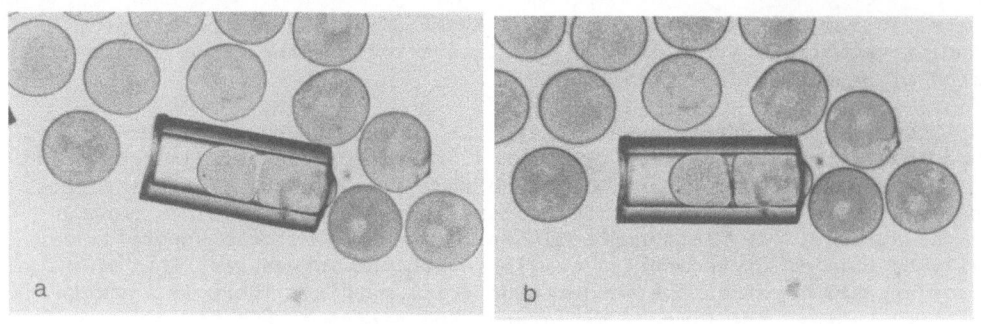

Fig. 1. Cleavage of a urethane-treated cell confined in an
80 μm i.d. short capillary tube. None of the spherical
treated cells in the chamber divide.
a. Cleavage beginning.   b. Cleavage completed.
Reprinted by permission from Journal of Experimental
Zoology, 231 (1984): 81-92, Ⓓ Alan R. Liss, Inc.

furrowing has been cited as supporting the idea of a specific role for
microtubules in the process, it must be pointed out that such substances
appear to affect the size of the mitotic apparatus as a whole, not just
the length of the microtubules (Hinkley, et al., 1982). Temperatures low
enough to interfere with microtubule polymerization do not prevent cleavage
furrow formation provided the mitotic apparatus is close to the surface
(Anstrom and Summers, 1983, a, b). Chambers (1917) found that asters and
the entire mitotic apparatus disappeared when an inserted needle was
vigorously stirred. The disappearance was accompanied by loss of ability
to carry out mitotic apparatus related activities such as cleavage.
Repeated displacement of the mitotic apparatus also reduces its size and
the visibility of its linear elements without abolishing its ability to
establish furrows (Rappaport and Ebstein, 1965; Rappaport, 1975, 1985).
It has also been shown that frequent stirring between the mitotic apparatus
and the surface does not prevent furrow formation (Rappaport, 1978).
Persistence of furrow formation activity despite these disruptive measures
suggests that this particular function of the mitotic apparatus does not
require its normal ultrastructure or precise arrangement.

Better understanding of the immediate, direct response of the surface
to the mitotic apparatus will help resolve many questions. The functional
response is the formation of an equatorial band of contractile material
that immediately relaxes when it is cut (Rappaport, 1966) or exposed to
localized concentrations of cytochalasin-D (Tanaka and Inoué, 1981). The
possibility that the characteristic circumferential orientation of equatorial microfilaments may be the consequence of local tensile force has
been discussed (Rappaport, 1975; Schroeder, 1975), but the reasons why
contractile activity is localized in that region are not completely
clear. In a normally spherical cell, the surface responds to the mitotic
apparatus by forming a furrow only when a pair of asters is present. The
apparent necessity for two has been attributed both to special geometrical
requirements and to simple quantitative factors.

One way to learn more about the immediate effect of the whole mitotic
apparatus on the surface is to determine whether a single aster can initiate
furrowing activity and if so, where. A single aster elicits no furrows
in a spherical sea urchin egg, although surface irregularities may appear.
(Hiramoto, 1971). But when an echinoderm egg is reshaped into a cylinder,
the presence of a single aster is sufficient to initiate furrowing
activity that is usually deepest in the plane of the astral center, provided that the aster is located some distance from the poles of the
cylinder (fig. 2) (Rappaport and Rappaport, 1985). When the aster

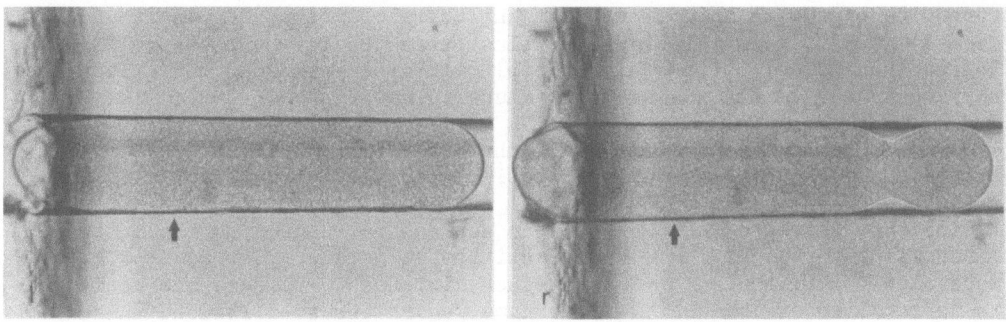

Fig. 2. Local surface constriction in the plane of the
center of an isolated aster. Linear elements
of the companion aster are visible in the left end
of the cell (arrowheads). Reprinted by permission
from the Journal of Experimental Zoology, 235 (1985):
217-226, © Alan R. Liss, Inc.

is near a pole, the contraction of the whole, hemispherical polar surface
is apparent. The first conclusion that can be drawn from these experiments
is that the establishment of localized furrowing activity does not
necessarily require the presence of two asters. The formation of a furrow
requires that some regions of the cell surface become more contractile
than others. In a spherical cell containing a single aster, the normal
astral expansion that occurs before division shifts its center until it
coincides with that of the cell. Because the effect of the aster on the
surface appears to be distance-related, it would in that circumstance be
expected to affect the entire cell surface uniformly and the degree of
localization of contractile activity necessary for furrowing activity
could not be achieved. In a cylindrical cell with a central aster, the
more affected surface would be bordered by two less affected surfaces and
the possibility of an annular region of greater contractility is realized.
The fact that the single aster of a cylindrical cell can be constricted
by the furrowing activity that it initiates indicates that the immediate
direct effect of the aster upon the surface is to cause contraction. If
the aster caused local surface relaxation, it would not be constricted by
the furrowing activity that it elicited. In normal cells and in cylindrical cells containing a single, central aster, the presence of less contractile surface zones allows the surface to yield to the extent that the
distortion of the cell requires. When the capacity to yield is blocked by
constraining surface movement close to the equatorial region, furrowing
activity stops (Rappaport and Ratner, 1967).

ESSENTIAL DISTANCE AND GEOMETRICAL RELATIONS BETWEEN THE MITOTIC
APPARATUS AND THE SURFACE

     In several respects the interaction between the mitotic apparatus
and the surface is distance related. Among normal eggs of different
species there are different degrees of nuclear eccentricity. In all
cases, furrowing activity begins first in the surface closest to the
mitotic apparatus. Depending on the degree of eccentricity, furrowing
on the distant margin may begin later or not at all. Cleavage of the
latter type has been termed "unilateral." That unilateral cleavage is
entirely a consequence of the position of the mitotic apparatus has been
shown in relocation experiments in which normally unilaterally cleaving
eggs were induced to form symmetrical furrows and normally symmetrically

furrowing eggs were induced to form unilateral furrows (Rappaport and Conrad, 1963). This relationship illustrates the idea that there is an "effective radius" beyond which the mitotic apparatus does not affect the surface to the degree necessary to cause furrowing.

In spherical echinoderm eggs, the effective radius is greater than the normal distance between the mitotic apparatus and the surface. In the range of distances between normal and the outer limit of the effective radius, the degree of response of the surface is related to the initial intervening distance. Furrows established at greater distances progress more slowly than those formed at lesser distances. Because the resistance to deformation of the cell was the same regardless of distance, these results were interpreted as implying that the force exerted by the slower furrow was reduced by reason of diminished intensity of interaction between the mitotic apparatus and the surface at greater distances (Rappaport, 1982). The previously described reversal of chemical inhibition by relocation of the surface or the mitotic apparatus is consistent with this interpretation, as it suggests that the treatment reduced the effective radius at the same time that it reduced the apparent size.

The information described above concerns the relationship between the mitotic apparatus as a whole and the surface. When the distances between different parts of the mitotic apparatus and different regions of the surface are considered, there is greater opportunity for complexity. In a normally spherical cell with large asters, the distance from the astral centers to the poles is less than the distance from the astral centers to the equator (fig. 3a). In view of the previously discussed distance-relatedness of the interaction between the mitotic apparatus as a whole and the surface, it is not illogical to suggest that the dimensional differences between different parts of the mitotic apparatus and different regions of the surface may be important in determining whether and where furrows develop.

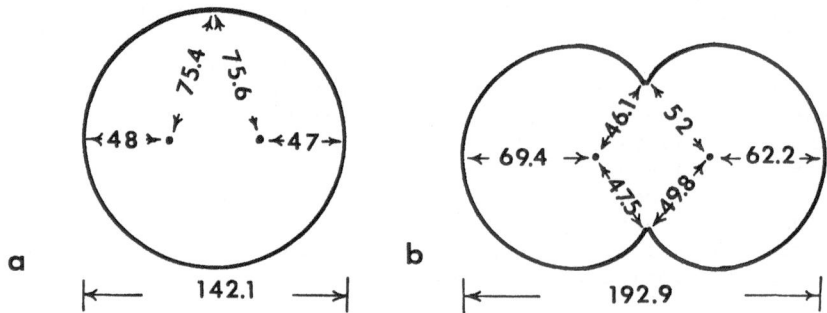

Fig. 3. Diagrams (to scale) of the positions of the astral centers of sand dollar eggs relative to the poles and the equator 5 min before cleavage, when furrow establishment takes place. The distances are in μm. Dots represent astral centers.
a. Normal spherical egg. Values are the means of measurements of 22 eggs. The distances from the astral centers to the equator were calculated.
b. Artificially constricted egg. The distances are the means of measurements made directly on 16 eggs.
Reprinted by permission from the Journal of Experimental Zoology, 231 (1984): 81-92. © Alan R. Liss, Inc.

The simplest and most direct method for determining whether certain dimensional relations are essential for furrow establishment is to alter the cell's shape during the period when the interaction takes place. When spherical echinoderm eggs are artificially constricted by a glass loop into two equal portions, the mitotic apparatus usually shifts as it expands so that it straddles the plane of the constriction. By adjusting the diameter of the loop relative to the diameter of the egg, artificial constriction may be used to reverse the normal distance relations between the astral centers and the polar and equatorial surfaces so that during constriction the distance to the equator is less than the distance to the pole (fig. 3 b). This reversal does not interfere with division; in fact, furrows usually appear in constricted cells before they appear in spherical controls (Rappaport and Rappaport, 1984). Furrow establishment does not appear to depend upon differences in distance between the astral centers and specific regions of the surface.

Reversal of chemical cleavage inhibition by manipulating the distance between the mitotic apparatus and the surface provides another opportunity to determine which distances are important in determining whether and where furrows develop. When the reduced mitotic apparatus of the treated cell is pushed closer to the surface it can establish a furrow in the presence of effective concentrations of the anaesthetic. It was postulated that the immediate effect of the treatment was to reduce the distance over which the mitotic apparatus could effect the surface and that the manipulation compensates for the deficiency by simply pushing the two closer together (Rappaport, 1971). The reduced mitotic apparatus of treated cells can also establish furrows in surfaces that are brought closer to it by re-shaping the cells into cylinders (Rappaport and Rappaport, 1984). Since this manipulation aligns the mitotic apparatus parallel to the cylinder axis, the polar and subpolar surfaces are displaced farther from the astral centers and the rest of the surface is brought closer (fig. 1). Because the deficiency is remedied by decreasing the distance, it was possible to determine by selective reduction which of the dimensions was most important in overcoming the effects of the treatment. Selective reduction was accomplished by inserting the treated cell in a glass loop with the same internal diameter as the cylinder that reversed the effect of chemical treatment. The mitotic apparatus was manuvered into various relations to the plane of the constriction, and it was found that only when the plane of artificial constriction fell between the astral centers was the effect of treatment reversed. The absence of furrowing when the mitotic apparatus was placed in other relations to the constriction indicated that neither the simple presence of the mitotic apparatus, nor the constriction were effective, and suggests that the interaction necessary to overcome the treatment takes place in the zone between the asters (fig. 4).

In cylindrical cells, isolated asters can cause visible contractile activity in adjacent cylindrical surface. Not only single asters remaining after aspiration of the rest of the mitotic apparatus, but also asters located farther apart than normal are effective. The plane of the constriction is usually located in the plane of the astral center (see above, and fig. 2). When the isolated asters of cylindrical cells (single or paired) are chemically treated to reduce their size, they do not cause surface constrictions. However, in cylindrical cells of the same dimensions exposed to the same treatment that blocks astral constriction, a furrow forms between pairs of asters that are as close together as they are in the normal mitotic apparatus. In these cells the astral centers are positioned on the cylinder axis. The only distances that decrease when they are moved together are the distance from center to center and the distance from the astral centers to a line on the surface equidistant

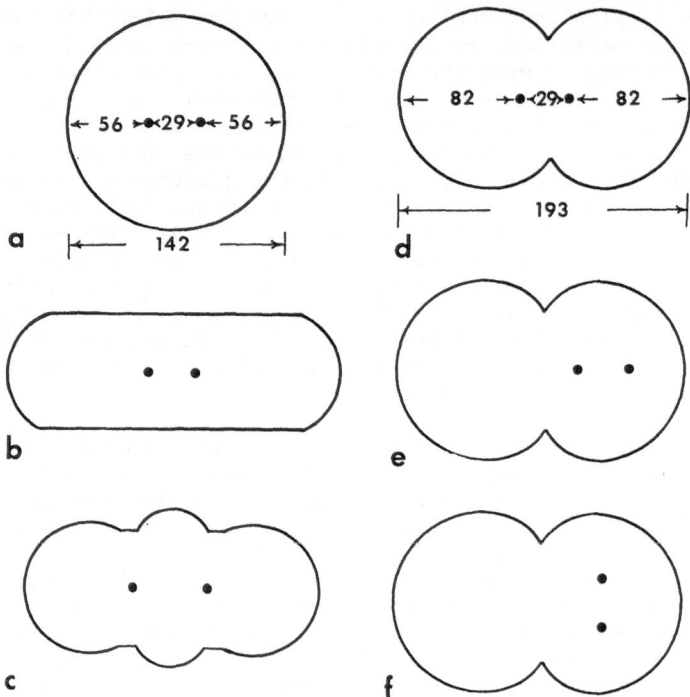

Fig. 4 Dimensions and relations between the astral centers and
        surface regions in urethane-treated sand dollar eggs.
        Distances are in μm. Dots represent astral centers.
        a. Positions of astral centers of sand dollar eggs 5 min
        before anticipated cleavage of controls (n=10). These eggs
        do not cleave. b. Positions of astral centers when treated
        cells are confined in 80 μm i.d. capillaries. These cells
        cleave. c. Positions of astral centers when treated eggs
        are constricted to 80 μm diameter bilaterally in the sub-
        polar and subequatorial regions by opposed pipets. These
        cells do not cleave. d. Diagram to scale of the positions
        of the astral centers relative to the poles in artificially
        constricted treated eggs. These cells cleave.
        e. Position of astral centers when the mitotic apparatus
        of treated artificially constricted cells was slid along
        the mitotic axis to the center of one of the cell halves
        formed by the constriction. These cells do not cleave.
        f. Position of astral centers when the mitotic apparatus
        of treated artificially constricted cells was slid along
        the mitotic axis to the center of one of the cell halves as
        in e, and then rotated 90 degrees. These cells do not cleave.
        Figures a and d reprinted by permission from the Journal of
        Experimental Zoology, 231 (1984), 81-92. © Alan R. Liss, Inc.
        Figures b, c, e, and f reprinted by permission from Inter-
        national Review of Cytology, in press. © Academic Press, Inc.

from the centers. The other distances remain the same or increase. It is apparent that in this circumstance a pair of asters can alter the surface to an extent that cannot be accomplished by either one alone. In view of the dimensional control that was exerted in these experiments, the results suggest that the closer proximity of the asters increases the intensity of the level of their interaction with the nearby surface, and that their effect is additive.

ORGANIZATION OF THE DIVISION MECHANISM

The preceding discussion has dealt with events that determine where and when the furrow develops. It is also appropriate to come back to the question of the basis of the unequal distribution of active contractile elements that is considered fundamental to the annular constriction that divides animal cells. Although the ultrastructural evidence of the contractile ring is localized in the equatorial cortex (Schroeder, 1972), normal cortical structure does not appear to be a prerequisite for subsequent division activity. Furrows appear in surfaces subjected to extreme stretching (Rappaport, 1976) and in the "precipitation membrane" that rapidly forms over the denuded endoplasm of portions of echinoderm eggs after the normal cortex is mechanically disrupted (Rappaport, 1983). It is possible that potential contractile elements are uniformly distributed under the surface and that only in a restricted area are they activated to exert tension. It is alternatively possible that localized contractile areas may result from localized accumulations of contractile elements. This mechanism was suggested by Lewis (1942) and more recently by Schroeder (1981) and by White and Borisy (1983). It has the advantage of hypothetically reducing the potential contractility of the area from which the element is moved at the same time that the contractility of the area to which it is moved is enhanced. The shifting of division-related contractile elements has not been documented but the results of some recent experiments suggest that it may occur. When the mitotic apparatus of a cylindrical cell is shifted after furrowing begins, the original furrow recedes as a new furrow develops in the normal relation to the new position of the mitotic apparatus. The method used to move the mitotic apparatus does not stress the cylindrical surface of the cell and it has been shown that, following its simple removal at that time, furrowing continues until the cell is permanently bisected (Rappaport, 1981). It was already pointed out that in cylindrical cells, isolated asters cause localized annular constrictions in the cylindrical surface so that it is highly unlikely in this circumstance that the asters of the shifted mitotic apparatus cause nearby localized surface relaxation (Rappaport and Rappaport, 1985).

When a sand dollar egg is confined in an 82 μm capillary, it is reshaped into a cylinder about 310 μm long. The way in which the original furrow recedes depends upon the distance that the mitotic apparatus is shifted. When the midpoint of the mitotic apparatus is shifted 45 μm, the constriction of the original furrow slides along the cylindrical surface until it reaches the midpoint of the mitotic apparatus in its new position where it resumes constriction and, provided the mitotic apparatus is not again shifted, completes division. When the mitotic apparatus is shifted 90 μm, the original furrow does not slide, and for a brief period the cell contains two furrows, the original and the one developing at the midpoint of the mitotic apparatus in its new location. As the new furrow deepens, the original furrow recedes and no further furrowing activity occurs in its vicinity unless the mitotic apparatus is returned to its original location (Rappaport, 1985). Since there are many well-known circumstances under which cells or cleaving eggs can simultaneously form more than the normal number of furrows (polyspermy, second cleavage following suppression of cytokinesis at first cleavage, etc.), it is not logical to argue that

9

the phenomenon is based upon the cell's inability to form more than one furrow at a time. It is tempting to speculate that the phenomenon is caused by a competition between active contractile systems for movable force producing units. If the contractile activity associated with the most recently formed furrow were the stronger, it could appropriate components of the original furrow and cause it to recede. Whatever the mechanism may be, it is apparent that similarly established contractile regions can affect each other at distances as great as 90 μm. If a subsurface lattice of potential contractile material exists (Schroeder, 1981), then its distribution can be affected by localized tension differences.

All of this serves to remind us that the parts of the complex process must be characterized and realistically evaluated by experimentation. Recent technical advances that improve the observability of structures and specific molecular species should, for instance, permit study of changes in their distribution and orientation that accompany division. But such demonstrations must be accompanied by experimentation. To try to reconstruct the mechanics of the process from appearances is to repeat old mistakes. It is sobering to consider that we are just now gaining a general understanding of the roles of some major cell structures that have been visible for more than a century, and the results were not anticipated by some of the best minds of the time. Experiments in this field are rarely heroic, but the accumulated results can reduce the number of possible alternative mechanisms. If the ultimate goal is to know how the cell actually accomplishes cytokinesis, then hypotheses and models must reflect the demonstrated properties and characteristics of the cell, and they must correctly predict the events of division under unusual imposed conditions.

We must resign ourselves to the possibility that when the real mechanism emerges from the shadows, it may not seem as sensible and protected from error as one we ourselves might have devised. For my own cells, I would prefer to see the mitotic apparatus mechanically involved in a way that did not require formation of a local contractile mechanism. The astral cleavage mechanism proposed by Chambers (1919) and by Gray (1924, 1931) has these characteristics. It is also consistent with all of the visible events and many of the experimental results; it contains fewer parts and constituent phenomena to malfunction. Unfortunately, it was eliminated by experimentation many years ago (Hiramoto, 1956).

ACKNOWLEDGEMENTS

The author's original research described in this paper was supported by grants from the National Science Foundation. Barbara N. Rappaport assisted in the carrying out of these experiments and the preparation of this manuscript.

LITERATURE CITED

Anstrom, J. A., and Summers, R. G., 1983a, A unilateral cleavage furrow in embryos of S.droebachiensis, J. Exp. Zool., 227: 395.
Anstrom, J. A., and Summers, R. G., 1983b, A morphological analysis of the first cleavage mitotic cytoskeleton isolated from sea urchin embryos (Strongylocentrotus droebachiensis), J. Morph., 177: 329.
Bütschli, O., 1876, Studien über die Ersten Entwicklungsvorgange der Eizelle, die Zelltheilung und die Conjugation der Infusorien, Abhandl. d. Senkenbergschen Naturf. Ges., 10: 213.

Chambers, R., 1917, Microdissection studies II. The cell aster: a reversible gelation phenomenon, J. Exp. Zool., 23: 483.

Chambers, R., 1919, Changes in protoplasmic consistency and their relation to cell division, J. Gen. Physiol., 2: 49.

Gray, J., 1924, The mechanism of cell division. I. The forces which control the form and cleavage of the eggs of Echinus esculentus, Proc. Camb. Philos. Soc. Biol. Series, 1: 164.

Gray, J., 1931, "A Textbook of Experimental Cytology," Cambridge University Press, Cambridge.

Hamaguchi, Y., 1975, Microinjection of colchicine into sea urchin eggs, Develop. Growth and Differ., 17: 111.

Harvey, E. B., 1935, The mitotic figure and cleavage plane in the egg of Parechinus microtuberculatus as influenced by centrifugal force, Biol. Bull., 69: 287.

Heilbrunn, L. V., 1928, "The Colloid Chemistry of Protoplasm," Borntrager, Berlin.

Hinkley, R. E., Webster, D. R., and Rubin, R. W., 1982, Further studies on dividing sea urchin eggs exposed to the volatile anesthetic halothane, Exp. Cell Res., 141: 492.

Hiramoto, Y., 1956, Cell division without mitotic apparatus in sea urchin eggs, Exp. Cell Res., 11: 630.

Hiramoto, Y., 1971, Analysis of cleavage stimulus by means of micromanipulation of sea urchin eggs, Exp. Cell Res., 68: 291.

Kawamura, K., 1977, Microdissection studies on the dividing neuroblast of the grasshopper, with special reference to the mechanism of unequal cytokinesis, Exp. Cell Res., 106: 127.

Lewis, W. H., 1942, The relation of the viscosity changes of protoplasm to ameboid locomotion and cell division, in: "The Structure of Protoplasm," W. Seifriz, ed., Iowa State College Press, Ames, Iowa.

McClendon, J. F., 1908, The segmentation of eggs of Asterias forbesii deprived of chromatin, Arch. Entwicklungsmech. Organismen, 26: 662.

Rappaport, R., 1961, Experiments concerning the cleavage stimulus in sand dollar eggs, J. Exp. Zool., 148: 81.

Rappaport, R., 1965, Duration of stimulus and latent periods preceding furrow formation in sand dollar eggs, J. Exp. Zool., 158: 373.

Rappaport, R., 1966, Experiments concerning the cleavage furrow in invertebrate eggs, J. Exp. Zool., 161: 1.

Rappaport, R., 1971, Reversal of chemical cleavage inhibition in echinoderm eggs, J. Exp. Zool., 176: 249.

Rappaport, R., 1973, On the rate of movement of the cleavage stimulus in sand dollar eggs, J. Exp. Zool., 183: 115.

Rappaport, R., 1975, Establishment and organization of the cleavage mechanism, in: "Molecules and Cell Movement," S. Inoué and R. E. Stephens, eds., Raven Press, New York.

Rappaport, R., 1976, Furrowing in altered cell surfaces, J. Exp. Zool., 195: 271.

Rappaport, R., 1978, Effects of continual mechanical agitation prior to cleavage in echinoderm eggs, J. Exp. Zool., 206: 1.

Rappaport, R., 1981, Cytokinesis: Cleavage furrow establishment in cylindrical sand dollar eggs, J. Exp. Zool., 217: 365.

Rappaport, R., 1982, Cytokinesis: The effect of initial distance between the mitotic apparatus and surface on the rate of subsequent cleavage furrow progress, J. Exp. Zool., 221: 399.

Rappaport, R., 1983, Cytokinesis: Furrowing activity in nucleated endoplasmic fragments of fertilized sand dollar eggs, J. Exp. Zool., 227: 247.

Rappaport, R., 1985, Repeated furrow formation from a single mitotic apparatus in cylindrical sand dollar eggs, J. Exp. Zool., 234: 167.

Rappaport, R., and Conrad, G. W., 1963, An experimental analysis of unilateral cleavage in invertebrate eggs, J. Exp. Zool., 153: 99.

Rappaport, R., and Ebstein, R. P., 1965, Duration of stimulus and latent periods preceding furrow formation in sand dollar eggs, _J_. _Exp_. _Zool_., 158: 373.

Rappaport, R., and Rappaport, B. N., 1974, Establishment of cleavage furrow by the mitotic spindle, _J_. _Exp_. _Zool_., 189: 189.

Rappaport, R., and Rappaport, B. N., 1984, Division of constricted and urethane-treated sand dollar eggs: A test of the polar stimulation hypothesis, _J_. _Exp_. _Zool_., 231: 81.

Rappaport, R., and Rappaport, B. N., 1985, Surface contractile activity associated with isolated asters in cylindrical sand dollar egg, _J_. _Exp_. _Zool_., 235: 217.

Rappaport, R., and Ratner, J. H., 1967, Cleavage of sand dollar eggs with altered patterns of new surface formation, _J_. _Exp_. _Zool_., 165: 89.

Ris, H., 1949, The anaphase movement of chromosomes in the spermatocytes of the grasshopper, _Biol_. _Bull_., 96: 90.

Schroeder, T. E., 1972, The contractile ring II. Determining its brief existence, volumetric changes and vital role in cleaving _Arbacia_ eggs, _J_. _Cell Biol_., 53: 419.

Schroeder, T. E., 1975, Dynamics of the contractile ring, _in_: "Molecules and Cell Movement," S. Inoué and R. E. Stephens, eds., Raven Press, New York.

Schroeder, T. E., 1981, The origin of cleavage forces in dividing eggs. A mechanism in two steps, _Exp_. _Cell Res_., 134: 231.

Tanaka, Y., and Inoué, S., 1981, Does cytochalasin-D induce reversible disruption of the cleavage contractile ring?, _Biol_. _Bull_., 161: 311.

White, J. G., and Borisy, G. G., 1983, On the mechanisms of cytokinesis in animal cells, _J_. _Theor_. _Biol_., 101: 289.

Wilson, E. B., 1901, Experimental studies in cytology. II. Some phenomena of fertilization and cell division in etherized eggs, _Arch_. _Entwicklungsmech_. _Organismen_, 13: 353.

Yatsu, N., 1912, Observations and experiments on the ctenophore egg. I. The structure of the egg and experiments on cell-division. _J_. _Coll_. _Sci_. _Tokyo_, 32: (Art. 3) 1.

Ziegler, H. E., 1898, Experimentelle Studien über die Zelltheilung. I. Die Zerschnürung der Seeigeleier. II. Furchung ohne Chromosomen, _Arch_. _Entwicklungsmech_. _Organismen_, 6: 249.

Ziegler, H. E., 1903, Experimentelle Studien über Zelltheilung. IV Die Zelltheilung der Furchungzellen bei Beroe und Echinus, _Arch_. _Entwicklungsmech_. _Organismen_, 16: 155.

MECHANICAL PROPERTIES OF THE PROTOPLASM OF

ECHINODERM EGGS AT VARIOUS STAGES OF CELL CYCLE

Yukio Hiramoto

Biological Laboratory, Tokyo Institute of Technology
Tokyo 152 and Department of Cell Biology, National
Institute for Basic Biology, Okazaki 444, Japan

INTRODUCTION

Investigation of the mechanical properties of the protoplasm is of primary importance in studies of cell division, since any motions or morphological changes in the protoplasm, including chromosome movement, spindle elongation and cytokinesis, result from structural changes of the protoplasm, which are accompanied by mechanical changes. Echinoderm eggs are ideal for studying cell division because many synchronously dividing cells can be obtained within an hour or so after insemination. Their relatively large size and regular shape are suitable for various biophysical studies, such as the measurement of mechanical properties of the protoplasm.

The echinoderm egg consists of three parts with different physical properties: the cell membrane, the cortex and the endoplasm. The present article reviews the results of some experiments on the mechanical properties of the endoplasm and of the cell surface (which are mainly due to the properties of cell membrane and the cortex) in eggs of the sea urchin and the starfish at various stages of maturation, fertilization and cleavage. These experiments were carried out or are in progress in my laboratory (Hiramoto, 1969 a, b; 1974; 1979; Nakamura and Hiramoto, 1978; Shôji et al., 1978; Kaneda, Kamitsubo and Hiramoto, unpublished).

MEASUREMENTS OF MECHANICAL PROPERTIES OF THE ENDOPLASM

Three methods were used in the measurement of mechanical properties of the endoplasm; the centrifuge method, the magnetic particle method and the capillary method.

Centrifuge method

The mechanical properties of the endoplasm were determined from the movement of particle(s) in it under a centrifugal field. The viscosity of the endoplasm (η) was determined from the velocity (v) and the radius of the particle (a), the difference between the density of the particle (D) and that of the endoplasm (D'), and the acceleration of the centrifugal field (A) following Stokes' formula, assuming that the endoplasm behaves as a Newtonian fluid.

13

$$\eta = \frac{2a^2(D-D')A}{9v} \qquad (1)$$

In the present study, we determined mechanical properties of the endoplasm from the movement of spherical gold particle(s) in the endoplasm under a centrifugal field. This method will be described elsewhere in detail (Kaneda, Kamitsubo and Hiramoto, to be published). Briefly, minute gold particles of various sizes with irregular shapes were obtained by grinding commercially-available gold powder with sucrose powder in a mortar. A small amount of the ground gold powder mixed with sucrose was put on a thin quartz plate with a drop of water, and then heated with a gas burner. Each gold particle melted and became completely spherical, while the sucrose was burnt out and the water was evaporated.

The particle was introduced into the cell with a glass micropipette manoeuvered by a micromanipulator. The cell containing the particle(s) was held between two thin glass plates in a centrifuge chamber filled with a mixture of sea water and isosmotic sucrose or Percoll solution at the same density as the protoplasm (1.07). The movement of the gold particle(s) in the endoplasm under the centrifugal field was observed and recorded with a centrifuge microscope combined with a video system designed and constructed by Kamitsubo and Kikuyama (unpublished). The movement of the particle under the centrifugal field could be repeated by turning the cell in the chamber after the particle had been centrifuged to the end of the cell. The magnitude of the centrifugal field used in the present study was in the order of 10^2 times gravity.

As shown in Fig. 1, the gold particle moved in the endoplasm at an uneven speed as if it was impeded by invisible meshwork structures in the endoplasm. Therefore, the "viscosity" calculated from Eq. 1 indicates merely the "mean viscosity" representing mean mechanical properties of the heterogeneous body.

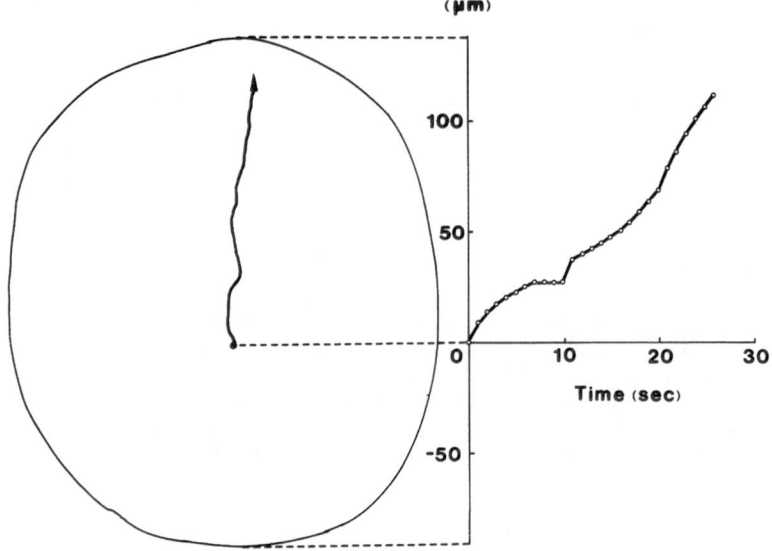

Fig.1   Movement of a gold particle in the endoplasm of a mature egg of the starfish, <u>Asterina</u> <u>pectinifera</u>. The particle moves discontinuously (at a variable velocity) in the endoplasm, suggesting a heterogeneous consistency. The magnitude of the centrifugal field was 106 xg.

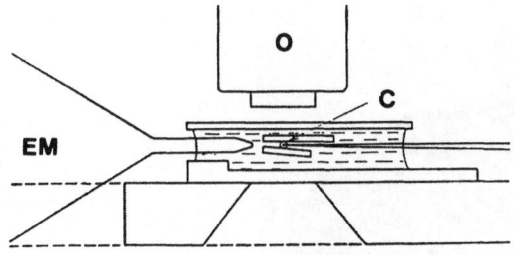

Fig.2 Experimental setup for the
       magnetic particle method. c,
       cell containing a magnetic
       particle; EM, pole of
       electromagnet; O, microscope
       objective.

Magnetic particle method

We determined mechanical properties of the endoplasm from the
movement of a spherical iron particle (5 to 7 μm in diameter), in the
endoplasm under a magnetic field obtained by switching on the circuit of
an electromagnet placed close to the cell (Hiramoto, 1969 a, b). The
particle was injected into the cell by applying a strong magnetic force
after bringing it close to the cell. The particle could be moved by
smaller magnetic forces if it was successfully introduced into the cell by
breaking the cell surface.

As shown in Fig. 2, the cell containing the particle was held between
thin glass plates in a chamber placed on the stage of a microscope. The
movement of the particle in the endoplasm before, during and after the
application of the magnetic field was recorded with a 16 mm cine camera
after an appropriate magnification with the microscope optics.

The image of the cell containing the particle was recorded on 16 mm
film and was projected with a cine projector through a narrow slit (s in
Fig. 3) onto a sheet of photographic paper moving in the direction at
right angles to the direction of the movement of the particle image, so
that the particle image moved along the slit. As shown in Fig. 3, the
particle moved toward the electromagnet at a decreasing rate during the
application of the magnetic force and it recoiled at a decreasing rate
after the application of the force.

The magnitude of the force applied to the particle depended on the
size of the particle, the electric current of the electromagnet circuit
and the distance between the electromagnet and the particle. The
dependence of the force on these factors was determined as follows. At
the end of each experiment the cell containing the particle was impaled
with a fine glass needle held with a micromanipulator and the glass plates
supporting the cell were removed so that the cell was held only with the
needle. When the circuit of the electromagnet was switched on, the needle
bent towards the electromagnet owing to the force applied to the particle
in the cell. Since the bending stiffness of the needle had been
determined beforehand, the absolute value of the force applied to the
particle could be determined. Such a determination was carried out for
every particle at various electromagnet currents and at various distances
of the particle from the electromagnet. The time courses of the applied
force during experiments were determined from the position of the particle
in the microscope field, which depended on the distance from the
electromagnet, and the current of the electromagnet, which was recorded
with an oscilloscope. As shown in Fig. 4, the time course of the

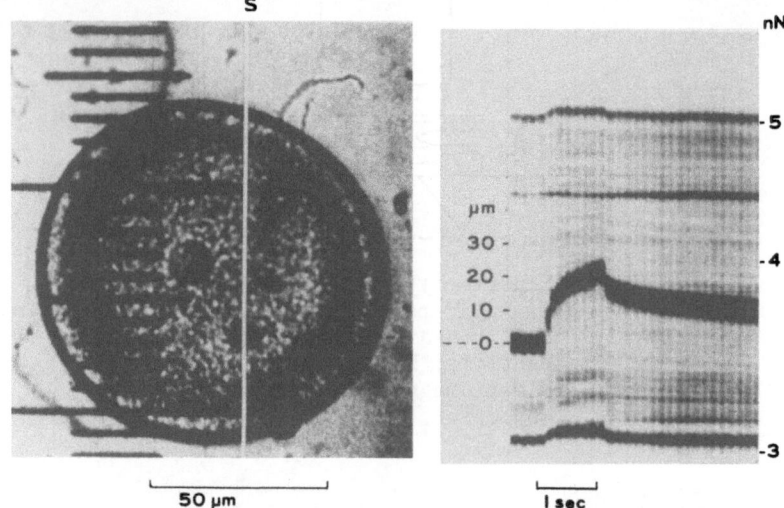

Fig.3 Photokymographic display of the movement of a
magnetic particle in the cell by application of a
magnetic field. A cinemicrographic image of the
particle in the cell (left figure) was projected
through a narrow slit (s) onto a sheet of
photographic paper moving in the direction at right
angles to the direction of movement of the image of
the particle, so that the particle image moving
along the slit could be recorded on the sheet of
photographic paper. The scale on the right side of
the figure represents the magnitude of the force
applied to the particle; this differed between
particles in the microscopic field, due to the
differences in distance from the electromagnet.

applied force was nearly square, though it was slightly deformed by the
presence of an inductance in the circuit and the change in the distance of
the particle from the electromagnet during the experiment.

The speed of the particle gradually decreased when the force was kept
constant or even slightly increased, and the particle recoiled with
decreasing speed after terminating the applied force. This result
indicates that mechanical properties of the protoplasm can be represented
by a visco-elastic model as shown in Fig. 5 a. Fig. 5 b shows the
movement of a spherical particle during the application of a constant
force in a fluid represented by this model. The visco-elastic constants of
the protoplasm, G, η_1 and η_2 in Eq. 2 were determined from observed
movement (x) of the particle with a radius a during the application of a
force F for time t (cf. Hiramoto, 1969 a).

$$x = \frac{F}{6\pi a G} \left(1 - e^{-Gt/\eta_1} + \frac{Gt}{\eta_2}\right) \qquad (2)$$

The consistency of the protoplasm was also semi-quantitatively represented
by F/ax_t where x_t is the displacement of the particle by application of a
force F for over a fixed period, e.g. 0.5 sec (cf. Hiramoto, 1969).

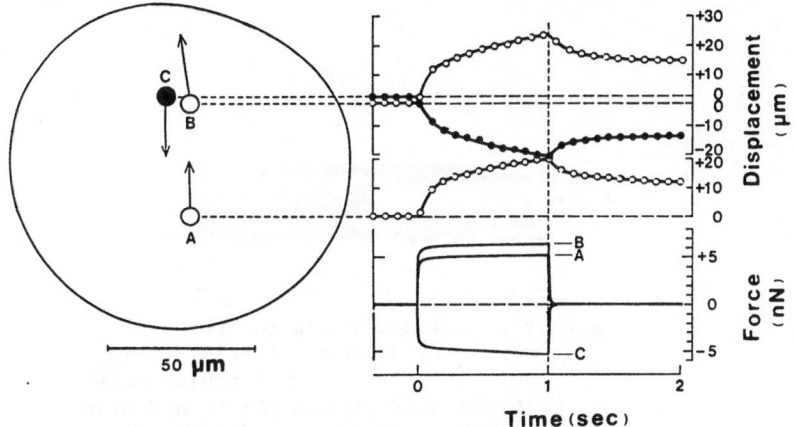

Fig.4   Movement of an iron particle in the endoplasm of
an unfertilized egg of the sand dollar,
Clypeaster japonicus under different magnetic
fields. The tracks of the particle in the
endoplasm are represented by the arrows in the
left figure. The time courses of the applied
force are represented by the continuous curves
(A, B and C) in the lower right figure. The
response curves of the particle are shown in the
upper right figure. After Hiramoto (1969 a).

Capillary method

        Mechanical properties of the protoplasm can be determined from the
rate of movement of the protoplasm in a capillary when it is forced to
move by a pressure difference applied between the two ends (Kamiya and

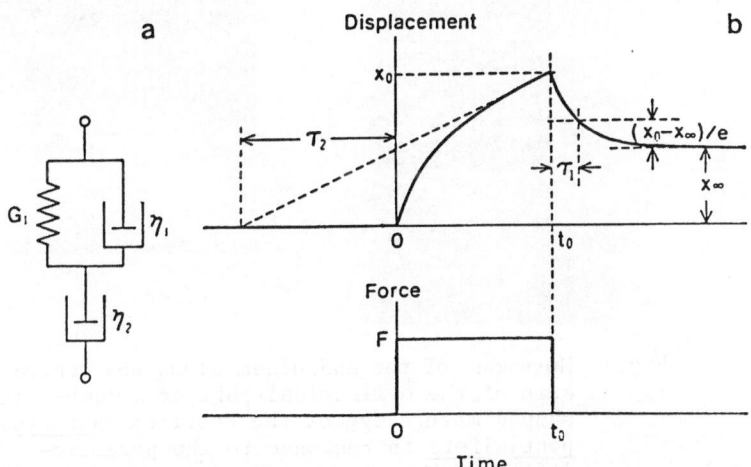

Fig.5   Visco-elastic model representing properties of
the endoplasm and the movement of a spherical
particle in the model by application of a
constant force. a, visco-elastic model. b,
response of a spherical particle to application
of a force (F). G, η_1 and η_2 are visco-elastic
constants. After Hiramoto (1969 a).

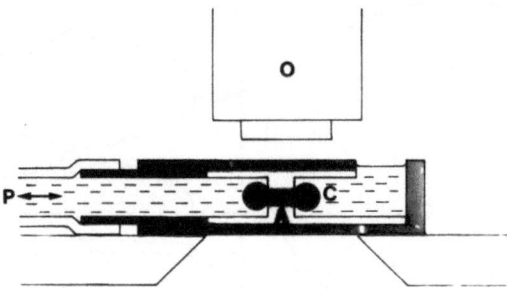

Fig.6 Experimental setup for the
capillary method. A cell (C) is
aspirated through a circular hole
in the agar septum (A) of a double
chamber, and forms a dumb-bell.
The mechanical properties of the
endoplasm are determined from the
movement of the endoplasm in the
cylindrical part of the cell in
response to a pressure-difference
applied between the two
compartments of the double chamber.
O, microscope objective.

Fig.7 Movement of the endoplasm along the central
axis of the cylindrical part of a dumb-bell
shaped mature egg of the starfish, Asterina
pectinifera in response to the pressure-
difference between the two compartments of
the chamber. Left, photomicrograph of the
cylindrical part of the cell. Right,
photokymographic record of the movement of
the endoplasm along the central axis of the
cylindrical part in response to a
constant pressure-difference of 10 sec in
duration.

Kuroda 1965, Shoji et al., 1978). In the present study (cf. Shoji et al., 1978), a part of the cell was deformed to a cylindrical shape by introducing the cell into a double chamber as shown in Fig. 6. Movement of the protoplasm during the application of a hydrostatic pressure difference between the two compartments of the double chamber was recorded by cinemicrography. The movement of the protoplasm along the central axis of the cylindrical part of the cell is displayed in Fig. 7. This was achieved by projecting the image of the cylindrical part of the cell, recorded on cine film, through a narrow slit on a sheet of photographic paper moving in the direction at right angles to the direction of movement of the image of the protoplasm, so that the image of the protoplasm at the central axis of the cylindrical part moved along the slit. The movement of the protoplasm in Fig. 7 indicates that the mechanical properties of the protoplasm can be represented by the visco-elastic model shown in Fig. 5 a. The visco-elastic constants G, η_1 and η_2 in Eq. 3 were obtained by fitting the observed movement of the protoplasm along the axis of the cylindrical part to the equation;

$$x^{\circ} = \frac{PR^2}{4LG} \left(1 - e^{-Gt/\eta_1} + \frac{Gt}{\eta_2} \right) \qquad (3)$$

in which x° is the displacement of the endoplasm by application of a constant pressure difference (P) between the two compartments of the double chamber of which L and R are the length and radius of the cylindrical part (cf. Shôji et al., 1978). The relative consistency was represented by PR^2/Lx_t°, where x_t° is the displacement of the endoplasm at the central axis of the cylindrical part during the application of a constant pressure difference (P) for a defined period, e.g. 10 sec (cf. Shôji et al., 1978).

MEASUREMENTS OF MECHANICAL PROPERTIES OF THE CELL SURFACE

We determined mechanical properties of the cell surface by the suction method (Mitchison and Swann, 1954; Hiramoto, 1986) and the magnetic particle method (Hiramoto, 1974).

Suction method

In this method, the cell surface is deformed by applying negative pressures of various magnitudes to a part of the cell surface through a micropipette whose tip is in close contact with the surface (cf. Fig. 8 a). The height of the bulge formed by a constant negative pressure gradually increases approaching a steady level during the application of a constant negative pressure as shown in Fig. 8 b. This indicates that the cell surface is visco-elastic.

The relationship between the applied negative pressure and the height of the bulge was found to be almost linear in sea urchin eggs (Mitchison and Swann, 1954) and starfish oocytes and eggs (Nakamura and Hiramoto, 1978). The slope of the negative pressure-height of the bulge relationship, which was defined as the stiffness of the cell surface by Mitchison and Swann (1954), depends both on the surface force (Cole, 1932), namely the tension at the surface (Harvey, 1931), and the elasticity of the cell surface (cf. Hiramoto, 1970; Nakamura and Hiramoto, 1978).

We determined changes in mechanical properties of the cell surface during maturation and cleavage by changes in the height of the bulge when a constant negative pressure was applied continuously or repeatedly at one or two parts of the cell surface (Nakamura and Hiramoto, 1978; Hiramoto, 1979; Ohtsubo and Hiramoto, 1985). In some experiments, changes in mechanical properties of the cell surface were represented by changes

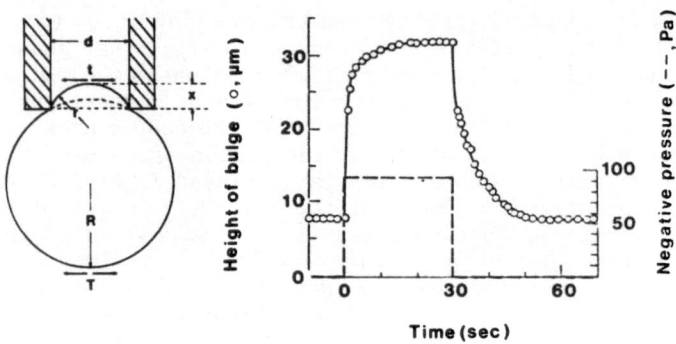

Fig.8   Suction method for determining the stiffness of
        the cell surface. a, principle of the method.
        b, change in the height (x) of the bulge (circles
        with continuous lines) formed by application of a
        constant negative pressure (broken line) in a
        primary oocyte of the starfish, <u>Asterina
        pectinifera</u>. Modified from Nakamura and Hiramoto
        (1978).

in the applied negative pressure required to keep the height of the bulge
constant (Hiramoto, 1979; Ohtsubo and Hiramoto, 1985).

<u>Magnetic particle method</u>

The stiffness of the cell surface was determined from the deformation
of the cell surface when a force was applied to an iron particle embedded
in the cortex of the cell (Hiramoto, 1974). The magnitude of the
stiffness was represented by the reciprocal of the height (x in Fig. 9) of
the bulge formed by a force of a defined magnitude.

CHANGES IN MECHANICAL PROPERTIES OF STARFISH OOCYTES DURING MATURATION

Starfish oocytes dissected from ovaries into sea water are at the
primary oocyte stage, having a large germinal vesicle. Maturation can be
induced by putting them into sea water containing 1-methyl adenine of an
appropriate concentration (Kanatani, 1973). In the present study, 1-
methyl adenine solution was applied to the oocyte of the starfish,
<u>Asterina pectinifera</u>, to give a final concentration of 10^{-6}M. Measurements
of mechanical properties of the endoplasm by the capillary method or of
the cell surface by the suction method were continued before and during
the process of maturation.

<u>Mechanical properties of the endoplasm</u>

Fig. 10 shows a series of measurements of the movement of the
endoplasm along the central axis of the cylindrical part of the cell
induced by application of a defined pressure difference (50 Pa in this
case) at various times after the application of 1-methyl adenine to a
starfish oocyte. Record a was taken after germinal vesicle breakdown and
before the formation of the first polar body; record b was taken during
the formation of the first polar body; record c was taken after the
formation of the first polar body and before the formation of the second
polar body; record d was taken during the formation of the second polar
body; and records e and f were taken after formation of the second polar
body.

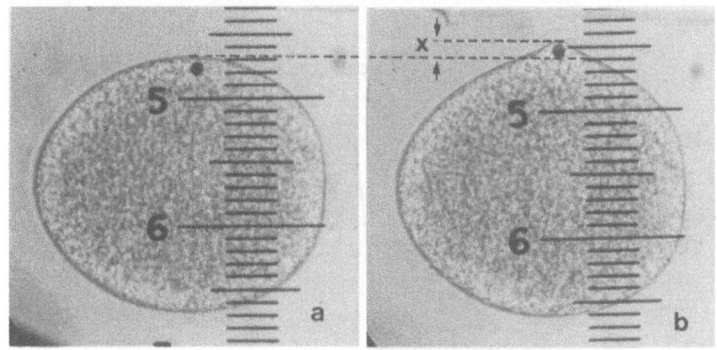

Fig.9 Deformation of an unfertilized egg of
       <u>Hemicentrotus pulcherrimus</u> by application of
       a force to a particle near the cell surface.
       a, in the absence of the force. b, during
       application of the force. Smallest division
       of micrometer scale indicates 5 μm.

The changes in consistency of the endoplasm during the process of
maturation are shown in Fig. 11, in which the consistency is expressed by
PR^2/Lx_t^o, described above, where the displacement of the endoplasm during
application of the pressure for 10 sec was used as x_t^o. It should be noted

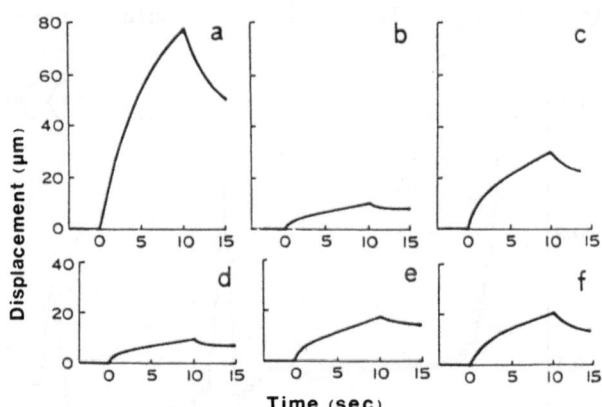

Fig.10 Measurements of consistency of the endoplasm of
       a starfish oocyte during maturation.
       Displacements of the endoplasm along the
       central axis of the cylindrical part of a dumb-
       bell shaped oocyte of the starfish, <u>Asterina</u>
       <u>pectinifera</u> were recorded before the formation
       of the first polar body (a), during the first
       polar body formation (b), after the first polar
       body formation and before the second polar body
       formation (c), during the second polar body
       formation (d) and after the second polar body
       formation (e and f). After Shôji, Hamaguchi
       and Hiramoto (1978).

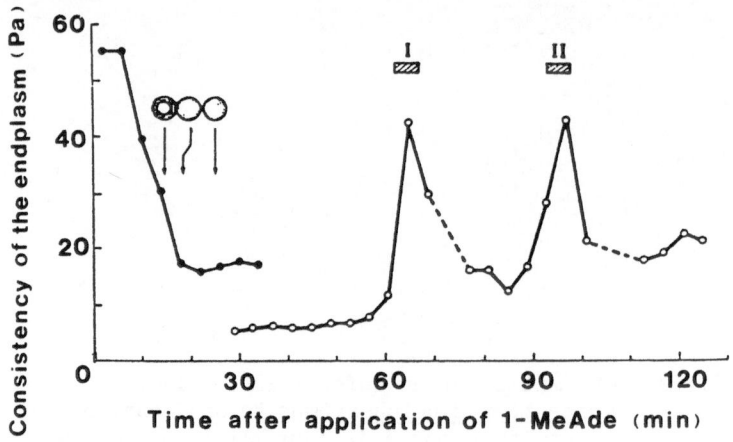

Fig.11 Changes in consistency during maturation in oocytes of the starfish, <u>Asterina pectinifera</u>, determined by the capillary method. Points connected with lines were obtained from the same oocyte. Stage of maturation and times of polar body formation are diagramatically shown at the top of figure. For details of the measurements, see text. After Shôji, Hamaguchi and Hiramoto (1978).

that the consistency decreases during the process of germinal vesicle breakdown and increases during the formation of the first and the second polar bodies (cf. Shôji et al., 1978). It should be noted that the change

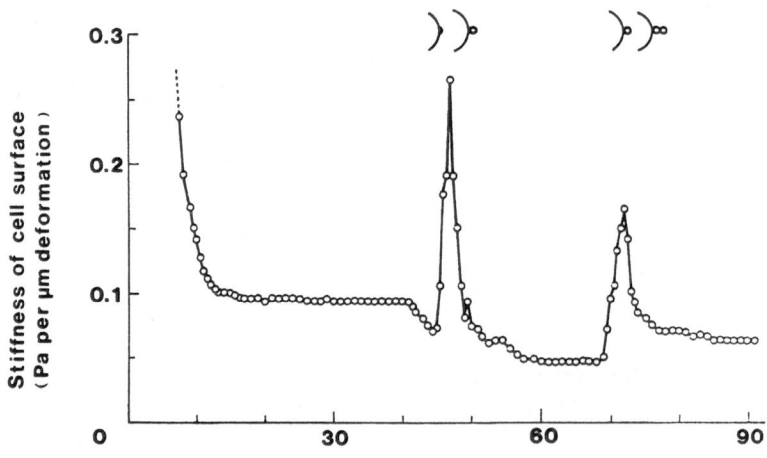

Fig.12 Changes in stiffness of the cell surface during maturation in an oocyte of the starfish, <u>Asterina pectinifera</u>, determined by the section method. Stages of polar body formation are diagramatically shown at the top of figure. After Nakamura and Hiramoto (1978).

in consistency occurs in the endoplasm in general while the meiotic spindle and the polar body are localized at a restricted region near the animal pole.

Changes in mechanical properties of the cell surface during maturation

Fig. 12 shows mechanical properties of the cell surface at various stages of maturation. The ordinates represent the stiffness of the cell surface calculated from the height of the bulge formed by a negative pressure applied to a part of the cell surface through a micropipette whose tip is in close contact with it (Nakamura and Hiramoto, 1978). Fig.

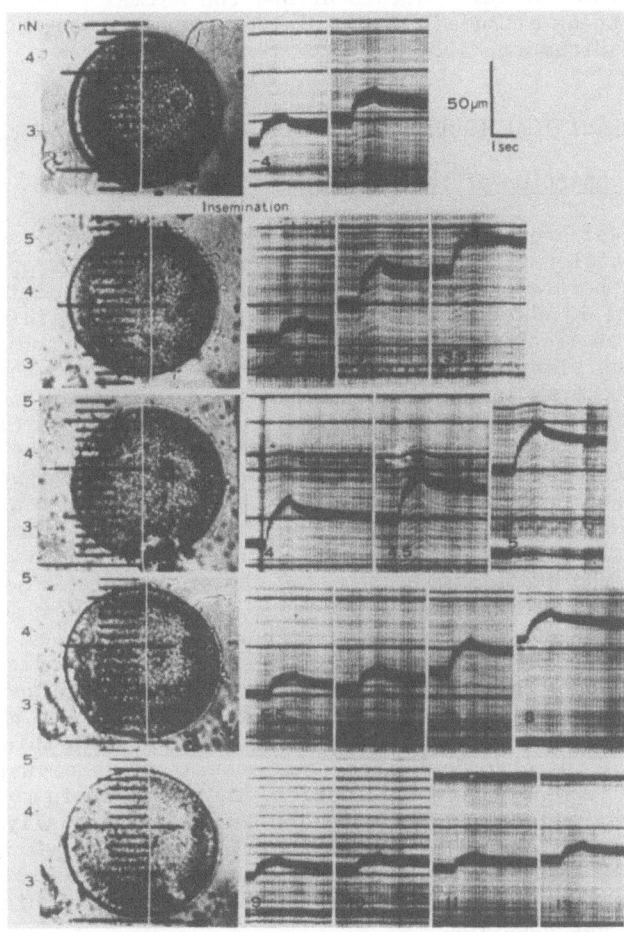

Fig.13 Photokymographic displays of the movement of an iron particle in response to application of a magnetic field in an egg of the sea urchin, Temnopleurus toreumaticus. The numeral in each record represents time in min after insemination. Scales on the left side represent the magnitude of the force applied to the particle which varies with the position of the particle in the microscopic field owing to the difference in distance from the electromagnet.

12 shows that the stiffness decreases within 20 min of the application of 1-methyl adenine, during which time the germinal vesicle breaks down, increases shortly before and during the formation of the first and the second polar bodies, and decreases towards the end of and after the polar body formation.

By using a pair of micropipettes of the same size for measuring the stiffness at two different parts of the cell surface simultaneously (Ohtsubo and Hiramoto, 1985), it was found that the stiffness of the cell surface increased almost in parallel over the entire cell surface before the onset of polar body formation. The change in stiffness occurred over the entire surface while the polar body was extruded at a restricted region around the animal pole of the cell. During polar body formation, the stiffness of the cell surface around the animal pole, where polar bodies were being extruded, was definitely larger than that of other parts (Ohtsubo and Hiramoto, 1985).

CHANGES IN MECHANICAL PROPERTIES OF SEA URCHIN EGGS UPON FERTILIZATION

Mechanical properties of the endoplasm

Mechanical properties of the endoplasm were determined by the magnetic particle method in eggs of the sea urchin, Temnopleurus toreumaticus before and at various stages after fertilization (Hiramoto, 1969 b). Fig. 13 shows a series of photokymographic displays of the movement of an iron particle obtained from cinemicrographic records in an egg at various times before and after insemination. In this experiment, forces of 3 to 5 nN in magnitude, depending on the position of the particle in the microscope field, were applied to the particle in the endoplasm of an egg. The movements of the particle in the endoplasm varied at different stages of fertilization. In this figure the response curves obtained from the egg before and at various times after insemination are similar to one another if the scales are adjusted to make the displacements the same. This may indicate that the visco-elastic constants of the endoplasm, G, η_1 and η_2 change in parallel upon fertilization and during the early one-cell stage, their relative values remaining unchanged. Therefore, the change in mechanical properties of the endoplasm can be represented by the change in a single constant, F/ax_t.

Fig. 14 shows the change in the consistency of the endoplasm represented by F/ax_t where x_t is the displacement of the particle during the application of a force F for 0.5 sec. As shown in this figure, the consistency of the endoplasm increases soon after insemination, reaches a maximum about 3 min after insemination, increases again to reach a second maximum about 9 min after insemination, and then gradually decreases. The increase in consistency 3 to 9 min after insemination may be due to the development of microtubule structures of the sperm aster, but no marked cytological change corresponding to the rapid change within 3 min after insemination was observed.

Mechanical properties of the cell surface

Fig. 15 shows change in stiffness of the cell surface upon fertilization in a sea urchin egg (cf. Hiramoto, 1974). The stiffness is represented by the reciprocal of the height of the bulge (μm^{-1}) formed by application of a force of a definite magnitude (2.7 nN in this case) of 1 sec in duration in the direction normal to the cell surface (cf. Fig. 9). The change in stiffness is similar to the change in consistency of the endoplasm shown in Fig. 14, though the stiffness level is always higher after fertilization than before fertilization, while the consistency is lower after fertilization (cf. Fig. 14).

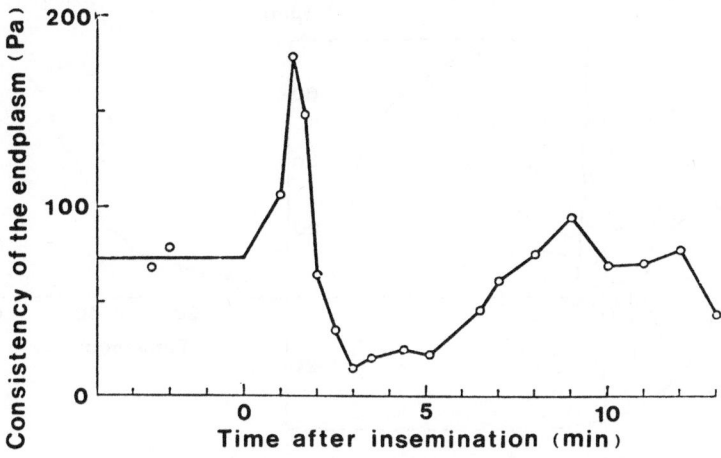

Fig.14 Changes in consistency of the endoplasm of an
egg of the sea urchin <u>Temnopleurus toreumaticus</u>
at fertilization determined by the magnetic
particle method. The consistency is
represented by F/ax_t where F is the applied
force, a is the radius of the particle, and x_t
is the displacement during the application of a
force 0.5 sec in duration. After Hiramoto
(1969 b).

The transient increase in the stiffness of the cell surface, which
occurs simultaneously with the increase in the endoplasmic consistency
within a few minutes of insemination, was reported by Mitchison and Swann
(1955) using the suction method and by Hiramoto (1963) using a method in
which the stiffness was determined by compressing the egg between
two parallel plates.

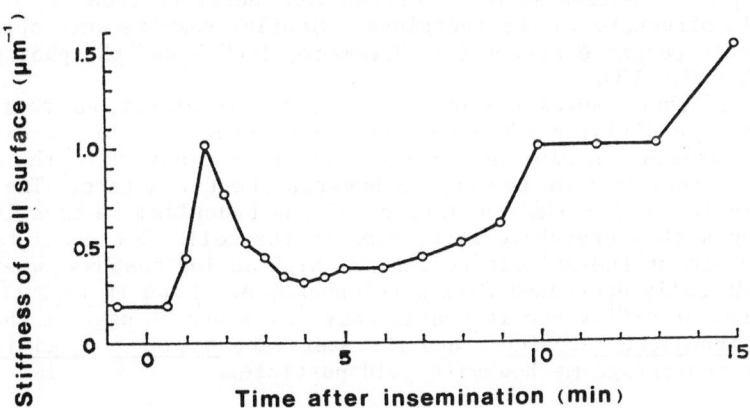

Fig.15 Changes in stiffness of the cell surface at
fertilization in an egg of the sea urchin
<u>Temnopleurus toreumaticus</u> determined by the
magnetic particle method. The stiffness is
represented by the reciprocal of the height of
the bulge (μm^{-1}) formed by application of a
force of 2.7 nN for 1 sec to an iron particle
embedded in the cortex. After Hiramoto (1974).

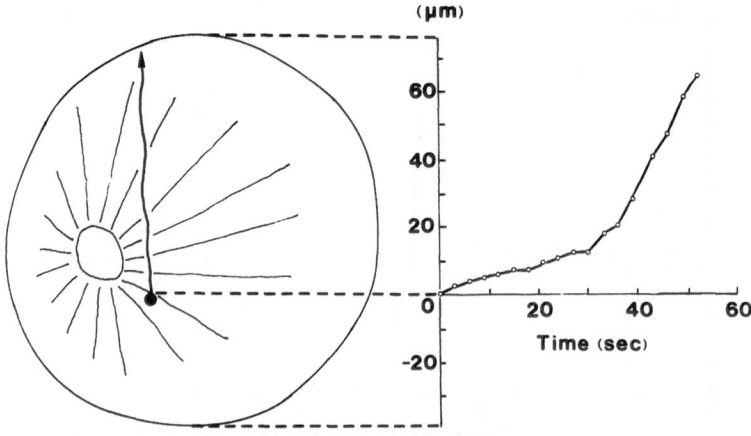

Fig.16 Movement of a gold particle in a fertilized egg
of the sand dollar Clypeaster japonicus under a
centrifugal field. The velocity of the
particle is different at different regions in
the endoplasm depending on its position
relative to the sperm aster. The magnitude of
the centrifugal field was 186 xg.

REGIONAL DIFFERENCES IN MECHANICAL PROPERTIES OF THE ENDOPLASM IN
FERTILIZED EGGS

     Fig. 16 shows the movement of a gold particle in the endoplasm of a
fertilized egg of the sand dollar, Clypeaster japonicus. The speed of the
particle was reduced at the region of the sperm aster and the particle
sometimes moved along the astral rays deviating from the direction of the
centrifugal field. These results indicate that the sperm aster,
consisting of a centrosome and microtubules radiating from it, may behave
as a rigid structure in the endoplasm. Similar results were obtained by
the magnetic particle method (cf. Hiramoto, 1969 b and the photographic
records in Fig. 13).
     Fig. 17 shows movements of an iron particle at various regions in the
endoplasm of a dividing sea urchin egg (Hiramoto, 1969 b). Clearly the
particle movement is slow near the centers of two asters of the diaster
while it is very fast at the region between the two asters. The above
results indicate that the consistency of the endoplasm is closely
correlated with microtubule structures in the cell. The consistency of
the endoplasm at the spindle region was high during anaphase whereas it
was dramatically decreased during telophase, as shown in Fig. 17.
     A similar difference in consistency was found in eggs of the sand
dollar, Clypeaster japonicus, and the starfish, Asterina pectinifera, by
using the centrifuge method with gold particles.

CHANGES IN CONSISTENCY OF THE ENDOPLASM FROM FERTILIZATION TO CLEAVAGE

     Fig. 18 shows the changes in the consistency of the endoplasm from
fertilization to cleavage as determined by the centrifuge method. In this
figure, the consistency is expressed by the viscosity calculated from the
mean velocity of a gold particle in the endoplasm under a centrifugal
field. The viscosity increases during the first half of the one-cell

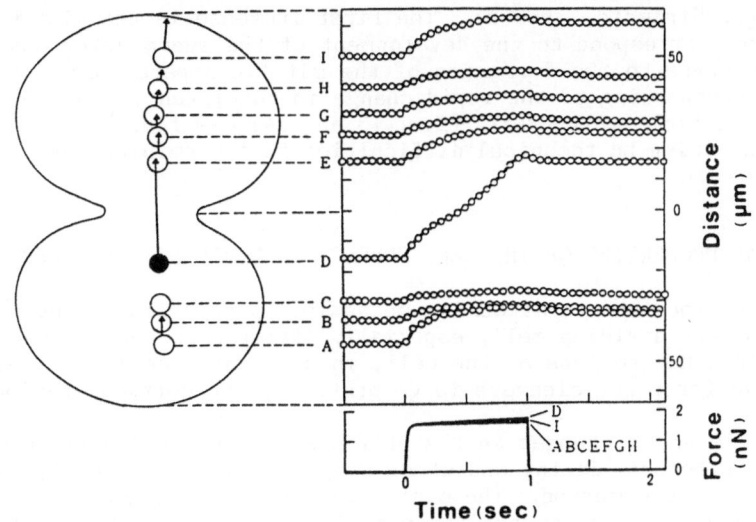

Fig.17 Measurements of visco-elasticity of a dividing
egg of the sea urchin <u>Temnopleurus</u> <u>toreumaticus</u>
determined by the magnetic particle method.
Curves, A-I on the upper right represent
movements of an iron particle at different
regions of the dividing egg, in response to a
magnetic field. The strengths of the magnetic
fields are shown by the continuous curves (A-I)
on the lower right. After Hiramoto (1969 b).

stage, decreases thereafter, and then increases before the onset of
cleavage. This change resembles the change in consistency of the
endoplasm during the one-cell stage as determined by the magnetic particle

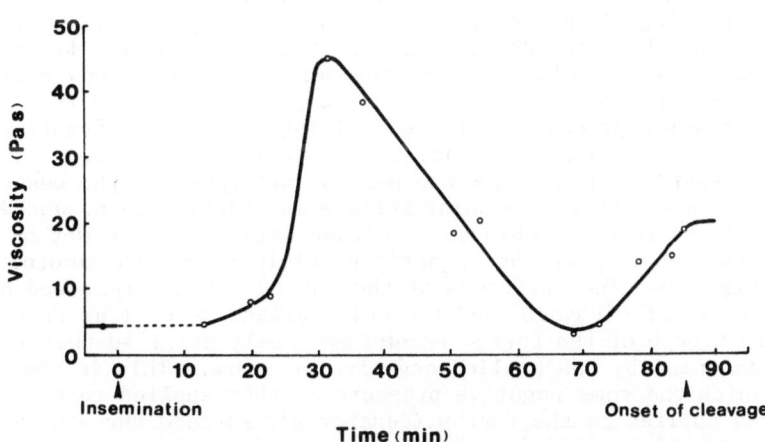

Fig.18 Changes in mean viscosity of the endoplasm
determined by the centrifuge method in an egg
of the sand dollar <u>Clypeaster</u> <u>japonicus</u>. The
mean viscosity was obtained from the mean
velocity of a spherical gold particle in the
egg before and at various stages after
fertilization.

method (cf. Hiramoto, 1969 b). The first increase in the viscosity in Fig. 18 may correspond to the development of the sperm aster and the second increase to the formation of the mitotic apparatus. No viscosity change corresponding to the rapid change in consistency shown by the magnetic particle method (cf. Figs. 13 and 14) was found in the this experiment owing to technical difficulties in determining rapid changes in the viscosity.

MECHANICAL PROPERTIES OF THE CELL SURFACE IN DIVIDING ECHINODERM EGGS

It is important to determine mechanical properties of the cell surface of the dividing cell, especially differences between the furrow region and other regions of the cell, in studies of cell division, because the motive force for cleavage is generated at the cortical region of the cell.

Fig. 19 shows changes in the stiffness of the cell surface of sea urchin eggs before, during and after cleavage as determined by the magnetic particle method. The ordinates represent the stiffness of the cell surface expressed as the reciprocal of the height of the bulge formed under forces of a defined magnitude of 1 sec duration. Two series of determinations of the stiffness of the cell surface at the region which will be the pole of the cell during cleavage are shown by the open circles connected by solid lines (A and B) in Fig. 19 a. As shown in this figure, two stiffness peaks, one 2 to 3 min before the onset of cleavage and the other 2 to 4 min after the onset of cleavage, are recorded. Fig. 19 b shows results from two examples (C and D) in which first the stiffness of the cell surface at the region expected to be the cleavage furrow during cleavage was determined (solid circles) and then the stiffness of the polar surface of the same cell after the start of cleavage was determined (open circles). The change in stiffness of the furrow surface resembles that of the polar surface, although it was not possible in this experiment to resolve whether or not a difference in stiffness existed between the polar surface in the same cell, owing to the variation of the stiffness values in individual cells.

Fig. 20 shows a series of determinations of the stiffness of the cell surface in fertilized eggs of the starfish, Asterina pectinifera, by the suction method. In these determinations, the height of the bulge at the polar surface (open circles in a) or the height at the furrow surface (closed circles in b) formed by a negative pressure applied through a micropipette was kept constant by controlling the applied negative pressure. In this case, the stiffness change is represented by the negative pressure change. Only one peak of stiffness at the onset of cleavage was observed at the polar surface in starfish eggs, whereas two peaks were found in sea urchin eggs, either using the same method (Ohtsubo and Hiramoto, 1985) or by the magnetic particle method (Hiramoto, 1974 and shown in Fig. 19). The stiffness of the furrow surface remained high after the start of cleavage, and the cell surface several μm on either side of the trough of the furrow became extremely stiff so that it was scarcely deformed by the applied negative pressure. This is shown in Fig. 20 c, in which the same negative pressure as that applied to the polar surface was applied to the furrow (equatorial) surface through the same size of micropipette. The height of the bulge at the furrow surface was more or less similar to the height at the polar surface before the onset of cleavage, but at the polar surface the height suddenly decreased as cleavage started, indicating an increase in the stiffness of the furrow. A similar increase in stiffness of the furrow surface at the onset of cleavage was found in sea urchin eggs by the same method but using two micropipettes.

The above results indicate that the mechanical properties of the cell surface change before the onset of cleavage almost synchronously over the

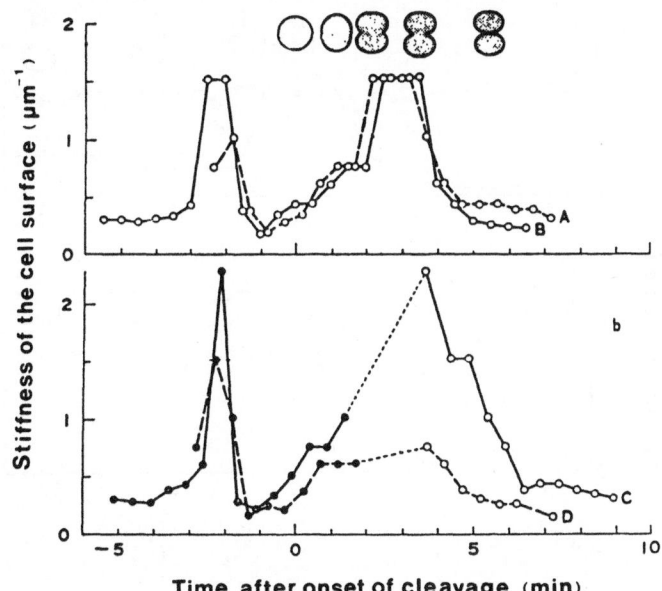

Fig.19 Changes in the stiffness of the cell surface
determined by the magnetic particle method in
eggs of the sea urchin <u>Temnopleurus</u>
<u>toreumaticus</u> before, during and after cleavage.
Stiffness is represented by the reciprocal of
the height of the bulge (x in Fig. 9) formed by
application of a defined force (8.5 nN in egg
A, 6.4 nN in B, 4.7 nN in C and 6.7 nN in D) of
1 sec in duration to the iron particle embedded
in the cortex of the egg. In eggs A and B, the
particle was embedded in the cortex at the
region which would form the pole during
cleavage. In eggs C and D, the particle was
first embedded in the cortex at the region
which would form the cleavage furrow during
cleavage and was moved to the cortex at the
polar surface a few min after the onset of
cleavage. Solid circles represent the
stiffness of the furrow surface while open
circles show the stiffness of the polar
surface. Stages of cleavage are represented
diagramatically at the top of the figure.
After Hiramoto (1974).

entire cell surface. The difference in mechanical properties between the
furrow surface and the polar surface occurs almost simultaneously with the
onset of cleavage. A similar change in mechanical properties of the cell
surface occurs during the process of polar body formation in starfish
oocytes, with the stiffness changing almost simultaneously over the entire
cell surface before the onset of the polar body formation. Judging from
the form of the cell surface under a negative pressure, the stiffness of
the cell surface at the base of the constriction for the polar body

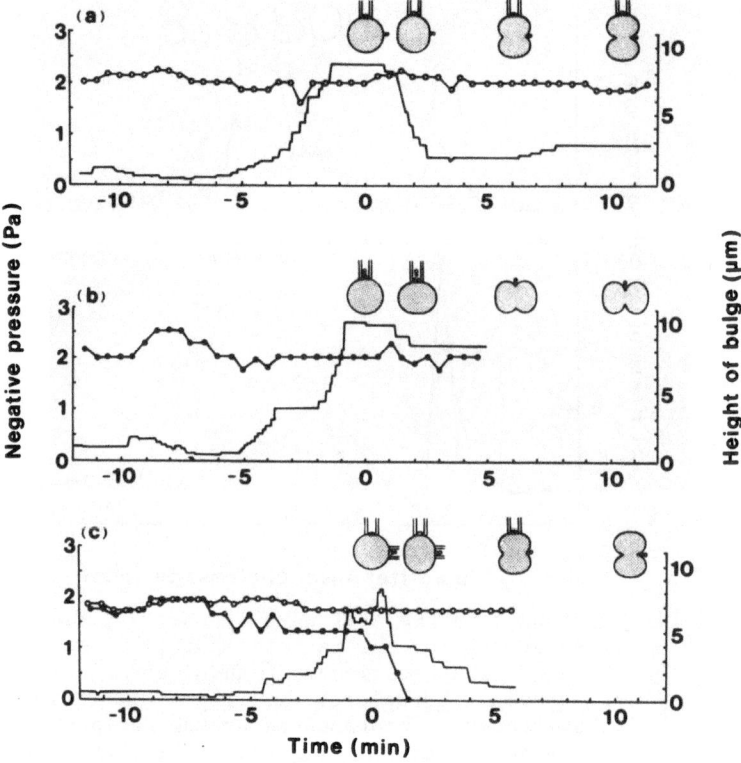

Fig.20 Changes in stiffness of the cell surface
determined by the suction method in eggs of the
starfish <u>Asterina</u> <u>pectinifera</u> before, during
and after cleavage. The stiffness is
represented by the negative pressure
(continuous lines) required to maintain the
height of the bulge formed at the cell surface
by application of negative pressure through a
micropipette closely attached to the cell
surface to be constant.   n a and c the height
of the bulge at the polar surface is kept
constant (open circles), while in b the height
of the bulge at the furrow surface is kept
constant (solid circles). Solid circles in c
represent the height of the bulge formed in a
micropipette attached to the furrow surface of
the cell under the same negative pressure as
that applied to the polar surface (cf. inset).
Stages of cleavage are represented on the top
of each figure. After Ohtsubo and Hiramoto
(1985).

formation increased dramatically when the polar body began to
be formed. Therefore, it may be concluded that two kinds of changes in
mechanical properties occur at the cell surface at the time of cell
division, both during cleavage and during polar body formation. One is
the change which occurs simultaneously over the entire surface and the
other is the change which occurs at the region which will form the
constriction for division. In the former, two peaks of stiffness were

found during cleavage of the sea urchin egg and one peak during polar body formation of the starfish oocyte and during cleavage in the starfish egg. The second kind of change may correspond to the formation of the contractile ring, consisting of aligned microfilaments, which starts almost simultaneously with the onset of division (cf. Schroeder, 1972). This indicates that the stiffness change and the contraction of the contractile ring during cell division may result from the same structural change at the cell surface.

CONCLUSION

From the measurements of mechanical properties of the endoplasm in echinoderm egg cells by three different methods (centrifuge method, magnetic particle method and capillary method), it is concluded that the cell displays characteristic changes in the consistency of the endoplasm at maturation, fertilization and cleavage.
At maturation, the endoplasmic consistency of the starfish oocyte decreases accompanying the breakdown of germinal vesicle, increases before the formation of the first polar body, decreases during and after the first polar-body formation, increases before the formation of the second polar body, and decreases during and after the second polar body formation. At fertilization, the endoplasmic consistency of the sea urchin egg displays a transient peak within a few minutes after insemination, and gradually increases accompanying the growth of the sperm aster after reaching a trough several minutes after insemination. The endoplasmic consistency decreases after it attains the second peak, and then increases accompanying the formation of the mitotic apparatus. The above changes and the regional difference in endoplasmic consistency within the fertilized egg strongly suggest that the consistency is closely correlated with microtubule structures formed in the endoplasm.
The stiffness of the cell surface, which is thought to be due to the tension and the visco-elasticity of the cortex and the cell membrane, also displays characteristic changes at maturation, fertilization and cleavage. At maturation, the stiffness of the starfish oocyte decreases accompanying the breakdown of the germinal vesicle, and transiently increases at the time of formation of the first and second polar bodies. At fertilization, the stiffness of the sea urchin egg displays a transient peak within a few minutes after insemination, then decreases and again increases after several minutes of postfertilization. The stiffness increases before the onset of cleavage over the entire surface. Stiffness peaks exist twice in sea urchin eggs at cleavage, one shortly before the onset of cleavage and the other at or shortly after the onset of cleavage, while the peak exists only once at the onset of cleavage in starfish eggs. The stiffness difference between the surface at the cleavage furrow and the surface at the pole develops simultaneously with the onset of cleavage, probably by the formation fo the contractile ring in the furrow cortex.

REFERENCES

Cole, K.S., 1932, Surface forces of the Arbacia egg, J. Cell. Comp. Physiol., 1:1.
Harvey, E. N., 1931, The tension at the surface of marine eggs, especially those of the sea urchin, Arbacia, Biol. Bull., 61: 273.
Hiramoto, Y., 1963, Mechanical properties of sea urchin eggs II. Changes in mechanical properties from fertilization to cleavage, Exp. Cell Res., 32: 76.
Hiramoto, Y., 1969 a, Mechanical properties of the protoplasm of the sea urchin egg I. Unfertilized egg, Exp. Cell Res., 56: 201.

Hiramoto, Y., 1969 b, Mechanical properties of the protoplasm of the sea urchin egg II. Fertilized egg, Exp. Cell Res., 56: 209.

Hiramoto, Y., 1970, Rheological properties of sea urchin eggs, Biorheology, 6: 201.

Hiramoto, Y., 1974, Mechanical properties of the surface of the sea urchin egg at fertilization and during cleavage, Exp. Cell Res., 89: 320.

Hiramoto, Y., 1979, Mechanical properties of the dividing sea urchin egg, in: "Cell Motility: Molecules and Organization", S. Hatano, H. Ishikawa and H. Sato, eds., Univ. Tokyo Press, Tokyo.

Hiramoto, Y., 1986, Determination of mechanical properties of the egg surface by elastimetry, in: "Methods in Cell Biology vol. 27, Echinoderm Gametes and Embryos, T. E. Schroeder, ed., Academic Press, New York.

Kamiya, N., and Kuroda, K., 1965, Rotational protoplasmic streaming in Nitella and some physical properties of the endoplasm, in: Proceedings of the 4th International Congress in Biorheology. Aug. 26-30 1963, John Wiley and Sons.

Kanatani, H., 1973, Maturation-inducing substance in starfishes. Intern. Rev. Cytol., 35: 253.

Mitchison, J. M., and Swann, M. M., 1954, The mechanical properties of the cell surface I. The cell elastimeter, J. Exp. Biol., 31: 443.

Mitchison, J. M., and Swann, M. M., 1955, The mechanical properties of the cell surface. The sea-urchin egg from fertilization to cleavage, J. Exp. Biol., 32: 734.

Nakamura, S., and Hiramoto, Y., 1978, Mechanical properties of the cell surface in starfish eggs, Develop. Growth and Differ., 20: 317.

Ohtsubo, M., and Hiramoto, Y., 1985, Regional difference in mechanical properties of the cell surface in dividing echinoderm eggs, Develop. Growth and Differ., 27: 371.

Schroeder, T. E., 1972, The contractile ring II. Determining its brief existence, volumetric changes, and vital role in cleaving Arbacia eggs, J. Cell Biol., 53: 419.

Shôji, Y., Hamaguchi, M. S., and Hiramoto, Y., 1978, Mechanical properties of the endoplasm in starfish oocytes, Exp. Cell Res.,117: 79.

ROLE OF PLASMA MEMBRANE FLUIDITY IN THE REGULATION OF CELLULAR ACTIVITIES

Giovanna Curatola and Giorgio Lenaz

Istituto di Biochimica, Facoltà di Medicina, Via Ranieri
60131 Ancona, °Istituto ed Orto Botanico, Facoltà di Scienze
Via Irnerio 42, 40126 Bologna (Italy)

INTRODUCTION

The plasma membrane, located at the interface between external
environment and the cell metabolic machinery, is the site of regulatory
events which control functional activities and cellular processes.
The involvement of the plasma membrane in cell growth, division and
differentation is well demonstrated in many different cell types, but a
detailed description at the molecular level as well as a unified framework
is lacking at present. In general it would be necessary to clearly
distinguish between plasma membrane properties which primarily influence the
progression of the cell cycle and the changes produced at the membrane level
as a passive consequence of morphologic and metabolic adaptation throughout
the different phases of the cell cycle.
Nevertheless, the interplay of these factors might be a specific feature of
each system analysed and of the events which induce the cell to leave the
resting state for a proliferative condition. Also intracellular membranes
have a crucial role in the regulation of cell activities, not discussed in
this report, which will try to outline a possible correlation between
changes in plasma membrane bulk properties and modifications of cell
functional activities occurring during cell division. A general view is
briefly summarized in Figure 1 and represents only a speculative model for
the purpose of discussion, showing that the role of the plasma membrane in
cell division is manyfold and involves both mechanical properties and
regulatory activities.
In discussing physico-chemical features of biomembranes, they are usually
viewed as a bidimensional fluid endowed with properties of continuum
structure in the molecular range; this approach is useful to obtain a
quantitative description and to predict overall behaviour, however in the
regulation of membrane functions local hindering of motion, segregation of
microdomains and specific interactions are extremely important on molecular
grounds. These factors are even more important if we want to relate membrane
functions with cellular activities specially during the cell cycle which is
characterized by changes in cell surface morphology and ultrastructure.

33

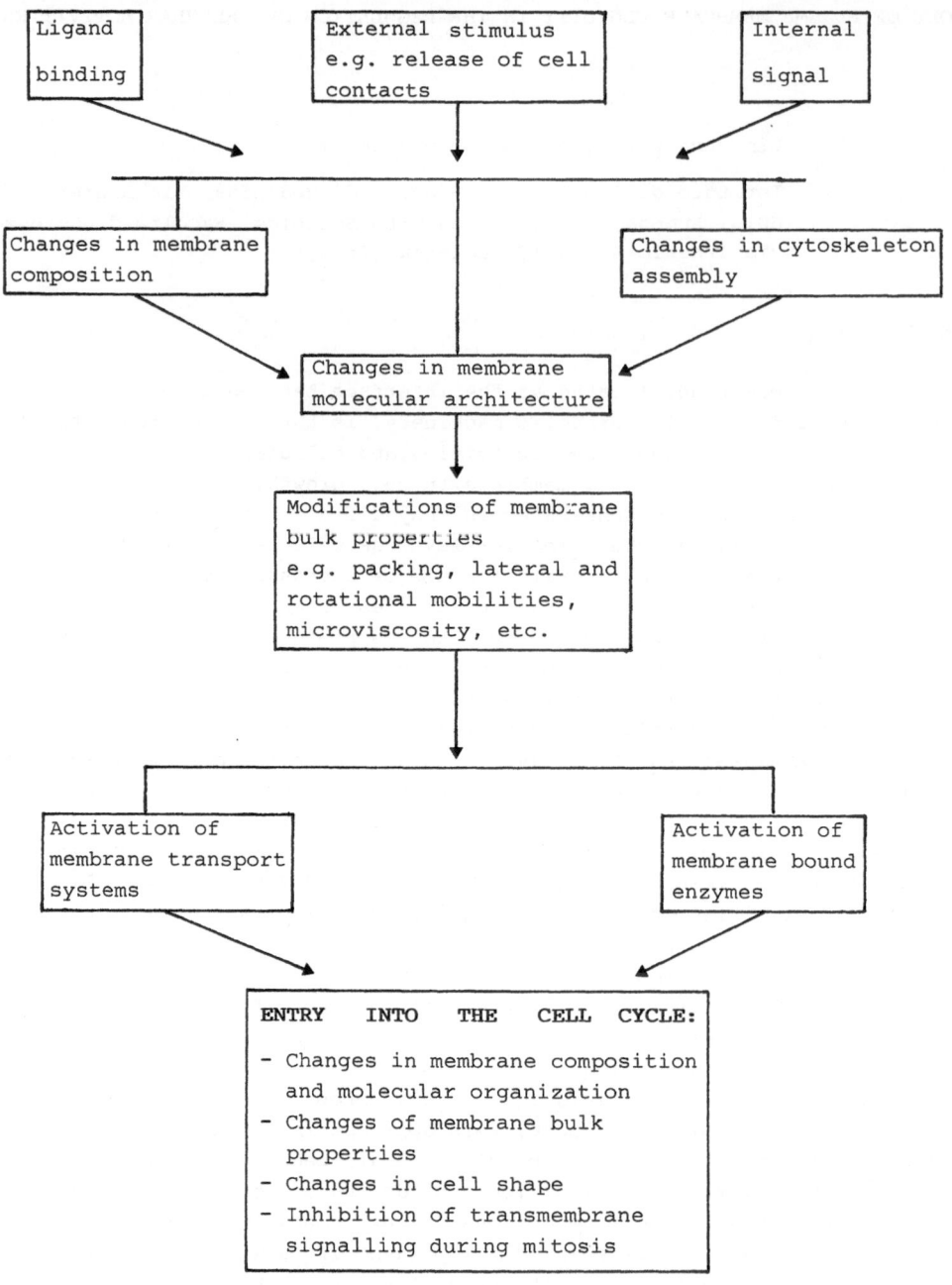

FIGURE 1. Plasma membrane involvement in cell activation and cell cycle.

Static and dynamic properties of biomembranes depend on the structural organization of membrane phospholipids.

Lipids, besides their general role as a bidimensional matrix in which proteins and other components are dispersed, have different functions related both to their bulk properties as liquid-crystalline solvent and to specific interactions with intrinsic or extrinsic membrane molecules[1-4] (Table I). For example, glycolipids are involved as binding sites and in the regulation of receptor clustering; lipid-protein interactions regulate membrane bound enzyme activities, and non-bilayer phases have been hypothesized to be required in complex membrane functions (e.g. fusion, pinocytosis etc.)

Thus phospholipid composition is the first determinant of biomembrane properties, as suggested by the observation that each membrane has its specific genetically determined composition of phospholipid headgroups and fatty acyl chains. Moreover, in each membrane phospholipids are asymmetrically distributed between the two monolayers, so that different phospholipid headgroups can be accomodated in the two membrane surfaces which have different curvature radii. The slow rates of phospholipid exchange (flip-flop) between the two leaflets imply that after a perturbation a biomembrane can relax to equilibrium very slowly unless a different structural organization or induced flip-flop is not acquired. Compositional changes, even if not relevant on a quantitative point of view, might modify membrane properties and membrane functional activities.

TABLE 1 - Role of lipids in membrane structure and functions

BULK	Relatively unspecific
	- Membrane assembly
	- General physical and chemical membrane properties (viscoelastic behaviour, microviscosity, permeability, surface potential, etc.)
SPECIFIC	Often only few molecules
	- Glycolipids as recognition sites
	- Binding sites for cytoskeleton attachment
	- Formation of non-bilayer structures
	- Regulation of membrane-bound enzymes and transport proteins (Solvent for lipid-soluble substrates or cofactors; compartmentation for vectorial reactions; anchoring device for 2 - dimensional collisions; to induce the optimal conformation for catalysis).

The knowledge of compositional changes in membranes, both plasmatic and intracellular, during the cell cycle is fragmentary at present although quantitative and qualitative modifications of many cellular components have been studied in detail. The data reported for different experimental systems do not offer any unified profile, suggesting some kind of specific behaviour for each cell type used.

In general, it is possible to distinguish compositional changes implicated in the mechanisms of external stimulus transduction during cellular activation and those related to biosynthetic cycling of the plasma membrane. The former are quantitatively small but elicit specific responses which influence the cell cycle activity and are related to membrane microdomain organization.

The study of plasma membrane biosynthesis throughout the cell cycle is particularly difficult due to the concomitant rearrangement of other membranes such as the nuclear envelope and endoplasmic reticulum in the late mitotic cell and to the necessity of establishing a correct relationship between time scale of cell cycle phases and half-life of phospholipid classes.

In this respect E.coli is a good model because it is devoid of intracellular membranes, the pool of lipid precursors is extremely small and the growth of the cell envelope may represent the essential process of cellular growth [5]. In all strains studied and under different growth conditions an asynchrony was observed in the rate of synthesis of membrane phospholipids, which increase two fold only 12-20 min after cell division, while cell replication initiates 5-10 min later [6].

During cell division a constant content of phospholipid phosphorus was observed in E.coli B/r CSH [7]. Moreover in this strain a considerable decrease of phosphatidylethanolamine and a simultaneous increase of cardiolipin content always occurs 15-20 min before the subsequent division when the chromosome replication is just completed. Similar patterns were obtained also in Saccharomyces cerevisiae [8] and in Bacillus megaterium [9].

Almost all dividing cells double their components and membranous structures prior to each mitosis; the biosynthesis of new molecules and their insertion can be simultaneous, so that no changes in composition and physico-chemical properties of plasma membrane have to be expected, or, due to a preferential synthesis during a specific phase of the cell cycle, one can observe modifications of the relative amounts of different membrane components[10,11].

In suspension cultured cells of mastocytoma P815Y Pasternak et al.[10] have shown a contemporary increase of phospholipids, cholesterol and protein between G_1 and G_2 without any changes of their relative concentrations, moreover the fatty acid composition of membrane phospholipids is unmodified throughout the cell cycle. During the interphase the increase of biosynthetic pathways is concomitant with a doubling of surface area and appearance of numerous microvilli. However the structural organization of plasma membrane observed as intramembranous particle distribution does not exhibit any substantial change also in microvillous regions.

On the other hand, the membranes of HeLa S_3 cells, grown in suspension culture, have a modified pattern of phosphatidylcholine (PC) synthesis[12]. In the recent mitotic cell the PC net synthesis, through cytidine-diphosphate-

choline is increased two or threefold. Moreover this biosynthetic
specificity introduces in the membrane more saturated fatty acids than the
PC biosynthetic pathway from phosphatidylethanolamine (PE) which
preferentially introduces polyunsaturated fatty acids. A large part of newly
synthetized PC is primarly localized in the nuclear envelope probably as a
step of its transport from the endoplasmic reticulum; the compositional
changes could be significant in the short time range of cell division due to
the short half-life of PC[13].

However in proliferating cells the maintenance of specific phospholipid
composition and cholesterol-phospholipid ratio independently of the growth
medium composition[14] indicates that the production of different membrane
phospholipid classes and their assembly into membranes is well coordinated
with DNA synthesis and cell-cycle. In human fibroblasts (WI-38) in culture,
the rate of lipid synthesis changes proportionally with the growth rate and
is equal for all lipid classes. The lipid synthesis decreases when cells are
arrested in G_1 due to an inhibition of cholesterol biosynthesis induced by
inhibitors of hydroxymethylglutaryl-CoA reductase[15] .

The availability of cholesterol both from exogenous sources or biosynthetic
origin is necessary for macromolecular synthesis which however declines
later than the lipid synthesis, indicating that a minimal sterol content is
necessary for cell survival in a resting state.

These observations have to be interpreted taking in account the role of
cholesterol as a primary regulator of membrane physico-chemical properties.
When the maintenance of proper membrane fluidity conditions is no longer
possible, it is communicated to the cell cycle control step.

To this respect the role exerted by the lipid protein ratio due to the
ordering effect of proteins on lipid bilayer is also revelant[16]; however
direct quantitative indications of membrane protein content changes
throughout the cell cycle are lacking.

Mouse neuroblastoma cells, clone Neuro-2A, during the different phases of
cell cycle, show changes of membrane ultrastructure which suggest specific
patterns of lipids and protein biosynthesis and spatial organization in
newly synthetized membranes, the surface of which increases gradually at
mid-S phase and roughly doubles before and around mitosis[11]. The appearance
of large domains devoided of Intra Membrane Particle (IMP) has been
interpreted as a preferential lipid over protein insertion from G_2
through M [17,18].

The discontinuos structure of these lipidic surfaces and the concomitant
presence of multilayer vesicles in the cytoplasm in continuity with internal
membrane surfaces demonstrate that mitotic membrane growth in neuroblastoma
cells is accomplished by incorporation and fusion of lipid membrane
vesicles. The insertion of membrane proteins is preferential from early G_1
until S phase as suggested by the gradual increase of IMP density which
reaches a maximum at mid-S[19]. Distribution of IMP occurs through the cell
cycle, with a random arrangement proceeding from G_1 to S when cells reattach
to substratum. The dynamic organization of membrane structure and the
asynchronous synthesis of membrane components have been correlated with
changes in their mobility and in membrane viscosity which will be discussed
later in this presentation.

Moreover in Chinese hamster ovary, incorporation of [^{35}S] sulfate into
glycosaminoglycans (GAG) is depressed during mitosis while it is stimulated

in early G_1 cells. The rate of GAG synthesis could be correlated with the process of detachement and adhesion of cells to substratum[20] confirming that specific components vary at selected times of cell-cycle[10].

In more complex systems, as regenerating liver, Bruscalupi et al.[21] have observed a marked decrease of cholesterol-phospholipid ratio, during G_1 phase, both in sham operated and in hepatectomized rats (Table 2). The value recovered to normal only for sham operated animals, whereas liver membranes from hepatectomized animals showed a decreased ratio throughout the S phase. Also the phospholipid-protein ratio was lower during liver regeneration at all times considered, whereas no significant change of the percentage distribution between different phospholipid classes was observed. The saturation index and the saturated/unsaturated fatty acid ratio appeared to be slightly modified by hepatectomy with an increase of saturation mainly due to a decrease of arachidonic acid more evident during G_1 phase. At the end of S phase the pattern was reversed and hepatectomized animals had a lower content of saturated fatty acids.

TABLE 2 - Changes in composition of rat liver plasma membranes during liver regeneration (SO sham operated, PH partially hepatectomized rats)

	G_1 phase		Early S phase		Late S phase	
	SO	PH	SO	PH	SO	PH
Cholesterol/Phospholipid*	0.15	0.16	0.18	0.13	0.22	0.17
Phospholipid/Protein*	0.31	0.23	0.34	0.26	0.33	0.25
Phospholipids**						
Lysophosphatidylcholine	–	–	1.88	2.73	2.37	2.00
Sphingomyelin	–	–	21.31	21.45	23.80	21.74
Phosphatidylcholine	–	–	35.55	37.50	37.68	38.87
Phosphatidylserine + Phosphatidylinositol	–	–	18.93	16.52	12.78	15.87
Phosphatidylethanolamine	–	–	22.30	21.70	23.30	21.31
Acyl groups						
Saturated/unsaturated fatty acid ratio	1.07	1.17	0.95	1.08	1.23	1.06
Unsaturation index***	1.31	1.16	1.40	1.27	1.21	1.29

*Values are expressed as a (w/w) ratio
**Values are expressed as percentage of total phospholipid phosphorus
***Average number of double bonds/fatty acid

In the same system no significant variations were detected in microsomal membranes obtained from hepatectomized animals except an increase of phosphatidylcholine and a concomitant decrease of phosphatidylserine plus phosphatidylinositol at the onset of S phase [22]. An increase of phosphatidylcholine was observed in microsomal membranes of hepatectomized rats[23], mainly due to an activation of transmethylating enzymes.
The results obtained on the composition of regenerating liver will be discussed in view of concomitant modifications of membrane fluidity and insertion of newly synthetized proteins[21].

Although fragmentary, the bulk of the observations on changes of membrane composition during cell cycle indicate the possibility that specific patterns accompany the different phases, suggesting a correlation between membrane component synthesis and assembly with the metabolic control of cell macromolecular synthesis. Modifications in membrane composition might be responsible for changes in membrane bulk properties; moreover specific compositional changes at the membrane surface might modulate some cellular properties.

TABLE 3 - Changes in composition of liver microsomes during liver regeneration (SO sham operated, PH partially hepatectomized rats)

	Early S phase	
	SO	PH
Cholesterol*	46.21	50.53
Total phospholipids*	682.30	778.01
Phosphatidylserine + phosphatidylinositol	17.55	13.85
Phosphatidylcholine	49.70	54.92
Phosphatidylethanolamine	22.97	21.63
Saturated/unsaturated fatty acid ratio	3.30	3.21
Unsaturation index**	1.31	1.25

*Cholesterol and total phospholipids are reported as μmoles/mg proteins; Phospholipid composition is expressed as percentage of total phospholipid phosphorus
**Average number of double bonds/fatty acid

The dynamic organization of membrane structure is determined by energetic and entropic factors which define the thermodynamics of membrane self-assembly and by molecular geometry[24,25], which dictates the modalities of membrane components interactions by steric constraints. The contributions of attractive and repulsive forces to the organization of the lipid bilayer breefly summarized in Figure 2, are different in the hydrophobic membrane core and in the hydrophilic headgroup region.

The interactions between phospholipid fatty acids packed in the two membrane dimensions are dominated by attractive van der Waals forces and anisotropic steric repulsions, as demonstrated by the increase and decrease of transition temperature from crystalline to liquid-crystalline state, increasing the chain length and the degree of unsaturation respectively. Above transition temperature the chain conformational disorder increases while below this temperature steric costraints prevail so that lipid physical properties are not symmetric around transition[25] . Following Marcelja[26], the molecular energy (E) can be described by the molecular orientation and ordering energy :

$$E = V_o \phi \left(\frac{3}{2} \cos^2\theta - \frac{1}{2} \right)$$

where ϕ indicates the strength of the molecular field, θ is the molecular orientation relative to the direction of the molecular field and V_o a coupling constant. The degree of interaction depends on the possibility of rotation around each C-C bond with two energetically favourable conformations. In the trans-state the packing is favoured while the gauche conformers, by bending fatty acyl chains, decrease van der Waals interactions[27] . Moreover segmental conformational motions in lipid bilayer have to be correlated with freedom of motion of the whole molecules as suggested by Vaz and Doane[28] .

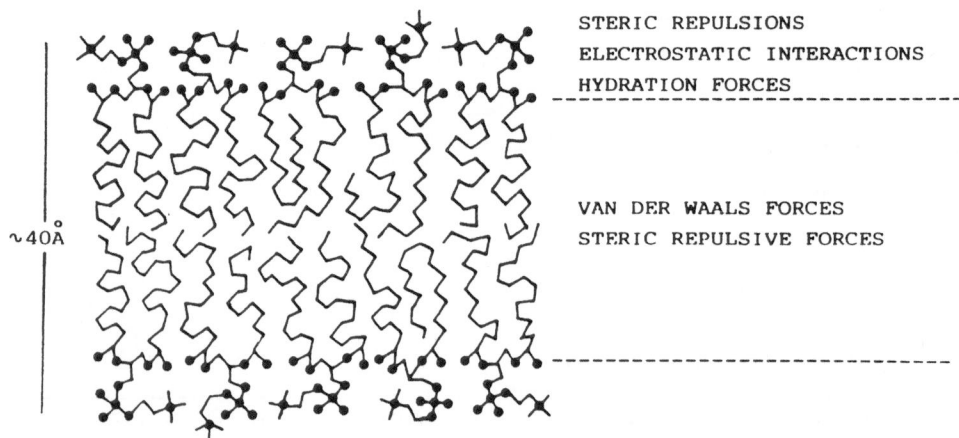

STERIC REPULSIONS
ELECTROSTATIC INTERACTIONS
HYDRATION FORCES

VAN DER WAALS FORCES
STERIC REPULSIVE FORCES

$\sim 40\text{Å}$

FIGURE 2. Forces contributing to the organization of lipid bilayer

In model membranes constituted by a single lipid species the head-group structure is similar to that of lipids in crystalline state[29] ; however in mixed systems NMR has shown an increased motion of phosphatidylserine in the presence of lecithin suggesting that the polar region structure strongly depends on mutual interaction between neighbouring molecules[30] .

Besides repulsive forces between homologous charges of phospholipids or of ions in the water phase or the dipolar molecules of water, strong repulsions can be represented by steric constraints which depend on molecular size and on the degree of water structuring at the bilayer surface. Attractive and repulsive electrostatic interactions are modified by the degree of charge dissociation or by the presence of counterions in the medium; however the value of surface potential (60-90 mV) does not vary substantially with membrane and electrolyte or solvent composition, indicating that the membrane surface is not fully ionized[31].

Acyl chain and head group structures are strictly related as it can be desumed by some observations discussed below. PE has a higher Tm than PC with the same fatty acid compostion due to the large hydrated choline headgroup which decreases hydrocarbon chains interactions[32] ; lecithin and other phospholipids during pretransition show tilted acyl chains which can be better packed than perpendicular structures in view of the presence of bulky headgroups [33].

In bilayers lipid chains have a more extended structure than in hydrocarbon systems suggesting an effect of lateral pressure[26] .

Energetic and entropic requirements dictate the mechanism of membrane assembly: following Tanford's approach[24] two opposing forces act at the interface between hydrophobic tails and the water phase. The molecular heagroup area, a , is determined by the interfacial tension forces (Figure 2).

The correlation between the mean free energy per molecule μ^0_N, the interfacial energy per unit area γ and the interfacial area per molecule (a), are represented in the following expression

$$\mu^0_N = \gamma_a + C/a$$

where γ_a and C/a are respectively the attractive and repulsive contributions. Israelachvili et al[34] , assuming both forces acting in the same plane, have calculated that the minimum free energy is given by μ^0_N (min) = $2\gamma_{a_0}$ where $a_0 = \sqrt{C/\gamma}$ identifies optimal surface area per molecule when the total interaction energy is at minimum.

The free energy can be expressed by

$$\mu^0_N = 2\gamma_{a_0} + \frac{\gamma}{a}(a - a_0)^2 ;$$

the effective area occupied and the headgroup optimal area identify the elastic contribution from molecular packing[35] defined as

$$\text{elastic energy} = \frac{k_c}{2a}(a - a_0)^2$$

where Kc is the elastic compressibility modulus.

The values calculated by Israelachvili et al[34] are in agreement with experimental results of Evans and Hochmuth[36] in lecithin bilayer and erythrocyte membranes.

Actually attractive and repulsive forces do not act on the same plane so that the total free energy depends on the distance between the center of applied forces .

When the center of repulsive forces is located at a distance D away from the interface, the mean free energy per molecule μ_N^0 becomes

$$\mu_N^0 = \gamma_a + \frac{C}{a\,(1 + D/R)}$$

where R is the surface curvature radius and C is a constant; correcting factors have to be taken into consideration to introduce optimal area and packing costraints for low and high curvature value[37] . Curvature elasticity has been described in detail using the continuum approach by Evans and Skalak[38].

In biological membranes the spontaneous curvature may be modified by the adsorbing and desorbing of ions and macromolecules on both surfaces or by changes in local pH and ionic strength so that elastic stress may be rapidly relaxed, as shown in monolayers, by rapid transfer of molecules between membrane and water phase [39]. Neglecting the contribution of shear deformation, a more simplified expression of changes in elastic energy associated with changes in spontanous curvature has been calculated by Helfrich[40] and Petrov [41]

$$\text{Elastic energy} = \tfrac{1}{2}\, k\, (C_1 - C_2 - C_0\,)^2 - k'C_1 C_2$$

where C_0 represents the spontaneous curvature and C_1, C_2 the membrane principal radii of curvature, k and k' are the curvature elastic constants. Changes in membrane elastic energy have been associated with fusion processes, internalization of coated pits, or accomodation into the bilayer of external molecules [39] .

Geometrical Constraints and Membrane Packing Properties
_____

The optimal head group area (a_o), the volume occupied by fatty acid chains (v) and the critical chain length (l_c), that is the chain length at maximum extension, characterize the geometric features of membrane phospholipids. All have been calculated in model systems[42] and of course are correlated with interacting forces and depend on conditions which can modify headgroup size and acyl chains conformations, for example changes in headgroup ionization or changes in degree of trans-gauche isomerization.

Using the three geometrical parameters a dimensionless value can be obtained, the packing parameter $[v/(a_o \cdot l_c)]$[43].

This parameter will determine the packing shape and the overall shape of the structure acquired by the assembling lipids. Following Israelachvili et al[34] ,at defined values of packing parameter a structure is acquired with minimum water hydrocarbon contacts, minimum steric constraints and the lowest free energy level (Table 4).

The influence of lipid geometrical features on membrane properties has been recognized of biological importance through the studies of Verkeley[44] and Cullis et al[45] . These authors have shown that in model systems, inverted micellar and hexagonal phases can be acquired by lipids normally present in biological membranes, such as cardiolipin following Ca^{2+} interaction or PE increasing the temperature. Non-bilayer phases could be involved in fusion processes, in controlling protein aggregation, and in ion transport[46] . Although the formation of hexagonal phase is difficult to demonstrate in biological membranes, indirect evidence points out its role in cytochrome P450 insertion into microsomal membranes[47]

TABLE 4 - Lipid packing properties and related structures

Packing parameter	<1/3	1/3-1/2	1/2-1	∿1	>1
Packing shape	Cone	Truncated cone or wedge	Truncated cone	Cylinder	Inverted truncated cone
Structures	Spherical micelles	Globular or cylindrical micelles	Flexible bilayer vesicles	Planar bilayers	Inverted micelles, hexagonal phase

Recently a model for a correlation between phospholipid fatty acid composition, lipid geometry and cell shape has been worked out by Kuypers et al[48]. The ratio GI between cross sectional area of hydrophobic tails and polar head defines a geometrical index of packing. Using the model of coupled bilayer[49], the membrane curvature (M_c) can be described by the following expression

$$M_c = \frac{\sum_{i=1}^{n} (GI)}{n} : \frac{m}{\sum_{i=1}^{m} (GI)}$$

where the first and the second terms refer to the average geometrical index and n and m to the number of molecules, respectively of external and cytoplasmatic leaflets; the value of 1 is acquired when the membrane is completely flat, while different values reflect an increase of membrane free energy level. Changing packing features of membrane phospholipids in absence of internal rearrangement by flip-flop, membrane structure relaxes modifying its curvature profile. The model fits well to the results obtained in erythrocyte membranes[50,51] in which changes in PC fatty acid composition induce shape changes (Figure 3).

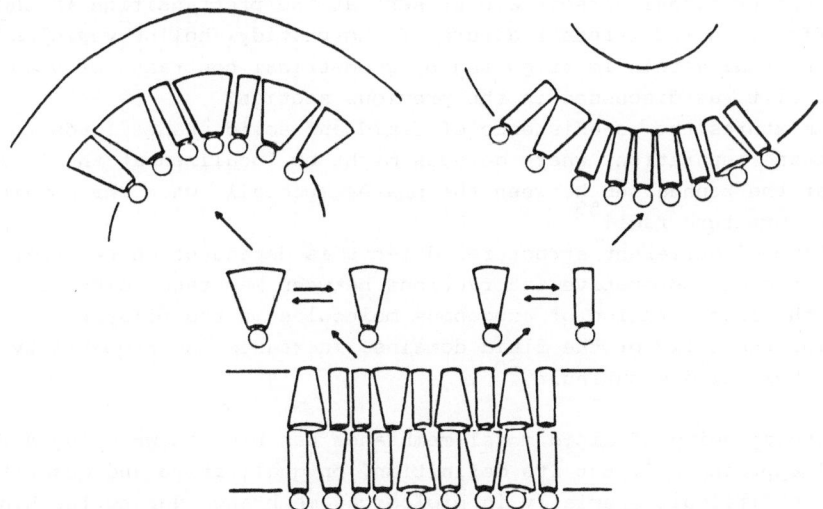

FIGURE 3. Influence of phospholipid molecular geometry on membrane curvature

The erythrocyte native PC species contains one saturated and one unsaturated fatty acid and have a moderate cone shape; the substitution with 1,2 diunsaturated species, which have a more pronounced cone shape causes an inwards bending of the outer layer.
On the contrary, the substitution with saturated PC species which have cylindrical shape, induces an outwards membrane bending.

Membrane Microheterogeneity

Membrane physical and chemical properties are usually measured by averaged values which describe the membrane as a whole, following the Singer and Nicolson model of a homogenous fluid matrix[52]. However the molecular mechanisms of many functional activities, specially related to signal transduction, imply a heterogenous distribution of membrane components, both lipids and proteins, with a mosaicism of membrane properties. Membrane heterogeneity can be described at different levels taking in account intrinsic and extrinsic factors which are involved in its regulation.

In analogy with lattice defects of crystals, structure imperfections are present also in single component bilayers: they have been described as point defects, line defects and grain boundaries [53]. Point defects in the hydrophobic region can be due the trans-gauche transition; the resulting free volume might be important for the diffusion of small molecules as defined in the random walk model[54]. The texture defects increase gradually near the main transition (Tm) when regions of disordered molecules are present in the context of ordered lipids[27], and thir presence explains an increased membrane permeability to organic molecules and ions[55,56].
In large vesicles, electric fields or osmotic pressure induce the formation of hydrophilic pores[57], associated with an increase of elastic energy due to the phospholipid reorientation[58] ; it is possible to find similar conditions in biological membranes during fusion processes[59]. In erythrocyte membranes the pore formation following lysolecithin action is due to the truncated cone shape of this phospholipid derivative which accumulates at the pore surface. Orientational defects are present at the pretransition as shown by X-ray diffraction and freeze fracture of phosphatidylcholine vesicles[53]; the bilayer internal strain is triggered by geometrical costraint of head groups and chain tilt, as discussed in the previous section.
In lipid mixtures the co-existence of fluid and solid phases leads to lateral phase separation, where defects might be localized at the dilatation regions at the boundaries between the phases specially when the domains have different curvature radii[59] .
The lifetime of different structural defects is dependent on temperature and on the degree of cooperative interactions between membrane molecules. Moreover the incorporation of exogenous molecules in the bilayer, stabilizing the solid or the fluid domains, increases the probability of membrane physical discontinuity[60] .

The heterogeneity of lipid model membranes has been shown using different technical approaches[61]. But its definition, on qualitative and quantitative grounds, is difficult specially in biological membranes due to the high compositional heterogeneity and complex intermolecular interactions.

Among many factors which take part in membrane lateral organization, lipid-protein interactions are of utmost importance on a functional point of view. The influence of proteins on bilayer structure are briefly schematized in Figure 4.

Extrinsic proteins following electrostatic interactions, might modify charge distribution (A) as in the case of the myelin basic protein binding with negatively charged phospholipids[62] or modify phospholipid order (B) and change local curvature (C)[63,59]. The association of proteins with a specific phospholipid contributes to the formation of compositional domains. The incorporation of integral proteins alters the structural order of lipid chains decreasing the rate of trans-gauche transition, and the rotational mobility (D)[64,65]. Moreover a change in the average lipid orientation respect to the membrane plane, producing a disordering effect, is also possible specially for proteins with noncylindrical shape (E).

Recently from data obtained with cytochrome P450[47] it has been hypothesized that an inverted micellar structure might surround an integral membrane protein with a distortion of the neighbouring bilayer (F); the dilatation regions located at the interface between the two different phases have to be endowed with peculiar permeability properties.

The presence of a slow exchangeable phospholipid pool (annulus) around membrane proteins is still debated.

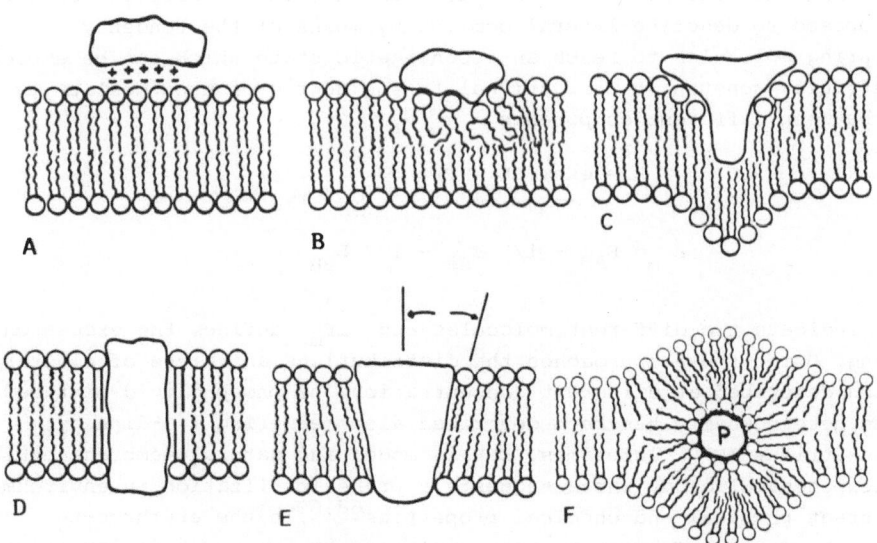

FIGURE 4. The influence of proteins on bilayer structure. In F, P indicates a protein incorporated by means of a non bilayer structure.

The immobilizing effect of proteins an membrane lipids, shown by Electron Spin Resonance[64] has been interpreted by Chapman et al[65] only as a consequence of increased lipid trapping at increasing protein concentration, so that local lipid mobility is restricted by a high environmental viscosity.

Besides the organizational defects induced by protein interactions with lipids, the presence of structural defects in bilayer matrix associated with elastic distortion energy might also contribute to protein distribution in the membrane plane with a preferential localization in the defect sites. The binding sites could be represented by membrane regions rich of gauche conformers; the protein incorporation might lead to an increase of trans to gauche ratio, minimizing the lateral compression.

As we have shown, a non-random distribution of membrane components is suggested by features intrinsic to the structure of phospholipid molecules, to the phospholipid-phospholipid interactions and to the interactions of lipids with other intrinsic and extrinsic membrane molecules. Uptodate three models have been advanced which take in account membrane domain organization; their specific aspects have been recently reviewed[61].

In the "plate model" proposed by Jain and White [53] the membrane is viewed as a mosaic of aggregates called plates or patches. The ordered domains are separated by fluid or disorganized regions; they exchange molecules directly and indirectly. Therefore their extension in the plane of the membrane, the composition and the physical properties might vary in the same membrane from plate to plate also with time. The domains could be made up of few molecules due to the possibility of a cooperative behaviour[66] or could be extended through the membrane by long range intermolecular interactions. To obtain direct evidence to support the plate model it is necessary to measure physical and chemical properties of different domains. Freire and Snyder[67] have proposed to describe lateral domains by means of the tendency of neighbouring molecules to reach an isoenergetic state which can be expressed by an affinity constant P proportional to the degree of non ideality in the mixing between different components

$$P = \exp\left(\Delta E_m / RT\right)$$

where

$$\Delta E_m = E_{AB} - 1/2\ E_{AA} - 1/2\ E_{BB}$$

A and B indicate two different molecules and ΔE_m defines the excess energy of mixing. Using these approaches the distributions and sizes of clusters have been described at different concentrations of phospholipid mixtures. Membrane heterogeneity has been described also measuring 1,6-diphenyl 1,3,5-hexatriene (DPH) lifetimes. DPH in model and natural membrane exhibits a multiexponential decay which suggests a probe localization in environments of different physical and chemical properties[68,69]. In the erythrocyte membrane the degree of heterogeneity evaluated by the width of DPH lifetime distribution is dependent on the cholesterol to phospholipid ratio, increasing at low cholesterol content[70].

Schiffmann advanced an alternative hypothesis of membrane structure in
different functional conditions based on theoretical considerations[71] and on
experimental observations, showing different distributions of IMP in the
resting with respect to the active state of crayfish gap junctions[72] . The
membrane structure fluctuates between two non equilibrium conditions, one
defined as the basal state with a random distribution of membrane components
corresponding to the Singer and Nicolson model and the other, triggered by
the binding of ligands, characterized by ordered molecular pattern defined
as lattice model. In this model the control of functional activities depends
on protein distribution.

Finally in another model Loor[73] takes in account the role exerted on
membrane domain mosaicism by the cytoskeleton connections with components of
biomembranes. Domains are divided in to three types. In the first class
domains the elements, as in fluid mosaic model, are free to move in the
membrane plane, they have a buffering effect compensating for quantitative
changes in the other two domain types. However they might segregate
following interactions with extrinsic proteins as fibronectin or cytoplasmic
proteins. Microfilaments control directly or indirectly the movements of
second class domain molecules, which are involved in microvilli expression
or in endocytosis or exocytosis events. Domains of the third type are
dependent on the linkages with microtubules, they are very stable and can be
modified only after microtubule depolimerization or dissociation; these
domains may be responsible for segregation of metabolic functions.
The role of cytoskeleton and cytomusculature in maintaining segregation of
membrane components has been demonstrated not only in polarized cells but
also in free cellular elements as erythrocyte and leukocytes where a random
distribution of membrane components could be expected.

Membrane Fluidity

The dependence of membrane protein activities and of the related
functional processes on membrane structural organization and physical
properties has been recognized although a detailed description at the
molecular level is still lacking. Among membrane properties, fluidity is
very frequently correlated with functional changes. This parameter does not
simply,indicate for membrane systems,the reciprocal of shear stress between
fluid layers or viscosity as in true fluids, expecially for the
bidimensional and anisotropic nature of bilayers. Fluidity is related to the
motional freedom of a segment or of the whole molecular structure; for
phospholipids organized in a membrane the possible motions are summarized in
the following:

a) Trans gauche transition. It has a frequency of $10^{-9} - 10^{-10}$ s and diffuses
   very rapidly (D = 10^{-5} cm^2/s) along molecular axis.

b) Segmental motion. Oscillation and flexing around an axis perpendicular
   to the bilayer plane. Time range $10^{-6} - 10^{-8}$ s. NMR results[74] indicate
   that the order parameter remains relatively constant along the chain
   decreasing only near the end of the chain. On the contrary spin label

technique shows a fluidity gradient with increased mobility suggesting an immobilizing effect by the headgroups. This controversial results suggest that separate order parameters are necessary to describe the motion of the molecular frame and segmental motion with respect to the frame. Headgroup flexibility has not been analyzed in physical details, different segments exhibit different motional properties, although the overall headgroup   rotational mobility is strongly restricted.

c) Rotational motion around an axis perpendicular to the membrane plane. It ranges from 10^{-8}s to 10^{-9}s in different systems.

d) Lateral diffusion in the membrane plane. The diffusion constant for lipid system ranges between 10^{-7}- 10^{-9}cm^2/s.

e) Translational motion from one monolayer to the other. It varies significantly from model systems to biological membranes in which transversal mobility is hindered by membrane component linkages to the cytoskeleton .

The phospholipid mobility is strongly dependent in biological membranes on interactions with other membrane molecules as proteins and cholesterol[75] .

Besides this dynamic aspect, fluidity implies a structural information related to the order of hydrocarbon chains and molecular packing, as discussed in previous sections; therefore different aspects are involved in the definition of membrane fluidity, which are strictly interrelated, as discussed by van der Meer[76].

In discussing the physical meaning of membrane fluidity it is important to realize that the different technical approaches, used to quantify it, give informations related to the range of motion (order) or the motional rate (mobility), or both, as shown in Table 5, so that the physical parameters measured might generate from an overlapping of dynamic and static contributions. Moreover the use of the same parameter in different techniques not necessarily implies physical equivalence[77] .

The technique most frequently used in the study of the relationship between membrane fluidity and membrane function has been fluorescence using DPH as molecular probe. The application in biological systems has been developed by Shinitzky et al[78] following the observation that fluorescence polarization is related to a submacroscopic property of the membrane $\bar{\eta}$ (the so called "microviscosity") in the following equation:

$$\frac{r_o}{r} = 1 + \frac{\tau}{\phi} = 1 + C(r)\ \frac{T \cdot \tau}{\bar{\eta}}$$

where r is the measured fluorescence anisotropy, r_o is the limiting value of r at infinite viscosity; T is the absolute temperature; τ is the excited state lifetime of the probe; ϕ is the rotational relaxation time; (C)r is a complex structural parameter. Therefore the steady state anysotropy can be resolved into a static part, related to order, and a dynamic part related to probe mobility.

48

The fluorescence lifetime is usually considered constant in biological membranes but the possibility to measure it with reliable and relatively easy techniques has recently shown that it is sensitive to the dielectric

TABLE 5 - Techniques used in membrane fluidity measurements

Technique	Physical parameters	
	Order	Mobility
NMR	S_{C-D}	τ_c
EPR	S	τ_c
FLUORESCENCE		
- time resolved	r_∞, θ_c, S	ϕ
- steady state anisotropy*	P, r_s, η	
- photobleaching (FRAP)	-	D_L
- excimer formation	-	D_L
RAMAN SPECTROSCOPY	S_{trans}	-
DIFFERENTIAL SCANNING CALORIMETRY	T_m, ΔH	

S_{C-D}, S, order parameter; r_∞ residual equilibrium anisotropy; θ_c, cone angle; P, polarization value; r_s, steady state anisotropy; η, microviscosity; τ_c, rotational correlation time; ϕ, rotational relaxation time; D_L, lateral diffusion constant; T_m, main transition temperature; ΔH, enthalpy changes. *Parameters relate to both range and rate.

gradient of the membrane bilayer and can be modified by changes of permeability of the membrane.
Therefore in the fluorescence polarization measurements both order parameter and rotational diffusion of the probe, which in turn is influenced by packing and permeability properties, are taken in account. These observations suggest that the technique could be very useful if the components of the physical parameters measured are properly analyzed.

The mobility of membrane-bound molecules appears essential for many biological functions[79]. The concept of membrane fluidity implies that all membrane constituents exist in a dynamic state of relatively free mobility of lipids and proteins. However, membrane fluidity is not distributed homogeneously but varies within lateral domains in the plane of the bilayer as well as transversally across the membrane; indeed it appears that the organizational problem most membranes must cope with, is not to maximalise motion of their component molecules, but to control it by restrictions allowing the motions to take place only for some molecules and at the times of specific physiological requirements.

General Theory

Mobility of membrane components includes both their lateral two dimensional displacement in the plane of the membrane (lateral diffusion, characterized by a lateral diffusion coefficient D_L) and rotational motion about an axis perpendicular to the plane of the membrane (rotational diffusion, characterized by a rotational diffusion coefficient D_R)[76]. Lateral and rotational diffusion will be treated together since they are closely related, being expression of the same general phenomenon; we will first illustrate some general useful parameters.

Brownian motion or diffusion is the random movement of a particle due to exchange of thermal energy with its environment, so that both its position and its orientation will exhibit noise. A rigid object in space has three positional and three angular coordinates, each being a randomly fluctuating function of time. For a spherical particle in a slow viscous motion in a three-dimensional viscous fluid we have:

$$D_L = \frac{KT}{6\pi\eta R} \qquad \text{and} \qquad D_R = \frac{KT}{8\pi\eta R^3}$$

where K is Boltzmann's constant, T is the absolute temperature, η is the viscosity and R is the particle radius. D_L has the dimension of cm^2/s and D_R of s^{-1}. The ratio between these two mobilities is indipendent of the viscosity:

$$\frac{D_L}{D_R} = \frac{4}{3} R^2$$

In biological membranes treatment of the mobilities is complicated by the reduction of dimensionality and by the anisotropic nature of the medium (the lipid bilayer).

A hydrodynamic model (Figure 5) for a membrane has been given by Saffman and Delbrück [80] for a cylindrical object embedded in a viscous continuum fluid sheet bounded by an aqueous fluid. Such a particle is restricted to move laterally in the x-y plane and to rotate around the z axis (D_R has two components, $D_{\parallel}$ and $D_{\perp}$, but only the former has been proven to take place, since rotation around the x or y axis involves an excessive energy requirement).

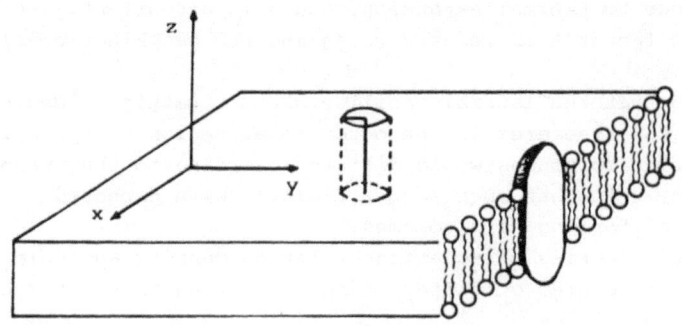

FIGURE 5. Hydrodynamic model for lateral diffusion of a membrane protein

Assuming the viscosity of the membrane η is much lower than the viscosity of the outer medium η', the following equations were derived:

$$D_R = \frac{K\,T}{4\,\pi\eta\,a^2\,h}$$

and

$$D_L = \frac{KT}{4\,\pi\,\eta\,h} \cdot \left(\log \frac{\eta h}{\eta'a} - \gamma \right)$$

where K and T have the meanings as before, h is the membrane thickness which as been equated to the height of the cylinder, a is its radius, and γ is Euler's constant (0.5772). The model assumes that the viscosities of the fluid bathing the membrane besides that of the lipid phase itself determine protein lateral diffusion, but not rotation. Therefore it is found that

$$\frac{D_L}{D_R} = a^2 \left(\log \frac{\eta h}{\eta'a} - \gamma \right)$$

The lateral diffusion of lipids or lipid-like molecules and of hydrophobic molecules embedded in the lipid bilayer, however, is not expected to obey to the Saffman-Delbrück model in depending on the viscous drag from the outer medium. In the case of a diffusant comparable in size to the solvent, the free volume theory of Cohen and Turnbull[81] and Montroll[82] seems to apply best [54,83]. According to this theory, the diffusion of a molecule in a fluid system may be divided into a three-step process: (a) creation of local free volume by density fluctuations which open up a hole within the cage where a solute molecule is situated; (b) the jump of the diffusing molecule into this hole, creating a void at the previous position; (c) the filling of the void by another solvent molecule. The lateral mobility of amphipatic molecules in a fluid lipid matrix will be determined by the free area according to

$$D_L = A_{exp} \left[\delta a^*/a_{g\,(T)} \right]$$

where a^* is close packed area per molecule and $a_{g\,(T)}$ is the mean free area per molecule at a given temperature T, calculated as the increase in free area at the phase transition plus the increase above the phase transition

temperature due to thermal **expansion**, and   is a constant. The
preexponential factor A is related to a* and the gas kinetic velocity of the
diffusant.

It is expected that the lateral mobility of amphipathic molecules will be
determined by the free area in the outer polar region of the bilayer,
whereas non polar molecules would diffuse according to the freedom of the
central region, which has much larger fluidity with expected higher
probabilities of forming void volumes.

Berg[84] provided a series of equations relating density and molecular
dimensions for molecules of different shapes moving in two dimensions (Table
6).

Lateral Diffusion

The first evidence on lateral diffusion of membrane proteins was obtained
in a classical experiment by Freye and Edidin[85]. The surface antigens of
two different cell lines were visualized by labelling with two types of
fluorescent antibodies and the cells were fused with inactivated Sendai
virus; after fusion the antigens became rapidly intermixed, and from the
mixing rate a D_L of $10^{-10} cm^2$/s was calculated.

The method of choice for measuring lateral diffusion of proteins has become
fluorescence recovery after photobleaching (FRAP)[79]. A method similar to
FRAP was first used for rhodopsin in the outer segment of rod disc membranes
[86], yielding D_L $3.5 \times 10^{-9} cm^2$/s.

TABLE 6 - Relationship of lateral diffusion with viscosity and molecular
          size

Sphere (Stokes-Einstein)	$D = \dfrac{KT}{6 \pi \eta r}$
Disk moving face on	$D = \dfrac{KT}{16 \eta a}$
Disk moving edge on	$D = \dfrac{3KT}{32 \eta a}$
Disk moving at random	$D = \dfrac{KT}{12 \eta a}$
Ellipsoid moving lengthwise	$D = \dfrac{KT}{4 \pi \eta a} \ln \dfrac{2a}{b} - \tfrac{1}{2}$
Ellipsoid moving sidewise	$D = \dfrac{KT}{8 \pi \eta a} \ln \dfrac{2a}{b} + \tfrac{1}{2}$
Ellipsoid moving at random	$D = \dfrac{KT}{6 \pi \eta a} \ln \dfrac{2a}{b}$
Cylinder with viscous drag from outer medium (Saffman and Delbrück)	$D = \dfrac{KT}{4 \pi \eta h} \ln \dfrac{\eta h}{\eta' a} - \gamma$

Since only few proteins contain natural chromophores, the method can be applied to nonchromophoric proteins by attaching suitable covalent probes to them[79,87,88]. The method involves photochemical bleaching of the chromophore in a small circular region of few microns of the membrane with a strong pulse of laser excitation transmitted through a microscope. As diffusion occurs the fluorescence intensity of the area, monitored with low level excitation from the laser, increases; the diffusion coefficients can be calculated by the recovery curves, since D_L is inversely proportional to the half-time of fluorescence recovery. The disadvantage of chemical modification of the diffusing molecule may be negligible for bulky proteins, but of major importance for smaller molecules.

A limitation of the method is that D_L measured by FRAP in nonplanar membrane surfaces, as is often the case in membranes with microvilli or invaginations, is underestimated, representing the mobility in projected flat planes of real nonflat membranes[89]. The method is unsuitable for membranes of small diameter, like subcellular organelles, unless they are modified to increase the size by fusion or other means.

As pointed out by Mc Closkey and Poo[90], the possibility that biomembranes are laterally inhomogenous on short distances[91] makes FRAP only suited for measuring long-range ($\geqslant 1\mu m$) lateral diffusion. Since chemical reactions and other collision-dependent interactions are probably more directly related to local than to long-range diffusion[90,92,93], this represents a real shortcoming of the FRAP method.

Protein diffusion is also detected by a method combining electrophoresis and freeze-fracture electron microscopy[94,95]; an electric current passed through microsuspensions of membranes causes the electrophoretic migration of the intramembrane particles in the membrane plane into a single patch at the anode; the membranes are quick-frozen at specific times after release of the electric field, while the particles are diffusing back to a random distribution. Calculation of D_L is then based on a theoretical model of particles diffusing laterally in a spherical membrane according to Fick's law.

The study of short-range lateral diffusion by other techniques widely used for lipids has not yet been extended to proteins, and the techniques seem rather unsuitable for the study of proteins. Such techniques involve EPR line broadening of spin labels[96], NMR[97], pyrene excimer formation[54], fluorescence collision quenching[98], triplet-triplet annihilation[99], and others[54,88]. The methods involve the determination of the rate of collisional encounters between two molecules, and therefore can only measure local short-range diffusion, as is involved in chemical reactions and molecular diffusion-dependent associations.

It is likely that dieletric spectroscopy[100] is able to measure both short-range and long-range diffusion, but the applicability of the method has yet to be defined to measure reliable diffusion coefficients.

Table 7 summarizes the diffusion coefficients for a variety of protein systems by different methods. The D_L of lipid usually range between $\geqslant 10^{-7}$ and $10^{-8}cm^2/s$[88]; they appear usually to fit the free-area theory as discussed before[83] and are slightly affected by the density of proteins in the membrane, only within the theoretical increase of viscosity induced by protein concentration.

TABLE 7 - Diffusion coefficients of some membrane proteins

Component	D (cm^2/s)	Method
Lectin receptors	1.1×10^{-11}	Photobleaching
Surface antigens (mast cells)	2.1×10^{-10}	Photobleaching
Surface antigens (mouse eggs,unfertilized)	1.6×10^{-9}	Photobleaching
Surface antigens (mouse eggs,fertilized)	10^{-11}	Photobleaching
Rhodopsin	2.6×10^{-9}	Photobleaching
Erythrocyte ghost proteins	3×10^{-12}	Photobleaching
Fibroblast proteins	2.1×10^{-10}	Photobleaching
Hormone receptors (fibroblasts)	$3-5 \times 10^{-10}$	Photobleaching
Surface antigens (mouse-human heterokaryons)	2×10^{-10}	Fusion
Erythrocyte ghost proteins	4×10^{-11}	Fusion
ConA receptors (embryonic muscle)	$4-7 \times 10^{-10}$	Electrophoresis
Mitochondrial proteins	8.3×10^{-10}	Electrophoresis
Integral proteins (normal erythrocytes)	4.5×10^{-11}	Photobleaching
Integral proteins (spherocytic erythrocytes)	2.5×10^{-9}	Photobleaching

For dilute suspension of noninteracting speres, Einstein[101] derived the relationship:

$$\frac{\eta}{\eta_o} \approx 1 + 2.5\ V_g$$

where is the actual viscosity, η_o is viscosity at zero concentration, and V_g is the volume fraction occupied by the spheres (i.e. the proteins). Contrary to lipids, the diffusion coefficients of membrane proteins usually range in a much broader field, between $\geqslant 10^{-9}$ and $\leqslant 10^{-12}$ cm^2/s , although there are reports of higher values[100,102], particularly in liposome reconstituted systems [90], and conversely of complete immobilization.

Rotational Diffusion

Spectroscopic methods to measure rotation[79] depend on photoselection whereby an oriented population of excited molecules is optically selected from an initially random distribution by excitation with plane-polarized light: only those molecules whose transition dipole moment for absorption is parallel to the electric vector of the incident light are excited. By excitation with a brief pulse of light (flash photolysis) the initial emission or absorption anisotropy decays as the molecules again become randomized. Since protein rotation is a relatively slow process compared to the lifetime of the excited state for normal fluorophores it is necessary to use a long-lived spectroscopic state; this is the case of the "triplet probes" which are attached covalently to the protein[79]. Another method for detecting slow rotational movements is saturation transfer EPR with covalently linked nitroxides[103].
By both such methods, the rotational mobility of several proteins was found to range between D_R of 20 µs for rhodopsin[104] in accordance to the very fluid nature of the retinal rod outer segment membrane [105] to complete

immobility of bacteriorhodopsin in the native purple membrane of Halobacterium halobium[106]. When proteins are mobile, their rotation usually conforms to the Saffman-Delbrück relation.

The reason for the immobilization detected in the rotational measurements of several proteins will be discussed in the next section.

Factors Restricting Protein Mobility in Natural Membranes

The Saffman-Delbrück model has been tested[93] by studying the mobility of bacteriorhodopsin in bilayers of dimyristoyl phosphatidylcholine, allowing to calculate a molecular radius near 20 Å and a membrane viscosity between 1.1 and 3.5 P at molar lipid:protein ratios between 210 and 90; at lower lipid protein ratios the D_L of the protein decreased more than expected by the theoretical viscosity increase[101], suggesting other complications such as the crowding or "archipelago" effect due to steric hindrance, by the concentrated protein dispersion, to long-range diffusion.

A further test of the Saffman-Delbrück relation has been the diffusion dependence on the viscosity of the outer phase: an increase from 0.76 to 9.54 cP by sucrose addition yielded a two-fold decrease of D_L, as expected[93]. On the contrary Vaz and Hallmann[107] have produced direct evidence against the applicability of the Saffman Delbrück model to lipid diffusion, since D_L was found independent of the height of the diffusing species.

All evidence points out that D_L depends largely on lipid viscosity[88], as also shown by the effect of pressure[108].

Also the 8-fold decrease of diffusion coefficient of a fluorescent lipid analogue accompanying water removal from phospholipid multibilayers is explained by the decrease of the area per lipid molecule and the increased proximity of lipid headgroups[109]. The lateral mobilities of phospholipids in the external and internal leaflets of erythrocytes can change independently and the temperature dependence of phospholipid diffusion is different[110], in accordance with the vertical asymmetry of lipids and its dependence on cytoskeletal influences[111].

In natural membranes the mobility of proteins, however, does not appear to follow the theoretical behaviour suggested by Saffman and Delbrück (Table 7). The observation that the mobility of membrane proteins appears largely dependent on the type of membrane and on the type of protein, and is usually lower than theoretically expected, suggests that mobility may be hindered by several physiological restrictions [112,79].

A possible restriction is the increase of membrane viscosity[113]. The lateral diffusion coefficents of proteins are usually decreased by two-to three orders of magnitude below the lipid phase transition[114]. However, the correspondence between viscosity and lateral diffusion of proteins is often quite poor. The changes of lipid microviscosity as measured by the rotational mobility of a fluorescent or paramagnetic probe do not necessarily correspond to changes in the lateral mobility of proteins embedded in the lipid bilayer[115]. This lack of correlation is consistent with the suggestion that lateral mobility of membrane proteins in situ is not modulated by the lipid viscosity but by the constraints from the aqueous matrix.

In certain instances the viscosity of the outer media (which can be different at the two membrane sides) may be responsible for the restriction.

At high outer medium viscosity(η')the Saffman Delbrück relation is not applicable; Hughes et al[116] have found that only values of $\eta'= 1-2$ P and η (membrane viscosity) = 0.1-0.2 P can give diffusion coefficients approaching the experimentally determined ranges of values for proteins.

More specifically, the elements of the cytoskeleton, in particular the microfilaments[117], or other peripheral proteins produce a marked reduction of the measured diffusion coefficients. Agents destabilizing the cytoskeleton also increase the lateral mobility of integral proteins[118]: the lateral mobility of fluorescein labeled membrane glycoproteins in erythrocytes is increased by polyphosphates (ATP and 2,3 diphosphoglycerate) and decreased by polyamines (neomycin and spermine) through their interactions with the cytoskeletal proteins[118]. Accordingly, lateral diffusion of integral membrane proteins but not of lipids is considerably lower in normal erythrocytes with respect to spherocytic cells which lack the principal components of the cytoskeletal matrix[119]. The effects of cytoskeleton on lateral diffusion appear to be dynamic and metastable, being a function of temperature and ionic strength[120].

Depolymerization of microtubules increases the motional freedom of spin label probes in plasma membranes of Chinese hamster ovary cells[121].

Other restrictions are represented by the regions of specific membrane junctions[122].

Another factor that appears to strongly modify protein diffusion is their concentration in the membrane. The dependence of the lateral distribution of membrane proteins on the protein:lipid ratio has been modeled by Monte Carlo calculations[123] and shown to vary from random to aggregated in a continous network. It was calculated that long-range diffusion is relatively sensitive to the area fraction of the impermeable patches, representing the membrane proteins, and, at a critical area fraction, diffusion is completely blocked[124]. It was shown in the reconstituted bacteriorhodopsin-lecithin system quoted previously[93] that the diffusion coefficient of bacteriorhodopsin decreased from 3.4 to 0.15 x 10^{-8}cm /s when the molar lipid protein ratio was decreased from 210 to 30, much more than expected from the increased viscosity.

In reconstituted membranes from E.coli over a range of protein concentrations of 0-60% by weight D_L for lipopolysaccharide decreased 10 fold, whereas D_L for phospholipid remained constant even in the long-range investigated by FRAP[125].

Although the lateral diffusion coefficients should be relatively insensitive to the size of the diffusing molecule[80], it has been shown that aggregation, by increasing the size of the mobile unit, resulted in lowered mobilities of membrane antigens.

A consequence of this is that measurements of D_L (and specifically of D_R) enable the protein diameter to be calculated and should therefore provide a demonstration of whether a given protein exists in a monomeric or oligomeric state in the membrane. It is also possible to measure D_L from measurement of D_R alone, provided the protein diameter can be estimated. This appear the only possible way up to now to calculate short-range diffusion for proteins. This may be of particular importance in collision controlled reactions[92,93].

Another factor that can affect the values for diffusion coefficients is the concentration gradient. By a careful study of the distribution of the intramembrane particles in the growing olfactory axons, Small et al[102] have derived diffusion coefficients in the range of 10^{-7}cm^2/s , which are moreover linearly dependent upon the inverse particle diameter in accordance with the Stokes-Einstein equation. Non-equilibrium processes depend on the rate of entropy production[126] and this is a major driving force of diffusion in a chemical gradient, as in the growing neuron.

The factors affecting diffusion of proteins and lipids are summarized in Table 8.

INVOLVEMENT OF MEMBRANE COMPONENT MOBILITY IN CELL FUNCTIONS, CELL BIOGENESIS AND MAINTENANCE OF STRUCTURE

As pointed out by Berg and Purcell[127], "In the world of a cell..., transport of molecules is effected by diffusion, rather than bulk flow, movement is resisted by viscosity, not inertia; the energy of thermal fluctation, KT, is large enough to perturb the cell's motion".

Adam and Delbrück[128] proposed that organisms resolve some of the problems of timing and efficiency of diffusion of certain molecules by reducing the dimensionality in which diffusion takes place from three-dimensional space to two-dimensional surface. Even if the efficiency of two-dimensional versus three-dimensional diffusion has been questioned[90], membrane-bound diffusion may well compete with transport inside soluble compartments of the cell, which is expected to be not so effective due to the high viscosity of the cytoplasmic matrix[129,130].

Cell function apparently requires that cell membranes have both dynamic and static properties; as pointed out by Peters[87], the insertion of newly synthetized proteins into the plasma membrane and their distribution on the cell surface, or the uptake of material by pinocytosis and the subsequent recycling of membrane components are examples of dynamic events, where diffusion contributes to homogenize the concentration differences of the newly assumed components; on the other hand, the maintenance of size, shape, polarity, and permanent surface specializations reflect membrane statics, where presumably free diffusion of components is hindered.

TABLE 8 - Factors affecting lateral diffusion coefficients in membranes

		Proteins		Lipids	Hydrophobic molecules
		Long range (>μm)	Short range (nm)	Short range (nm)	Short range (nm)
Theoretical (25°C)	η(P)	$\geqslant 1$	$\geqslant 1$	$\leqslant 0.1$	
	η'(P)	0.01	−	−	
	D(cm^2/s)	10^{-8}-10^{-9}	10^{-7}-10^{-8}	$\leqslant 10^{-7}$	
Increased protein concentration		↓	(↓)	(↓)	(↓)
Increased size (protein aggregation)		↓	↓	−	−
Viscosity of outer medium		↓	↓	0	0
Viscosity of membrane		↓	↓	↓	↓
Temporary protein association		↓	(↓)	−	−
Protein concentration gradient		↑	(↑)	0	0

One major function of diffusion is certainly related to membrane biogenesis: the newly synthetized proteins and lipids distribute in the membrane after synthesis and insertion, and move to their specific sites by lateral diffusion. Particularly fascinating is the problem of the very long distances travelled by membrane components in the elongated neuronal cells. Small et al[102] have shown that intramembrane particles of the plasmalemma of growing olfactory axons form density gradients, with density decreasing with increasing distance from the perikarya, with a slope that depends upon particle size[131]. The diffusion coefficients calculated range from 0.5 to $1.8 \times 10^{-7} cm^2/s$ for particles between 14.8 and 3.6 nm diameter respectively. The rates in a concentration gradient appear considerably enhanced with respect to iso-concentration system, assuring a fast distribution of the protein in the growing axon.

Synaptogenesis involves a highly localized concentration of acetylcholine receptors at the postsynaptic membrane. It has been shown that receptors in vitro cultures of embryonic neurons and myotubes move laterally from non aggregated regions toward the site of the developing synapse[132,133,134]. Once gathered at the synapse, receptors become immobile, as the result of some cytoplasmic anchoring[135]. This is but one example of diffusion-mediated trapping, by virtue of which mobile molecules diffuse to a trap region where they become immobilized. This appears to represent a major control of cell surface topography[90].

Receptor-mediated endocytosis involves the capture by coated pits of diffusible ligand-receptor complexes from adjacent regions of the membrane followed by intracellular formation of coated vesicles which deliver the ligands to the cytoplasmic compartments[136].

From the D_L value of $2-3 \times 10^{-11} cm^2/s$ measured for LDL receptors[137,138] it was concluded that the diffusional trapping mechanism must operate close to the diffusion-limited rate. McCloskey and Poo[90] consider that the diffusion-collection process may be more efficient than having receptor sites permanently inserted in the coated pits. Theory dictates that maximal efficiency is assured by a number of receptors not exceeding the ratio of cell diameter to receptor diameter[127], the best arrangement being maximum dispersal of any different receptor system. Since several types of ligands can enter the cell through a single coated pit[139], there may be insufficient room inside a single coated pit to house several copies of all the different receptor types.

Diffusion followed by trapping cannot be the only mechanism by which potentially mobile proteins are fixed in specialized regions of the membrane. The presence of contractile elements of the cytoskeleton is certainly directly participating in creating and maintaining the localization of membrane proteins.

ATP-driven directed motion of protein species attached to the cytoskeleton, to specialized zones of the cell surface, has often been proposed to account for localizing movement[140]. The influence exerted by the cytoskeleton on redistribution of immunoglobulins on the B lymphocyte surface upon binding with antibodies has received large attention[113]. The initial phase after binding involves redistribution of immunoglobulins into small patches; this process is independent on metabolism and cytoskeletal integrity. At 37°C in the presence of metabolic energy and microfilaments, the patches collect to a cap at one pole of the cell. Patching and capping have been observed in a variety of receptors. The mechanism of capping is not completely understood, and is ascribed either to the hydrodynamic drag on membrane proteins induced by lipid flow[141] or to the contraction of microfilaments[140]. Studies with a model system involving interaction of stearylated dextrans on 3T3 cell plasma membranes[142], confirmed that the patching process need not require cell metabolism or coupling with cytoskeleton. The slow diffusion of the stearyl dextrans in the cell plasma membrane in contrast with model lipid membranes, suggests binding to some immobile structure; in such simplified systems cap formation was shown not to be a diffusion process but to be rather preceded by cessation of diffusion, and appears to involve the cytoskeleton in some way. Since stearyl dextrans penetrate only half of the bilayer, the cytoskeleton can be associated only indirectly through some other membrane macromolecule.

Clearly in the case of epithelial cell monolayers carpeting the lumen of organs like intestine, kidney tubules etc., the properties of the luminal and basolateral surfaces of the plasma membranes are maintained by blocking unrestricted lateral motion. It is likely that tight junctions preserve lateral asymmetry by blocking the translational diffusion of membrane components between luminal and basolateral surfaces[143]. However, other mechanisms must work in concert with tight junctions to form the initial pattern[90].

Closely related to these problems are those of self-assembly and sorting of cellular components; mechanisms must exist to overcome the tendency of diffusion to randomise molecular distributions.

A complex example of this interplay is found in eukaryotes where the Golgi complex is engaged in sorting components received from the endoplasmic reticulum to lysosomes, secretory vesicles, plasma membrane etc.: these proteins must undergo lateral separation in the membrane plane. How this is achieved is however largely unknown[90]

Diffusion Control of Chemical Associations in Two Dimensions

All association processes in solution are ultimately limited by the time it takes to bring reactants together by diffusion. Most macromolecular reactions also require that the molecules attain a correct mutual orientation so that potentially reactive groups are properly aligned: usually the molecules have to collide many times before the reaction takes place. If the rate of a chemical reaction is limited by the time it takes to bring the reactive groups together via diffusion, the reaction is said to be diffusion-controlled; on the contrary, if subsequent chemical processes are limiting, the rate is reaction-controlled[144]. Diffusion limited reactions are viscosity dependent and have weak temperature coefficients. According to Smoluchowski[145] the bimolecular association rate constant, k_a, for two spherical molecules A and B is

$$k_a = \pi (D_A + D_B) \cdot (r_A + r_B)$$

where D_A and D_B are the diffusion constants and r_A and r_B the radii of the two molecules. Since macromolecules are not reactive over their entire surfaces, but on restricted active sites, a full description of the diffusion-limited association process must consider the establishment of the relative positions and orientations of the molecules that are needed for the reaction to occur[146]. It can be assumed that a molecule, owing to the erratic nature of the diffusional path (random walk), will come close to its starting point a large number of times prior to achieving an appreciable separation from its origin (microcollisions as distinguished from macrocollisions)[144]; molecules that have come together by diffusion will experience a large number of microcollisions with changes in orientation, facilitating the occurrence of the useful collision. For two spherical molecules, assuming one molecule to be completely reactive, and the other having a reactive patch over its surface, limited by an angle δ_A with the center of the molecule, the diffusion-limited association will be roughly proportional to $\sin \delta_A$ (Figure 6)

$$k_A = \pi (D_A + D_B) (r_A + r_B) \sin(\delta_A/2)$$

If the steric constraints are severe (i.e. δ_A is very small), there can be a difference of orders of magnitude with respect to the simple Smoluchowski relation. In real macromolecular associations, however, it is likely that long range and short range interaction forces will facilitate and prolong the collisions, giving the molecules ample opportunity to seek out orientations for reaction. Thus electrostatic and hydrophobic interactions are sources of useful adhesion interactions[147].

It has been proposed[128] that reduction of dimensionality from three-dimensional space to two dimensional surface, as it usually happens in membrane-mediated reactions, will enhance the rate constants by facilitating collisional encounters. There is some empirical evidence in favour of guided diffusion by reduced dimensionality[148,149,150], but the rate constants for two-dimensional diffusion have yet to be rigorously defined[90]. The equation developed by Hardt[151], relating reaction rates with diffusion in two dimensions, and often taken as a basis for calculation of diffusion-limited rate constants in membranes, has been severely criticized[90] in that it does not take into account that the rate "constants" for two dimensional diffusion controlled reactions decline continuosly with time[152].

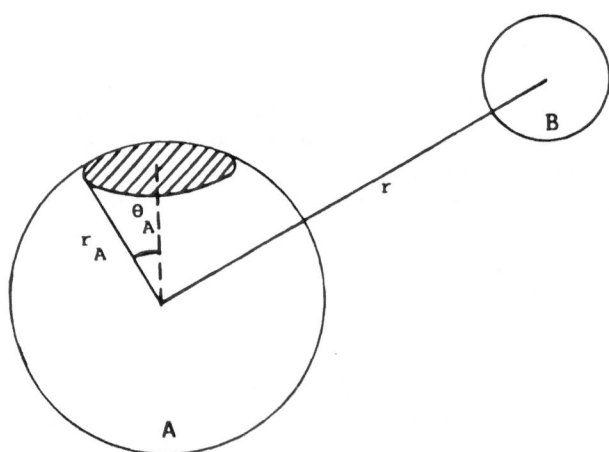

FIGURE 6. Reactive-patch geometry for the association of molecules with heterogenous reactivity.

By using the expressions of Torney and McConnell[153], Mc Closkey and Poo[90] have calculated the two-dimension vs three-dimension efficiency of a reaction such as that of cytochrome c with cytochrome c oxidase in mitochondria.

The ratio of diffusion controlled reaction rates in 2D vs 3D would be:

$$R_2/R_3 \approx 83(D_2/D_3) \ln (AD_2t)$$

where A is of the order of $10^{14} cm^{-2}$.

A membrane protein diffuses with D_2 $5 \times 10^{-9} cm^2/s$; a similar protein would diffuse in solution with D_3 of $10^{-7} cm^2/s$; with these values, R_2/R_3 would not be enhanced, but actually slowed down ($R_2/R_3 \approx 0.3$).

There are however some possible reasons favoring bidimensional reactions in membranes.

First of all, most cellular compartments are crowded with proteins making up a gel-like viscous network where diffusion could be severely limited, thus increasing theoretical D_2/D_3 and therefore R_2/R_3. In some cases diffusion coefficients in membranes can be very fast, as is the case of relatively small hydrophobic molecules. It is considered that electron transfer in mitochondria between the flavin dehydrogenases and the cytochrome bc_1 complex is coupled to the diffusion of ubiquinone[154]; the diffusion coefficient of ubiquinone has been shown to be as high as $10^{-6} cm^2/s$ by a fluorescence quenching method[155,156] although measurement by FRAP of long range diffusion of a fluorescent short chain analog of the quinone has given a lower value of $3 \times 10^{-9} cm^2/s$[157,158].

Since it is rationalized that short-range diffusion is more meaningful for chemical associations than long range diffusion, it is believed that diffusion within the low viscosity hydrophobic interior of the membrane can produce preferential paths for highly efficient collisional encounters. This is why in most instances it is shown that diffusion of ubiquinone is not rate-limiting for electron transfer turnover[112].

There is another reason why reduction of dimensionality can be of great advantage for useful collisions to occur.

Large amplitude rotational motion of integral membrane proteins is restricted to an axis normal to the membrane plane, thus aligning the reactive groups in a way that is impossible for soluble macromolecules; this orientation would provide a rate advantage in reactions involving large proteins. For two proteins each of radius a with a circular reactive patch of radius b, the probability ratio in two-dimensions vs three-dimensions would be:

$$P_2 / P_3 \approx (a / b)^2$$

This means that the reduction of dimensionality of rotational diffusion (3D to 1D) may provide more of a rate advantage than reduction of dimensionality of translational diffusion (3D to 2D)[90].

The lateral translational rate of a diffusing molecule is usually calculated by the Einstein-Smoluchowski relation

$$d^2 = 4 Dt$$

for a bidimensional path, where d is the distance seen by the diffusing molecule and t is time. However, the mean time τ required to reach a small target of radius r in the middle of a cell of radius L (L>>r) is[144] :

$$\tau = (L^2/3D) \ (L/r) \qquad \text{in three dimensions}$$

and

$$\tau = (L^2/2D) \ \ln \ (L/r) \qquad \text{in two dimensions}$$

TABLE 9 - Lateral displacements calculated at different diffusion coefficients

D (cm^2/s)	Time to run 30nm (ms) (Eq1) *	Time to run 30nm (ms) (Eq2) **	Assumed enzyme turnover	ms/turn-over	Collisions/turnover	
					Eq1	Eq2
10^{-5}	0.000225	0.004	5,000	0.2	889	50
			500	2	8,889	500
			50	20	88,889	5,000
10^{-6}	0.00225	0.04	5,000	0.2	89	5
			500	2	889	50
			50	20	8,889	500
10^{-7}	0.0225	0.4	5,000	0.2	9	0.5
			500	2	89	5
			50	20	889	50
10^{-8}	0.225	4	5,000	0.2	0.9	0.05
			500	2	9	0.5
			50	20	89	5
10^{-9}	2.25	40	5,000	0.2	0.09	0.005
			500	2	0.9	0.05
			50	20	9	0.5
10^{-10}	22.5	400	5,000	0.2	0.009	0.0005
			500	2	0.09	0.005
			50	20	0.9	0.05

*Eq.1: $t = d^2/4D$;
**For diffusion of a particle to a small target of diameter r, the displacement d from a diffusion coefficient D is given by Eq. 2: $t = (d^2/2D) \ [\ln \ (d/r) - 3/4]$. The value of 30 nm has been chosen, representing an average distance between different complexes of the mitochondrial electron transfer chain[127].

Thus, the diffusional search for a small target is much more efficient in two dimensions than in three, assuming D to be of comparable magnitude in three and two dimensions (Table 9).

The transmembrane control of cellular activity involves the transfer of information from the environment to the cell via reception of the signal at the membrane surface, followed by coupling to an internal effector system, and subsequent discharge of a second messenger into the cell interior. The eukaryotic adenylate cyclases are integral membrane enzymes controlled by hormones and neurotransmitters. One class of hormones is known to activate the enzyme, while another class is inhibitory[159].

The coupling of hormone binding to cAMP production requires three units: the catalytic unit (adenylate cyclase proper), the hormone receptor, and the GTP binding protein (G-protein). The bulk of evidence suggests that collisional -encounters between catalytic units and receptor-hormone complexes produce the catalytically active state. Hormone-stimulated cyclase activity has been restored in heterokaryons from cell partners lacking functional catalytic units and hormone receptor, respectively[160]. Tolkowsky and Levitzki[161] observed that progressive inactivation of the β -adrenergic receptor by a specific affinity-label caused a decrease of the maximal binding capacity of the receptor and a proportional decrease of the rate of activation, but no change in the maximum level of activity. Conversely, progressive inactivation of the cyclase by hydroxymercuribenzoate was found not to change the rate of activation nor the capacity of the receptor to bind hormone, but to decrease the maximal level of activation.

These data strongly suggest that receptor and cyclase are separate units that can homogeneously collide in the plane of the membrane.

The collision coupling mechanism[161] involves the following steps:

$$ H + R \underset{K_H}{\rightleftharpoons} HR + E \xrightarrow{k_2} HRE \xrightarrow{k_3} HR + E' $$

where K_H is the hormone-receptor dissociation constant, k_2 is the molecular rate constant governing the formation of HRE, and k_3 is the rate constant governing the activation of the enzyme.

The role of the G protein in the collision coulping mechanism is not clear; Arad et al[162] have provided evidence that the stimulatory GTP regulatory unit Ns is tightly associated with adenylate cyclase, favoring the following refinement of the model:

$$ R + N_S (GDP) \cdot C \xrightarrow{H,GTP} HR \cdot N_S (GTP) \cdot C' \xrightarrow{} HR + N'_S (GTP) \cdot C' $$
$$ \downarrow Pi $$
$$ N_S (GDP) \cdot C $$

where C is inactive cyclase and C' is the active form of the enzyme. In accordance with the model, inhibition of the lateral diffusion of membrane proteins also inhibited hormone-dependent adenylate cyclase activity[163].

Hanski et al[164] observed that insertion of cis-vaccenic acid into the membrane enhances the rate of adenylate cyclase activation though the β -receptors by an order of magnitude. This treatment was shown to enhance the rotational motion of the fluorescent probe DPH within the membrane.

The results would suggest that the coupling of receptor and cyclase is diffusion-controlled. However a FRAP study[165] of a fluorescent phospholipid analog showed that lateral diffusion was only slightly enhanced by cis-vaccenic acid incorporation; moreover the mobile fraction of the probe was also not significantly increased.

This study would show that the increased activation of adenylate cyclase following membrane fluidization is not simply due to increased long range lateral motion but rather to specific effect on domains containing the adenylate cyclase components (or to direct effects on the coupling between the catalytic unit and the β-receptor).

The lateral mobility of β-receptors as measured by FRAP appears quite slow[166]; Henis and Elson[167] observed that in human liver cells β-receptors are immobilized, but agonist binding causes dispersal and mobilization of the receptors: the time course of mobilization is however much smaller than that of cyclase activation.

It can be concluded that although the collisional coupling model for cyclase activation appears very plausible, the molecular details may still be obscure, and long range lateral diffusion may not be necessarily involved. Another possible interpretation is that binding of agonists can induce local increases of fluidity, as shown hy Hirata and Axelrod[168] as a consequence of β-agonist- induced phosphatidylethanolamine methylation and by Curatola et al[169] after incubation of erythrocyte ghosts with S-adenosylmethionine. The increased lipid fluidity may increase the frequency of encounters between receptors and catalytic units.

The problem is further complicated by the possible involvement of cytoskeletal regulation. Microtubule disassembly can increase hormonally stimulated cyclase activity[170]; cytoskeleton-disrupting drugs enhance the G unit mediated activation of adenylate cyclase[171], suggesting that it is the ability of the G unit to diffuse laterally in the membrane to represent the limiting factor in cyclase activation.

Clearly, further studies are needed to completely clarify the details of the hormonal activation, also in other systems.

MEMBRANE FLUIDITY IN THE REGULATION OF MEMBRANE LINKED ENZYMES

Lipids are required for the activity of membrane-linked enzymes[3]. As a bulk phase, their physical properties are involved while their chemical nature may be of relatively minor importance. In this respect, the bilayer represents a dispersing medium and also represent a way for interfacial regulation[172,4]. It is not surprising that under certain conditions such role can be exerted by detergents[173]; detergents of course cannot maintain the bidimensional structure of the membrane, but disperse the protein in a tridimensional solution, so that their action can only mimic that of a real membrane for non-integrated and non-vectorial enzymatic reactions. It is also understandable how lipid fluidity can be involved in the regulation of enzymic activity[174].

On the other hand, specific interactions of lipids with membrane proteins seem necessary for catalytic events, possibly as part of the active site or for keeping the enzymes in their optimal conformation. Perhaps the most investigated property of membrane-bound enzymes in relation to membrane lipid fluidity has been the presence of breaks or discontinuties in their Arrhenius plots[175,3,176]. The break is well defined in thermodynamic terms according to the transition state theory; when a break is found in an Arrhenius plot of an enzymic activity (Figure 7), the usual presence of two

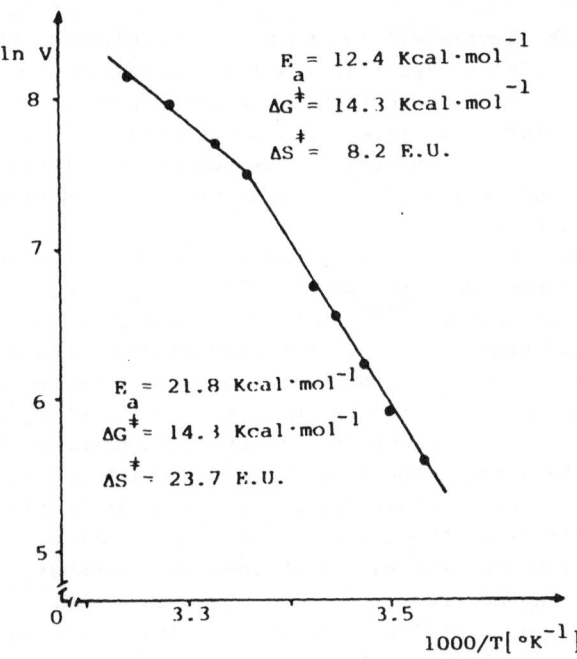

$E_a = 12.4 \text{ Kcal} \cdot \text{mol}^{-1}$

$\Delta G^{\ddagger} = 14.3 \text{ Kcal} \cdot \text{mol}^{-1}$

$\Delta S^{\ddagger} = 8.2 \text{ E.U.}$

$E_a = 21.8 \text{ Kcal} \cdot \text{mol}^{-1}$

$\Delta G^{\ddagger} = 14.3 \text{ Kcal} \cdot \text{mol}^{-1}$

$\Delta S^{\ddagger} = 23.7 \text{ E.U.}$

$1000/T[\,^\circ \text{K}^{-1}]$

FIGURE 7. Arrhenius plot of mitochondrial ATPase

intersecting lines indicates that at the point of intersection the catalytic constants are identical above and below the break, so that $\Delta G^{\ddagger}$ of activation must be the same; since however the two processes above and below the break have different activation energies and hence $\Delta H^{\ddagger}$, it follows that the change of $\Delta H^{\ddagger}$ requires a corresponding and compensating change in $\Delta S^{\ddagger}$ to maintain a constant $\Delta G^{\ddagger}$

There are several reasons that may cause nonlinear Arrhenius plots. A reason for non linearity could be merely kinetic if in a composite reaction the rate limiting step changes with temperature. Another cause for a nonlinear Arrhenius plot is a temperature-dependent conformational change of the enzyme[178]. Evidence for a structural transition coincident with a break was found in soluble aminoacid oxidase[179] in myosin subfragment 1 complexing with ATP and in isolated F_1 ATPase[180]. Another cause for a nonlinear Arrhenius plot is that the solvent structure undergoes a structural transition[176]. The breaks in Arrhenius plots of membrane enzymes appear strongly affected by the physical properties of lipids, often independently of the enzyme species considered; moreover the same enzyme in the same organism can exhibit different temperature dependencies in relation to different lipid composition. These observations led Raison[175] to postulate that the breaks are the result of a phase change in lipids. Following this reasoning it can be speculated that a constraint by freezing of the lipid could limit essential conformational changes between different structural states involved in catalysis; more simply the rigid state could prevent such motions of the enzyme and/or substrate are needed for useful collisions to occur.

A correlation between break in Arrhenius plot and phase transition of the surrounding lipids has been found e.g. in sarcoplasmic reticulum ATPase reconstituted with dipalmitoyl lecithin[181] and in mitochondrial F_1F_0-ATPase reconstituted with dimyristoyl lecithin[182]; accordingly, the enzyme reconstituted with dioleyl lecithin (having a transition temperature at -21°C) exhibited no clear discontinuity in the Arrhenius plot. Unfortunately the precise correlation between breaks and lipid phase transitions represents the exception; the calorimetric phase transitions in membranes usually fall below 0°C[183] whereas breaks in enzymic activities usually range between 15-25°C [3].

On the other hand, the mobility of paramagnetic and fluorescent probes in the membrane often undergoes changes in slope at the same temperatures where enzymatic breaks take place[184,185]. Such breaks are often ascribed to phase transitions of some kind occurring in microdomains of the protein-rich lipid bilayer or to redistribution of such microdomains with temperature[91,186]. The presence of an excess lipids in the form of bilayer appears necessary for the break to occur, as shown in[187] in the mitochondrial bc_1 complex; the nature of the lipid affects the temperature at which the break occurs, in the sense that increased rigidity elevated the break temperature[188]; the effect of increased rigidity on rising the temperature of occurrence of the break is substantiated by the fact that increased pressures increase the temperature at which the break occurs by about 25°C/kbar[189]. In many cases cholesterol was found to modify biphasic Arrhenius plots into linear plots, as in ATPase from Mycoplasma mycoides var capri[190], in microsomal drug monooxygenation[191], in mitochondrial F_1F_0-ATPase[192] and ubiquinol cytochrome c reductase[193] (Figure 8).

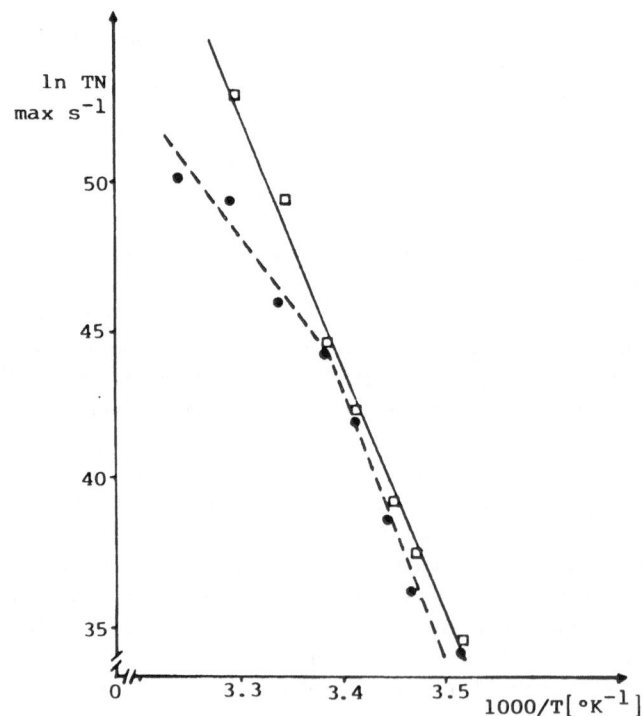

FIGURE 8. Arrhenius plot of ubiquinol-2cytochrome c reductase in phospholipid vesicles.

The bulk of the data reviewed suggest that a property of the lipids affected by composition, temperature and pressure, may be responsable of the catalytic changes. One such property may be viscosity. The catalytic properties of protein can indeed be affected by the viscosity of the medium by the hindering the transition between two conformational states involved in catalysis, as elegantly shown by Beece et al[194] for the binding of O_2 and CO to myoglobin in many different solvents by flash photolysis.

Swann[195] investigated the temperature dependence of brain (Na^+-K^+)ATPase, an enzyme showing two conformational states E_1 or E_2 during catalysis; a break was found at 17°C with the E_2 conformation more stable below the break and the E_1 more stable above the break. The increased order (low entropy) accompanying E_2 may be favoured by increased lipid viscosity.

To this purpose the explanation by Hidalgo et al.[181] for sarcoplasmic reticulum ATPase with lipids below the phase transition could be applied to lipids reaching a critical viscosity. When lipids assume a limiting viscosity value, a less efficient catalysis, shown by a higher positive ΔH^*, is accompanied by a greater loss of order during activation (higher positive ΔS^*); phospholipids in a more viscous state stabilize a ordered preactivation state of low entropy content by the increased rigidity around the enzyme, so that the transition state is reached with a higher positive ΔS^*. Accordingly $\Delta G^*$of activation is independent of variation of the physical properties of the lipids.

An understanding of the role of viscosity on enzymic functions can be obtained by the view that discrete protein molecules, considered singly in a microscopic system, undergo considerable fluactions in thermodynamic properties, whereas time-averaged studies of ensambles of proteins yield a static view of these molecules[196,197]. According to Somoggi et al[198] proteins in solution exist as equilibrium mixtures of different conformational states having the same total free energy. The activity of an enzymic protein will depend upon fluctuations between such states as a device to obtain the appropriate enthalpy/entropy balance for the transition state. The bulky protein matrix functions as a fluctuating free energy reservoir in the generation of the catalytic conformation.

Enzymes effect catalysis by generating local events having high free energy at the active site. The mode of linkage between environment and active site has been defined in several models, such as surface hydration, mobile cooperative defects, local electric fields and transient strain due to thermal collisions and viscosity-dependent.

The exchange of energy at the protein surface by collisions with solvent molecules occurs with characteristic frequencies of $\approx 10^8$ Hz, in contrast with the turnover times of typical enzymes around 10^{-3}sec; therefore, whilst an enzyme is indeed a highly stochastically fluctuating molecule during its catalytic cycle it is in thermodynamic equilibrium with its surroundings[199]. The enzyme may "borrow" heat energy from its surroundings but must "return" it after using it in its enzymatic cycle.

According to Somoggi et al[198]

$$k_{cat} = (\frac{A}{\eta^\epsilon} + A') \exp(-\frac{\Delta}{kT}) .$$

where Δ stands for the activation energy (enthalpy) η is the solvent viscosity, A is a function of structural parameters characterising the potential-energy profile, the exponent ϵ (where $0 < \epsilon < 1$) relates to the attenuating influence of the protein matrix on solvent viscosity, and A' is a viscosity-independent empirical parameter.

Fluctuations propagate as quasi-particles (zymons) strongly dependent on medium viscosity. At high viscosity, k_{cat} is not linear with A/η^ϵ and depends on intrinsic motions of the protein only (A').

At difference with what considered by Somoggi et al[198], such theory might be considered valid also for membrane enzymes, provided that membrane viscosities be not extremely high. The viscosity of membranes is usually considered as on average to be >1 P[200], but the midplane of the bilayer experiences much lower viscosities of few cP[201].

Viscosity increased by lowering the temperature or other means could hinder that part of the catalytic mechanism that is a consequence of thermal fluctuations, allowing a different mechanism to emerge.

According to Kell[202], membrane bound energy-conserving enzymes are capable to store free energy over periods of seconds in a dispersionless fashion without losing energy by thermal exchange.

Non-thermally excited conformational states (high-energy states) of proteins can carry energy for long times and distances in the form of solitary excitations (solitons) in such a dispersionless fashion at a level greater than expected for Boltzmann distribution of vibrational energies.

These ideas appear particularly important in the mechanism of energy conservation.

CHANGES IN MEMBRANE FLUIDITY RELATED TO CELL CYCLE

Conflicting reports are present in the literature on membrane fluidity changes in different cellular systems during the cell cycle. Although the dissimilar results could reflect specific properties inherent to the biological models, it is necessary to take in account the features of the different techniques used to measure membrane fluidity, as previously discussed. The DPH fluorescence is the most widely used method in this type of studies but the rapid DPH localization also in intracellular membranes makes it an unsuitable probe for intact cellular systems.

Using (4-trimethylammonium)-6-phenyl 1,3,5 - hexatriene (TMA-DPH), a DPH derivative, which remains localized on plasma-membrane, we have observed an increase of membrane fluidity in intact polymorphonuclear leukocytes during cellular activation. The change was evident only after a preliminary treatment with colcemid, a microtubule disrupting drug[203] On the contrary, in both cases DPH did not evidentiate any modifications while an increase of DPH fluorescence has been observed in isolated membranes.

A decrease of microviscosity measured by DPH fluorescence has been observed in cultured lymphocytes after mitogenic stimulation with succynil-CoA and ConA ; the effect is evident at 51 h when the DNA synthesis is maximal[204] ; using the frequency distribution of the fluorescence polarization values of single cells, the stimulated cultures contained more cells with lower microviscosity respect to the control unstimulated cultures; therefore the data can be correlated with blast formation. Similarly, contact inhibited 3T3 fibroblasts after serum stimulation exhibit an increase of membrane fluidity more evident at the transition between G_1 and S phases.

No changes of membrane fluidity accompany the cell cycle in murine leukemic cells L1210 suggesting that some properties involved in the control of membrane fluidity are lost during transformation.

Similar results were obtained also in lymphocytes isolated from human blood and stimulated with mitogenic lectins while the non mitogenic lectins do not affect the membrane fluidity. The effect is temporary and desappears 1 h after stimulation and is not accompanied by changes in phospholipid and cholesterol level and probably has to be related to perturbation of membrane structure after the binding of lectins to membrane receptors[205].

Changes of membrane microviscosity correlated to changes in membrane composition and membrane structural organization have been shown in mouse neuroblastoma cells C1300 by de Laat[206] and summarized in table 10. Membrane viscosity increases during cell division being maximal during mitosis and early G_1 phase. Such increase is parallel with a decrease of membrane protein and phospholipid lateral mobilities; moreover, after mitosis, protein mobility rapidly increases at the beginning of G_1 phase, suggesting that in this phase a specific mechanism controls protein diffusion. Membrane ultrastructural analysis demonstrates changes in intramembrane particle density which is minimal at early G_1 phase and reaches a maximum abruptly during mitosis. These results have been interpreted as consequence of a preferential insertion of phospholipids respects to the protein during cell division and subsequent formation of membrane macrodomains.

In regenerating liver we have shown a correspondence between compositional changes and modifications of membrane fluidity[21]. In hepatocyte plasma membrane a decrease of 5-doxylstearic acid mobility (Figure 9) is more evident at the beginning of S phase, suggesting an increase of fluidity at the membrane surface without significant changes in the membrane hydrophobic core measured by 16-doxylstearic acid (Figure 9).

TABLE 10 - Plasma membrane changes during the cell cycle of Neuro-2 A cells

Cell cycle phases	Microviscosity S*	PL** mobility	Protein*** mobility	IMP**** density
Mitosis	0.64	$3.1^{\pm}0.3$	$1.3^{\pm}0.2$	1100
Early G_1	0.59	$4.2^{\pm}0.3$	$3.1^{\pm}0.3$	700
Late G_1	0.55	$5.6^{\pm}0.2$	$2.7^{\pm}0.3$	1100
S	0.50	$6.2^{\pm}0.3$	$1.8^{\pm}0.1$	1720

*S,DPH order parameter; **PL mobility measured as Diffusion coefficient of fluorescein-labeled ganglioside (F-GM1) ($D \times 10^{-9} cm^2/s$); ***Protein mobility measured as diffusion coefficient of conjugated antibody against EI_4 cells ($D \times 10^{-10} cm^2/s$); ****IMP density is expressed as IMP_s per μm^2 Data are taken from [207,208,11].

The fluidity increase can be correlated with changes in cholesterol /phospholipid ratio during G_1 and early S phase, and phospholipid / protein ratio at the end of S phase, already discussed. The loss of cooperativity shown by 5'-nucleotidase in the presence of ConA can be ascribed to the membrane fluidity modification.

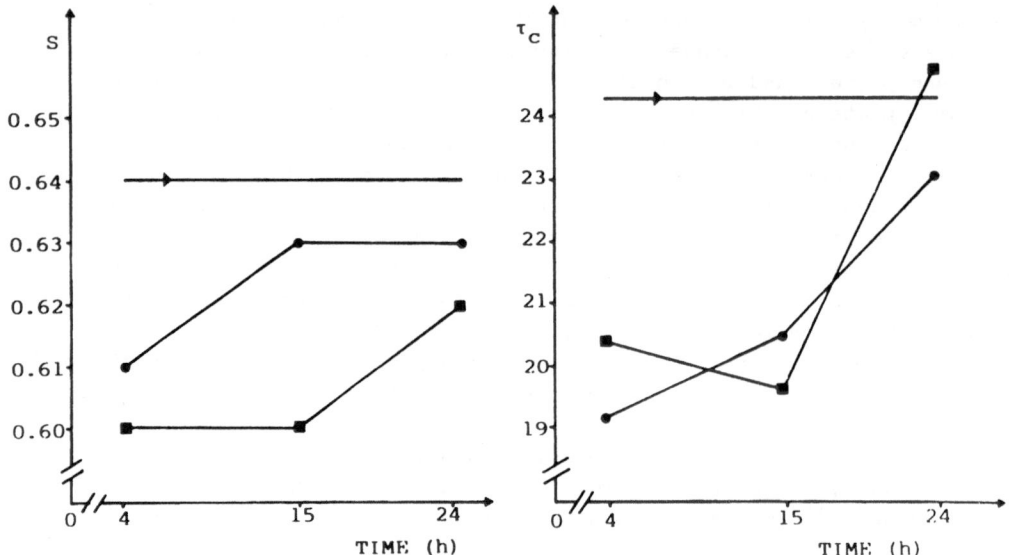

FIGURE 9. Mobility of doxyl derivatives of stearic acid in isolated plasma membrane obtained from liver of →— control, ●——● sham operated and ■——■ partially hepatectomized rats. The mobility are expressed as order parameter (S) for 5-doxylstearic acid, and correlation time (τ_c x10^{-10}s) for 16-doxylstearic acid

In the same system the fluidity is not modified during early S phase at the surface of endoplasmic reticulum [22] but partial hepatectomy disorganizes the deep portion of the bilayer as shown by the temperature dependence of the τ_c of 16-doxylstearate (Figure 10). An increase of the ratio between weakly (W) immobilized and strongly bound (S) protein SH groups, demonstrates a relative increase of SH group population localized in a more fluid environment (Figure 10). Therefore membrane physical changes are sensed by proteins as also suggested by changes in membrane bound enzymes activities.

The relations between membrane fluidity and the progression of cellular activities through the cell cycle open exciting developments and may represent a fruitful field of investigation in the near future.

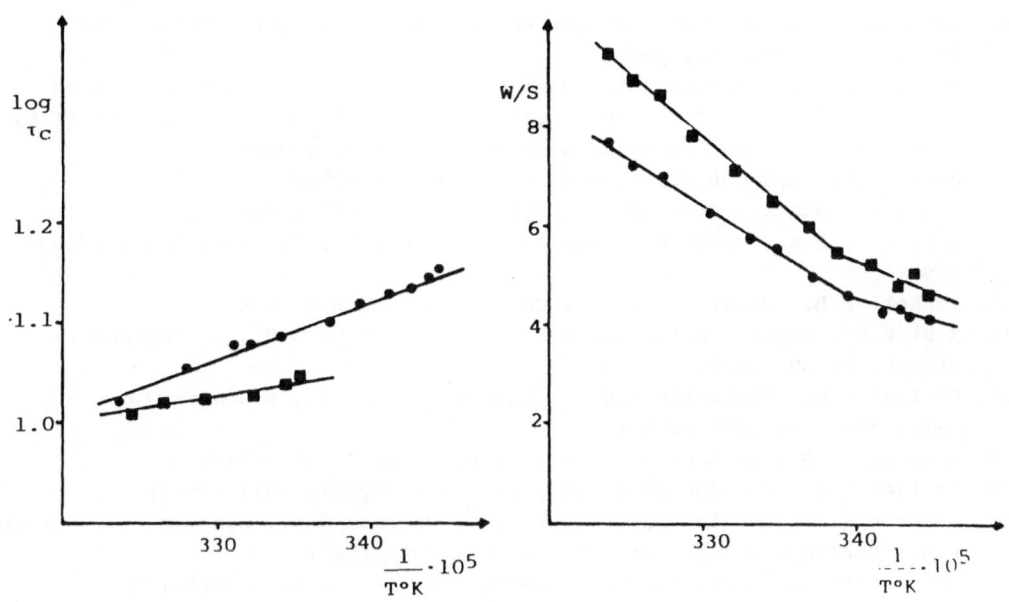

FIGURE 10. Effect of liver regeneration on temperature dependence of correlation time ($\tau_c \times 10^{-9}$s) of 16 doxylstearic acid, and on the W/S ratio of maleimide spin labeled SH groups. Liver microsomal membranes from ●——● sham operated, and ■—■ partially hepatectomized rats.

ACKNOWLEDGMENT

We would like to thank Mrs. P. Ricciotti for typing the manuscript and Mr. G. Grilli for drawing the original figures.

REFERENCES

1. Fourcans B., Jain M.K. Advanc. Lipid Res. 12:147 (1974)
2. McElhaney R.N. Curr. Topics Memb. Transp. 17:317 (1982)
3. Lenaz G. Subcell. Biochem. 6:233 (1979)
4. Wrigglesworth J.M. in "Structure and properties of cell membranes" vol.1 GH Benga ed., CRC Press, Boca Raton FL, pag.137, (1985)
5. Ledue M., Schaechter M. J. Bacteriol. 133:1038 (1978)
6. Joseleau-petit D., Kepes F., Kepes A. Eur. J. Biochem. 139:605 (1984)
7. Mozharov A.D., Shchipakin V.N., Fishov I.L., Evtodienko Y.V., FEBS Lett. 186:103 (1985)
8. Cottrell S.F., Getz G.S., Rabinowitz M., J. Biol. Chem. 256:10973 (1981)
9. Daniels M.J., Biochem. J. 115:697
10. Pasternak C.A. in "The cell surface and the cell cycle" vol.IV CRC Press, Boca Raton FL, pag.1
11. De Laat S.W., Bluemink J.G., Boonstra J., Mummery C.L., Van der Saag P.T., Van Zoelen E.J.J., in "Physiology of membrane fluidity" Vol.II M. Shinitzky ed., CRC Press, Boca Raton FL, pag 21 (1984)
12. Henry S.M., Hodge L.D., J. Cell Biol. 97:166 (1983)
13. Van den Bosch H., Biochim. Biophys. Acta 604:191 (1980)
14. Horwitz A.F.A., Wight P., Ludwig P., Cornell R., J. Cell. Biol. 77:334 (1978)
15. Cornell R.B., Horwitz A.F., J. Cell. Biol. 86:810 (1980)
16. Jost P.C., Capaldi R.A., Vanderkooi G., Griffith O.H., J. Supramol. Struct. 1:269 (1973)
17. De Laat S.W., Tertoolen L.G.J., Van der Saag P.T., Bluemink J.G., J. Cell. Biol. 96:1047 (1983)
18. Bluemink J.G., De Laat S.W. Cell. Surf. Rev. 4:403 (1977)
19. De Laat S.W., Van der Saag P.T., Int. Rev. Cytol. 74:1 (1982)
20. Preston S.F., Regula C.S., Sager P.R., Pearson C.B., Daniels L.S., Brown P.A., Berlin R.D., J. Cell. Biol. 101:1086 (1985)
21. Bruscalupi G., Curatola G., Lenaz G., Leoni S., Mangiantini M.T., Mazzanti L., Spagnuolo S., Trentalance A., Biochim. Biophys. Acta 597:263 (1980)
22. Bruscalupi G., Curatola G., Leoni S., Mangiantini M.T., Spagnuolo S., Trentalance A., Zolese G., Ital. J. Biochem. 32:391 (1983)
23. Jaiswal R.K., Rama Sastry B.V., Landon E.J., Pharmacology 24:355 (1982)
24. Tanford C. in "The hydrophobic effect: formation of micellse and biological membranes". John Wiley, New York (1973)
25. Nagle J.F., J. Chem. Phys. 58:252 (1973)
26. Marcelja S. Biochim. Biophys. Acta 367:165 (1974)
27. Lee A.G. Biochim. Biophys. Acta 472:285 (1977)
28. Vaz N.A.P. and Doane J.W. Phys. Rev. A. in the press
29. Pullman B., Berthod H. and Gresh N., FEBS Lett. 53:199 (1975)
30. Browning J.L., Seelig J., Biochemistry 19:1262 (1980)
31. Eisenberg M., Gresalfi T., Riccio T., McLaughlin S. Biochemistry N.Y. 18:5213 (1979)
32. Mabrey-Gaud S. in:"Liposomes: from physical structure to therapeutic applications" C.C. Knight ed., Elsevier Amsterdam, p. 105 (1981)
33. Jahning F., Harlos K., Vogel H., Eibl H., Biochemistry 18:1459 (1979)
34. Israelachvili J.N., Marcelja S., Horn R.G., Quart. Rev. Biophys. 13:121 (1980)

35. Mitchell D.J., Ninham B.W. (1981) Chem. Soc. Faraday Trans. II 77, 601
36. Evans E.A., Hochmuth R.M., in "Current topics in membrane and transport" vol. X, Kleinzeller A. and Bronner F. eds., Academic Press, New York, pag. 1 (1978)
37. Mitchell D.J., Ninham B.W. J. Chem. Soc. Faraday Trans. II (in the press)
38. Evans E.A., Skalak R., CRC Critical Revs. Bioeng. 3:181 (1979)
39. Sackmann E. in "Biological Membranes" vol. 5, Academic Press, London p.105 (1984)
40. Helfrich W., E. Naturforsch 28:693 (1973)
41. Petrov A.G., and Pavloff Y.V., J. Physique 40:455 (1979)
42. Tanford C., J. Phys. Chem. 76:3020 (1972)
43. Israelacvili J.N., Mitchell D.J., Ninham B.W., J. Chem. Soc. Faraday Trans. II 72:1525 (1976)
44. Verkeley A.J. Biochim. Biophys. Acta 779:43 (1984)
45. Cullis P.R., De Kruijff B., Hope M.J., Verkleij A.J., Nayar R., Farren S.B., Tilcock C., Madden T.D., Bally M.B. in "Membrane fluidity in Biology" vol. 1 Academic Press, New York p.39 (1983)
46. Rilfors L., Lindblom G., Wieslander A., Christiansson A. in "Biomembranes" vol. 12, Plenum Press, New York, p.205 (1983)
47. Stier A., Finch S.A.E., Bosterling B., FEBS Lett. 91:109 (1978)
48. Kuypers F.A., Roelofsen B., Berendsen W., Op Den Kamp J.A.F., Van Deenen L.L.M., J. Cell. Biol. 99:2260 (1984)
49. Shertz M.P., Singer S.J., Proc. Natl. Acad. Sci. USA 71:4457 (1974)
50. Christiansson A., Kuypers F.A., Roelofsen B., Op Den Kamp J.A.F., Van Deenen L.L.M., J. Cell. Biol. 101:1455 (1985)
51. Op den Kamp J.A.F., Roelofsen B., Van Deenen L.L.M. TIBS, 320 (1985)
52. Singer S.J. and Nicolson G.L., Science 175:720 (1972)
53. Jain M.K., White III H.B. Adv. Lipid. Res. 15:1 (1977)
54. Galla H.J., Hartmann W., Theilen U., Sackmann E. J. Membrane Biology 48:215 (1979)
55. Marsh D. Watts A., Knowles P.F. Biochemistry 15:3570 (1976)
56. Nagle J.F., Scott H.L., Biochim. Biophys. Acta 513:236 (1978)
57. Kinosity K, Tsong T.Y., Nature 272:258 (1978)
58. Petrov A.G., Mitov M.D., Deuzhaniski A. in "Advances in liquid crystal research and application" Bata L. ed., Pergamon Press, Budapest, pag. 695 (1980)
59. Sackmann E. in "Biological Membranes" vol. 5, Academic Press, London, pag. 105 (1984)
60. Klausner R.D., Kleinfeld A.M. in "Cell surface dynamics", Perelson A.S., Delisi C., Wiegel F.W. eds., Marcel Dekker, New York, 23 (1984)
61. Curatola G., Bertoli E. in "Biomembrane and receptor mechanisms" E. Bertoli and D. Chapman eds., Liviana Springer Verlag-Press, in press
62. Boggs J.M., Wood D.D., Moscarello M.A., Papahadjopoulos D., Biochemistry 16:2325 (1977)
63. Lenaz G., Mascarello S., Landi L., Cabrini L. Pasquali P., Parenti-Castelli G., Sechi A.M., Bertoli E. in "Membrane Bioenergetics" L. Packer ed., Elsevier Amsterdam, pag. 189 (1977)
64. Jost P.C., Griffith O.H., Capaldi R.A., Vanderkooi G., Biochim. Biophys. Acta 311:141 (1973)
65. Chapman D., Gomez-Fernandez J.C., Goni F.M., FEBS Lett. 98:211 (1979)
66. Yellin N., Levin I.W., Biochim. Biophys. Acta 468:490 (1977)

67. Freire E., Snyder B., Biochim. Biophys. Acta 600:643 (1980)

68. Parasassi T., Conti F., Glaser M., Gratton E., J. Biol. Chem. 259:14011 (1984)

69. Gratton E., Fiorini R.M., Valentino M., Wang P., Biochem. J. in press

70. Curatola G., Fiorini R.M., Gratton E., Biochim. Biophys. Acta submitted

71. Schiffmann Y., Prog. Biophys. Molec. Biol. 36:87 (1980)

72. Peracchia C., Dulhunty A.F., J. Cell. Biol. 70:419 (1976)

73. Loor F. in "Cytoskeletal elements and plasma membrane organization" C. Poste and G.L. Nicolson eds., Elsevier, North Holland, pp. 253 (1981)

74. Seelig J., Browning J.L., FEBS Lett. 92:41

75. Lenaz G. in "Biomembrane" Burton R.M., Guerra F.C. eds. Plenum Press, N.Y. pag. 111 (1984)

76. Van der Meer W. in "Physiology of membrane fluidity" vol. 1, M. Shinitzky ed., CRC Press, pag.53 (1984)

77. Stubbs C.D., Essay in Biochem. 19:1 (1983)

78. Shinitzki M., Yuli I., Chem. Phys. Lipids 30:261 (1982)

79. Cherry R.J. Biochim. Biophys. Acta 559:289 (1979)

80. Saffman P.G., Delbrück M., Proc. Natl. Acad. Sci. USA 72:3111 (1975)

81. Cohen M.H., Turnbull D., J. Chem. Phys. 31:1164 (1959)

82. Montroll E.W., J. Math. Phys. 10:753 (1969)

83. Vaz W.L.C., Goodsaid-Zalbuonds F., Jacobson K. FEBS Lett. 174:122 (1984)

84. Berg H.C. in "Random walks in biology" Princeton University Press (1983)

85. Freye L.D., Edidin M., J. Cell. Sci. 7:319 (1970)

86. Poo M. and Cone R.A. Nature 247:438 (1974)

87. Peters R. in "Structure and properties of cell membranes" vol. 1, Benga G.H. ed., CRC Press, Boca Raton FL, pp. 35 (1985)

88. Wade C.G. in "Structure and properties of cell membranes" vol. 1, Benga G.H. ed., CRC Press, Boca Raton FL, pp. 51 (1985)

89. Aizenbud B.M., Gerston N.D. Biophys. J. 38:287 (1982)

90. Mc Closkly M., Poo M.M., Int. Rev. Cytol. 87:19

91. Jain M.K. in "Membrane fluidity in biology", Aloia R.C. ed., Academic Press, New York pp. 1, (1983)

92. Kawato S., Kinosita K., Biophys. J. 36:277 (1981)

93. Peters R., Cherry R.T., Proc. Natl. Acad. Sci. USA 79:4317 (1982)

94. Poo M., Ann. Rev. Biophys. Bioenerg. 10:245 (1981)

95. Sowers A.E., Hackenbrock C.R., Proc. Natl. Acad. Sci. USA 78:6246 (1981)

96. Marsh D., Watts A. in "Liposomes: from physical structure to therapeutic applications" Knight C.G. ed.) Elsevier, Amsterdam pp.139 (1981)

97. Cornell B.A., Poge S.M., Chem. Phys. Lipids 27:151 (1980)

98. Lakowicz J.H., Hogen D., Chem. Phys. Lipids 26:1 (1980)

99. Razi-Nagvi K., Behr J.P., Chapman D., Chem. Phys. Lett. 26:440 (1974)

100. Kell D.B., Harris C.M., J. Biophys. 12:181 (1985)

101. Einstein A., Ann. Physik. 19:286 (1906)

102. Small R.K., Blank M., Ghez R., Pfenninger R.H., J. Cell. Biol. 98:1434 (1984)

103. Hyde J.S., Dalton L.R. in "Spin labelling II: theory and applications" Berliner L. ed., Academic Press, New York, p.1 (1979)

104. Cone R.A. Nature New Biol. 236:39 (1972)

105. Daemen F.J.M., Biochim. Biophys. Acta 300:255 (1973)

106. Razi-Nagvi K., Gonzales-Rodriguez J., CHerry R.J., Chapman D., Nature New Biol. 245:249 (1973)

107. Vaz W.L.C., Hallmann D., FEBS Lett. 152:287 (1983)
108. Muller H.J., Galla H.J., Biochim. Biophys. Acta 733:291 (1983)
109. Mc Cown S.T., Evans E., Diehl S., Wiles H.C., Biochemistry 20:3134 (1981)
110. Rimon G., Meyerstein N., Henis Y.I. Biochim. Biophys. Acta 775:283 (1984)
111. Lubin B., Chin D., Bastacky J., Roelofsen B., J. Clin. Invest. 67:1643 (1981)
112. Lenaz G., Fato R., J. Bioenerg. Biomemb. 18:365 (1986)
113. Nicolson G.L. Biochim. Biophys. Acta 457:97 (1976)
114. Vaz W.L.C., Kaptza H.E., Stumpel J., Sackmann E., Jovin T.M., Biochemistry 20:1392 (1981)
115. Kleinfeld A.M., Dragsten P., Klausner R.D., Pjura N.S., Matayoshi E.D., Biochim. Biophys. Acta 649:471 (1981)
116. Hughes B.D., Pailthorpe B.A., White L.R., Sawyer W.H., Biophys. J. 97:673 (1982)
117. Smith B.A., Clark W.L., Mc Connell H.M., Proc. Natl. Acad. Sci. USA 76:5641 (1979)
118. Schinder M., Koppel D.E., Sheetz M.P., Proc. Natl. Acad. Sci. USA 77:1457 (1980)
119. Koppel D.E., Sheetz M.P., Schindler M., Proc. Natl. Acad. Sci. USA 78:3576 (1981)
120. Golan D.E., Veath W. Proc. Natl. Acad. Sci. USA 77:2537 (1980)
121. Aszalos A., Yang G.C., Gottesman M.M., J. Cell. Biol. 107:1357 (1985)
122. Edidin M., in "Membrane and transport" vol. 1, Martonosi A.N. ed., Plenum Press, New York, pp.141 (1982)
123. Freire E., Snyder B., Biophys. J. 37:627 (1982)
124. Saxton M., Biophys. J. 39:165 (1982)
125. Schindler M., Osborn M.J., Koppel D.E., Nature 283:346 (1980)
126. Onsager L., Physiol. Rev. 37:405 (1931)
127. Berg H.C., Purcel E.M., Biophys. J. 20:193 (1977)
128. Adam G., Delbrück M. in "Structural chemistry and molecular biology" Rich A., Davidson N. eds., Freeman, San Francisco p.198 (1968)
129. Keith A.D., Snipes W., Science 183:666 (1973)
130. Keith A.D., Mastro A.M. in "Membrane fluidity in biology" vol.2, Aloia R.C. ed., Academic Press, New York, p.237 (1983)
131. Small R. and Pfenninger K.H., J. Cell. Biol. 98:1422 (1984)
132. Axelrod D., J. Memb. Biol. 75:1 (1983)
133. Cohen M.W., J. Exp. Biol. 8:43 (1980)
134. Cohen M.W., Weldon P.R., J. Cell. Biol. 86:388 (1980)
135. Tank D.W., Wu E.S., Webb W.W., J. Cell. Biol. 92:207 (1982)
136. Goldstein J.L., Anderson R.GW., Brown M.S., Nature 279:679 (1979)
137. Barak L.S., Webb W.W., Biophys. J. 33:749 (1981)
138. Goldstein B., Wofsy C., Bell G., Proc. Natl. Acad. Sci. USA 78:5695 (1981)
139. Carpentier J.L., Gordon P., Anderson R.G.W., Goldstein J.L., Brown M.S., Cohen S., Orci L., J. Cell. Biol. 95:73 (1982)
140. Oliver J.M., Berlin R.D., Int. Rev. Cytol. 74:55 (1982)
141. Bretscher M.S., Nature 260:21 (1976)
142. Wolf D.E., Henkart P., Webb W.W., Biochemistry 19:3893 (1980)
143. Evans E.A., Biochim. Biophys. Acta 604:27 (1980)

144. Berg O.G., Van Hippel P.H., Annu. Rev. Biophys. Chem. 14:131 (1985)
145. Smoluchowski M., Z. Phys. Chem. 92:129 (1985)
146. Sole K., Stockmayer W.H., J. Chem. Phys. 54:2981 (1971)
147. Chou K.C., Zhon G.P., J. Am. Chem. Soc. 104:1409 (1982)
148. Welch G.R., Gaertner F.M., Proc. Natl. Acad. Sci. USA 72:4218 (1975)
149. Mosbach K. FEBS Lett. 62:80 (1976)
150. Overfield R.E., Wraight C.A., Biochemistry 19:7328 (1980)
151. Hardt S.L., Biophys. Chem. 10:239 (1979)
152. Emeis C.A., Fehder P.L., J. Am. Chem. Soc. 92:2246 (1970)
153. Torney D.C., Mc Connell H.M., Proc. R. Soc. (London) 387:147 (1983)
154. Kroger A., Klingenberg M., Eur. J. Biochem. 34:358 (1973)
155. Fato R., Battino M., Parenti Castelli G., Lenaz G., FEBS Lett. 179:238 (1985)
156. Fato R., Battino M., Degli Esposti M., Parenti Castelli G., Lenaz G., Biochemistry 25:3378 (1986)
157. Gupte S., Wu E.S., Hoechli L., Hoechli M., Jacobson K., Sowers A.E., Hackenbrock C.R., Proc. Natl. Acad. Sci. USA 81:2606 (1984)
158. Hackenbrock C.R., Gupte S.S., Chazotte B. in "Achievements and perspectives of mitochondrial research" vol.1, Quagliarello E., Slater E.C., Palmieri F., Saccone C., Kroon A.M. eds., Elsevier Amsterdam p.83 (1985)
159. Levitzki A., J. Receptor Res. 4:399 (1984)
160. Orly J., Schramm M., Proc. Natl. Acad. Sci. USA 73:4410 (1976)
161. Tolkowsky A.M., Levitzky A., Biochemistry 17:3795 (1978)
162. Arad H., Rosenbusch J.P., Levitzki A., Proc. Natl. Acad. Sci. USA 81:6579 (1984)
163. Atlas D., Volsky D.J., Levitzki A., Biochim. Biophys. Acta 597:64 (1980)
164. Henski E., Rimon G., Levitzki A. Biochemistry 18:846 (1979)
165. Henis Y.I., Rimon G., Felder S., J. Biol. Cell. 257:1407 (1982)
166. Bakardjieva A., Peters R., Hekman M., Horning H., Burgmeister W., Helmreich E.J.M. in "Metabolic interconversion of enzymes" Holzer H. ed., Springer Verlag, New York p.378 (1981)
167. Henis Y.I., Elson E.L. Proc. Natl. Acad. Sci. USA 78:1072 (1981)
168. Hirata F., Axelrod J., Nature 275:219 (1980)
169. Curatola G., Ferretti G., Zolese G. Arch. Biochem. Biophys. submitted
170. Insel P.A., Kennedy M.S., Nature 273:471 (1978)
171. Rasenick M.M., Stein P.J., Bitensky M.W., Nature 294:560 (1981)
172. Sandermann H., FEBS Lett. 21:254 (1980)
173. Dean W.L., Tanford C. Biochemistry 17:1683 (1978)
174. Lenaz G. in "Biomembrane and receptor mechanisms" Chapman D. and Bertoli E. eds., Liviana Springer Verlag -press, in press
175. Raison J.K. J. Bioenerg. 4:559 (1972)
176. Lenaz G. in "Biomembranes: dynamics and biology" Guerra F.C., Burton R.M. eds., Plenum, N.Y., pp. 111 (1984)
177. Dixon M., Webb E.C. Enzymes, Longmans, London (1964)
178. Massey V., Curti B., Ganther H.A. J. Biol. Cell. 241:2347 (1966)
179. Biosca J.A., Travers F., Barman T.G. FEBS Lett. 153:217 (1983)
180. Baracca A., Curatola G., Parenti-Castelli G., Solaini G. Biochem. Biophys. Res. Commun. 136:891 (1986)
181. Hidalgo C., Ikemoto N., Gergely J. J. Biol. Cell. 251:4224 (1976)
182. Solaini G., Baracca A., Parenti-Castelli G., Lenaz G. J. Bioenerg. Biomemb. 16:391 (1984)

183.Hackenbrock C.R., Hoechli M., Chan R.M. Biochim. Biophys. Acta 455:466 (1974)

184.Raison J.K., McMurchie E.J. Biochim. Biophys. Acta 363:135 (1974)

185.Lenaz G., Curatola G., Mazzanti L., Zolese G., Ferretti G. Archives 223:369 (1983)

186.McMurchie E.J., Gibson R.A., Albywardena M.Y., Charnock J.S. Biochim. Biophys. Acta 727:163 (1983)

187.Battino M., Castelluccio C., Pilarska M., Sarzala M.G., Lenaz G. Cell. Biol. Int. Reports 10:154 (1986)

188.Bertoli E., Finean J.B., Griffiths D.E. FEBS Lett. 61:163 (1976)

189.Heremans K. Ann. Rev. Biophys. Bioeng. 11:1 (1982)

190.Rottem S., Cirillo K.P., De Kruiiff B., Shinitzky M., Rasin S. Biochim. Biophys. Acta 323:509 (1973)

191.Duppel W., Ulrich W. Biochim. Biophys. Acta 426:399 (1976)

192.Lenaz G. in "Biomembrane and Receptor Mechanisms" Chapman D., Bertoli E. ed. Liviana Springer Verlag-press in press

193.Lenaz G., Battino M., Degli Esposti M., Fato R., Parenti-Castelli G. in "Biomedical and clinical aspects of coenzyme Q" Folkers K., Yanamma Y. eds., Elsevier, Amsterdam pp.73 (1986)

194.Beece D., Gisenstein L., Frananfelder H., Good D., Manden M.C., Keinisch L., Reynolds A.H., Porensen L.B., Yue K.T. Biochemistry 19:5147 (1980)

195.Swann A.C. Archives 221:148 (1983)

196.Cooper A., Proc. Nat. Acad. Sci. 73:2740 (1976)

197.Cooper A., Progr. Biophys. Mol. Biol. 44:181 (1984)

198.Somogyi B., Welch G.R., Damyanovich S. Biochim. Biophys. Acta 766:81 (1984)

199.Careri G., Fasella P., Gratton E. Annu. Rev. Biophys. Bioeng. 6:69 (1979)

200.Hughes B.O., Pailthorpe B.A., White L.R., Sawyer W.H. Biophys. J. 97:673 (1982)

201.Seelig J. and Seelig A. Quart. Rev. Biophys. 13:19 (1980)

202.Kell D.B. in "Collective phenomena in physics of biophysics" Pohl H.A. ed., in press (1986)

203.Valentino M., Governa M., Fiorini R.M., Curatola G. Biochem. Biophys. Res. Commun. submitted

204.Coolard J.G., Dewildt A., Comen-Meulemans E.P.M., Smeekens J., Emmelot P., Inbar M. FEBS Lett. 77:173 (1977)

205.Toyoshima S., Osawa T. Exp. Cell. Res. 102:438 (1976)

206.De Laat S.W., Bluemink J.G., Boonstra J., Mummery C.L., Van der Saag P.T., Van Zoelen E.J.J. in "Physiology of membrane fluidity" vol. 2, (M. Shinitzky ed.) CRC Press, pag.21-51 (1981)

207.De Laat S.W., Van der Saag P.T., Shinitzky M. Proc. Natl. Acad. Sci. USA 74:4458 (1977)

208.De Laat S.W., Van der Saag P.T., Elson E.L., Schlessinger J. Proc. Natl. Acad. Sci. USA 77:1526 (1980)

ON THE IMPLICATIONS OF LATERALLY MOBILE CORTICAL TENSION ELEMENTS FOR CYTOKINESIS

J.G. White and A.A. Hyman
Medical Research Council Laboratory of Molecular Biology
Hills Road, Cambridge CB2 2QH, England

The processes of mitosis and cytokinesis work in concert to produce two cells from a single progenitor, each of which has a complete complement of chromosomes. These two processes have been extensively studied since it was first appreciated that all organisms are cellular, yet neither process is completely understood at the present time. Cytokinesis in animal cells occurs by the formation of a furrow which typically appears as an annular constriction at the equator of the cell. Cytokinesis in plant cells seems to be a completely different process that does not involve an active constriction; this mode of cytokinesis will not be discussed here.

The position of the constricting cleavage furrow in a dividing animal cell is such that the mitotic apparatus (MA) is bisected after anaphase has occurred (and the chromosomes have migrated to the two poles), thus ensuring the equal partitioning of the two sets of chromosomes into the daughter cells. In order to understand the process of cytokinesis it will be necessary to have explanations as to the nature and genesis of the contractile ring, and the timing of furrowing relative to the mitotic activity of the cell. In this chapter we will briefly review the current state of knowledge on cytokinesis. We will then discuss a theory that attempts to explain many of the aspects of cytokinesis on the basis of the behaviour of laterally mobile tension elements in the cortex of a cell. Finally we would like to describe some unusual cleavage configurations that occur in the eggs of Caenorhabditis elegans, and examine the relationshhip between the MA and the cleavage furrow in these situations.

THE RELATIONSHIP BETWEEN THE MITOTIC APPARATUS AND THE CLEAVAGE FURROW

The cleavage furrow nearly always bisects the mitotic spindle, even in the case of asymmetric divisions, where the spindle is eccentrically positioned within the cell. This generalisation also holds in the extreme case of polar body exclusion, where a small meiotic spindle is situated adjacent to the plasma membrane prior to cleavage (Odor & Renninger, 1960). Such behaviour suggests that there may be a causative link between the position of the MA (which develops well before the appearance of a furrow) and the position of the cleavage furrow. This has been verified by displacing the MA from its normal position by a variety of methods, such as centrifugation (Conklin, 1917) or micromanipulation (Rappaport & Ebstein, 1965). When these manipulations are done early enough the cleavage furrow is always seen to bisect the MA in its new orientation, even when this orientation is quite different from that which would have occured in the unperturbed state. No cleavage furrow forms if the formation of the MA is blocked by colchicine (Beams and Evans, 1940). Taken together these observations, along with many others, strongly implicate the MA as being necessary for cleavage furrow formation and also for specifying its orientation. However, several experiments have shown that the MA is not necessary for furrowing. The MA can be completely removed just before the onset of furrowing, yet a furrow will appear in a location that would have bisected the MA, were it in its normal location (Hiramoto, 1971). Thus the MA appears to be necessary to set up the conditions necessary for furrowing without playing a part in the actual process of furrowing.

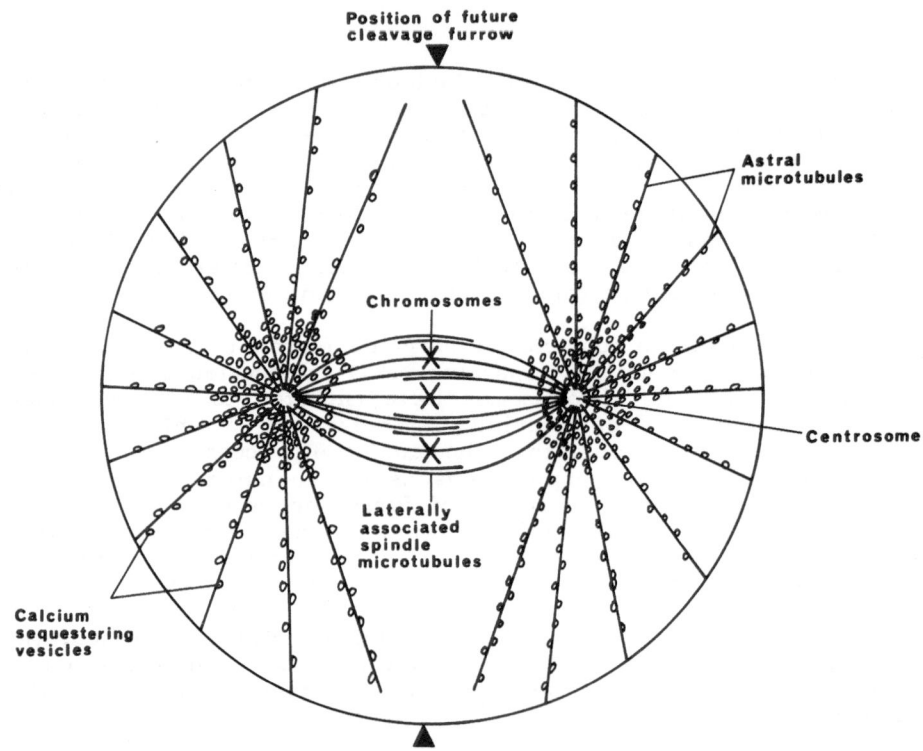

FIGURE 1.   Schematic drawing of the mitotic apparatus of a typical animal cell at metaphase. Microtubules radiate out of two centrosomes. Some are attached to to the kinetochores of chromosomes and others become laterally associated with microtubules from the opposite aster, making up the body of the mitotic spindle. Astral microtubules away from the spindle radiate out to the cortex of the cell. Large accumulations of membrane-bound vesicles are associated with the astral microtubules, some of which probably have a calcium sequestering activity. Furrowing occurs at the equator (arrowed) after the chromosomes have separated and migrated to the poles. From White, 1985.

At metaphase a MA typically consists of two centrosomes out of which radiate many microtubules (the asters). A central spindle is located between the centrosomes onto which the chromosomes are attached (Fig. 1). Given that a cleavge furrow bisects the MA, and also that the chromosomes lie on this future bisection plane at metaphase, it might be thought that the chromosomes or some closely associated entity might be the agents that specify the furrow location. In an elegant experimeent Rappaport (1961) demonstrated that this is not the case. A cell was made to take up a toroidal configuration by deeply indenting it with a small blunt object (Fig. 2). The first cleavage cut the torus, bisecting the MA but did not divide the cell into two. At the second metaphase two MAs formed in the single cell and a cleavage furrow bisected each as expected. There was, however, an additional cleavage furrow between the two asters from the adjacent MAs. This experiment neatly demonstrated that, at least in the case of the sand dollar, the apposition of two asters was a sufficient condition for the induction of a cleavage furrow, even if there is no interconnecting spindle. Further evidence that none of the nuclear components are necessary for furrowing has come from Hiramoto (1971), who demonstrated that the nucleus may be removed without affecting furrowing, provided that the asters stay separated. In cases of polyspermy, where there are more than two asters in a single cell, furrows are seen between all combinations of adjacent pairs of asters (Sugiyama, 1951). This suggests that all asters are equivalent i.e. they do not exist as complementary pairs.

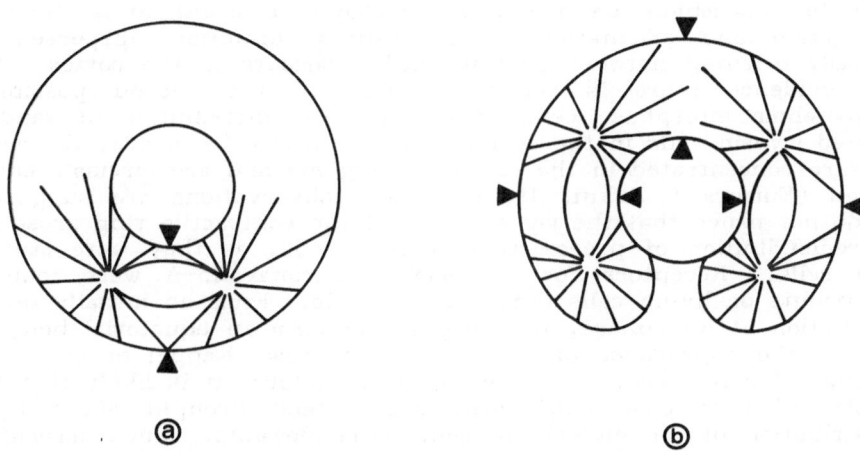

(a)   (b)

FIGURE 2. When a sand dollar egg is constricted into a toroidal configuration by pressing a blunt needle down in the middle of the egg, a binucleate cell is produced after the first cleavage (a). When this cell cleaves (b), furrows (arrowed) form to bisect the two mitotic spindles, and an additional furrow also forms between the apposed asters of the adjacent spindles. From White, 1985; redrawn from Rappaport, 1961

Asters appear to act cooperatively over a limited range as, in a given cell type, there is a maximum astral separation and maximum distance of the asters from the cortex beyond which furrowing will not occur (Rappaport, 1969).

The general conclusions that one may derive from these observations are: that asters are the cell organelles that induce furrowing, that they work in pairs and that, furthermore, one aster is very much like another.

THE NATURE AND LOCATION OF THE FORCE GENERATING ELEMENTS

Expriments have shown that although the presence of a MA is necessary to initiate furrowing, it can be removed along with much of the cytoplasm, just prior to the appearance of a furrow, without inhibiting cytokinesis (Hiramoto, 1956). These observations strongly suggest that the forces for furrowing must be generated within the cortex of the cell. The cortex is the region that is immediately subjacent to the plasma membrane. This region contains a network of filaments that are randomly oriented (Yomemura & Kinoshita, 1986; Yumura & Fukui, 1985). At the apex of the cleavage furrow there is a high density of such filaments; here they they are oriented circumferentially, making up a structure that is generally referred to as the contractile ring (Schroeder, 1975). The contractile ring has been shown to contain actin (Schroeder, 1975) and myosin (Yumura & Fukui, 1985). Furrowing requires ATP (Landau et al., 1955) and is blocked by the applicaton of antibodies to myosin (Mabuchi & Okuno (1977). It therefore seems probable that the forces for furrowing are generated by an acto-myosin system. The equatorial location of the contractile ring, together with the circumferential orientation of its component filaments, suggests that it operates to constrict the equator by a sliding filament mechanism.

One of the central questions that must be answered in order to understand cytokinesis is how the contractile ring arises in the cortex in its stereotyped location relative to the two apposed asters of the MA. It could be assembled de novo, or perhaps it could arise from the aggregation and orientation of an initially uniformly dispersed and randomly oriented network of contractile filaments in the cortex. What little evidence there is seems to point to the second possibility. Dictyostelium interphase cells have a uniform distribution of randomly oriented myosin rods in the cortex. When these cells cleave, the myosin rods are concentrated in the equatorial regions and are circumferentially aligned (Yumura & Fukui, 1985). These observations are suggestive, but do not prove that the myosin rods of the contractile ring arose from the redistribution of pre-existing cortical rods, as they were made on fixed cells. Receptors for the lectin Concanavalin-A were found to redistribute on living cells during cytokinesis. From an initially uniform distribution at interphase, they migrate to form an equatorial belt, just prior to the appearance of the cleavage furrow (Kappel et al., 1982). Although Con-A receptors are on the cell surface, it is likely that their redistribution is coordinate with, and indeed brought about by, a redistribution of the underlying contractile elements. Such associations are seen to occur when Con-A receptors are crosslinked and as a consequence cap (Berlin & Oliver, 1978).

CORTICAL ACTIVATION PRIOR TO CLEAVAGE

It is a common observation that cells tend to round up prior to cleavage. At this time cortical stiffness (Mitchison & Swann, 1955) and internal hydrostatic pressure (Hiramoto, 1963) increase, but start to fall off just prior to the onset of furrowing. These changes are likely to be brought about by a generalised activation of the cortical contractile apparatus. This activation has been shown to be independent of the presence of an intact MA, as cyclic changes in the cortical tension still occur in enucleated cytoplasts from amphibian eggs (Hara et al., 1980). The period of these cycles corresponds to the period of the normal mitotic cycle. The cytoplasts do not cleave however, presumably because they lack a MA.

Thus there are to be two seperable aspects to cytokinesis: an inital cortical activation which is independent of the presence of the nucleus and the MA, and an interaction of the activated cortex by the MA to produce a cleavage furrow (Schroeder, 1981). These two activities are independent but are coordinated in cells undergoing mitosis.

SPECULATIONS AS TO THE NATURE OF THE STIMULUS FROM THE MITOTIC APPARATUS

Given that the asters of the MA are strongly implicated as the effectors of the furrowing stimulus, it is worth considering whether their location or structure gives any clue as to the nature and distribution of the the stimulus. Astral microtubules originate from the centrosome and generally extend right out to the cortex (Fig 1 & Fig. 7b), suggesting that they could interact with the cortical apparatus. However, neither tubulin nor microtubules have been demonstrated to be involved in the direct regulation of actomyosin systems. The stimulus could be some factor that is released from the asters or it could equally well be some factor that is removed by the asters (Conrad & Rappaport, 1981). Dense accumulations of membrane bound vesicles are associated with asters (Fig. 1). These vesicles have been shown to have an ATP activated calcium seqestering activity (Silver et al., 1980). They are presumably derived from the endomembrane system of the cell which breaks down and vesiculates on entry into metaphase. During the early stages of mitosis vesicles have been seen to travel along astral rays and accumulate at the centrosomes in time-lapse video recordings of the early divisions of <u>Caenorhabditis</u> <u>elegans</u> (J.G. White, unpublished observations).

It is tempting to speculate that the stimulus from asters involves a local sequestration of calcium mediated by the calcium sequestering vesicles associated with the asters. Such a notion is attractive in that it may also suggest explanations for other aspects of mitosis. As has been pointed out, the contractile apparatus in the cortex seems to be uniformly activated prior to cytokinesis. A global release of calcium could achieve such an activation. Intracellular injections of calcium ions (Gingell, 1970) or the the surface application of calcium ionophores (Schroeder & Strickland, 1974) elicit cortical contractions, suggesting that the contractile apparatus may be calcium regulated, in common with many other acto-myosin systems. Furthermore, these and other experiments (Bluemink, 1972) demonstrate that any part of the cortex may be induced to contract, suggesting that functional contractile elements exist all over the cortex. High levels of calcium are inimical to the stability of microtubules (DeBrabander & DeMay, 1980), and indeed the interphase microtubule array breaks down on entry to mitosis. It is likely that calcium sequestering vesicles have some motor attached to them which drives them along astral microtubules towards the centrosome. Newly nucleated microtubules from the centrosomes may be able to survive in high levels of calcium if they are surrounded by such calcium sequestering vesicles which act to keep the concentration of calcium low in their immediate vicinity (Schatten et al., 1980).

DISTRIBUTION OF MICROTUBULES PRIOR TO CLEAVAGE

If microtubules are the direct or indirect (ie via associated calcium sequestering vesicles) mediators of cortical stimulation, then a knowledge of their distribution around the cortex may give some indication as to the distribution of the stimulus on the cortex. If one assumes for simplicity that the asters of the MA are spherically symmetric structures, then simple geometric arguments demonstrate that, in an initially spherically symmetric cell, the lowest density of microtubules at the cortex will always be in the equatorial regions (White & Borisy, 1983). These estimates probably over-estimate the density of equatorial microtubules, because the assumption of spherical symmetry is probably not valid. In a real MA it is likely that many of the microtubules become captured by the kinetochores of chromosomes to form chromosomal microtubules (Euteneur & McIntosh, 1981). Other microtubules from the apposing asters probably become laterally associated and make up the body of the spindle (McIntosh et al., 1979). Thus there is likely to be a paucity of free microtubules on the side of the aster which is adjacent to the spindle (Fig. 1 & 7b & d). A better approximation to an aster may therefore be a hemisphere. In this case there will clearly be an even lower density of cortical microtubules in the equatorial regions than is the case for spherically symmetric asters. An electron microscopic study has been made of the distribution of astral microtubules on the cortex of a dividing cell, and it was found that few if any microtubules were present in the equatorial regions (Asnes & Schroeder, 1979). It therefore seems likely that if the stimulus is borne to the cortex via the astral microtubules, and if it also bears some simple relation to microtubule density, then such a stimulus has to have its lowest value at the equator.

EQUATORIAL VS. ASTRAL STIMULATION

We have put forward arguments to suggest that the stimulus emanates from the asters and acts on the contractile apparatus of the cortex and therefore has to be lowest in the equatorial regions, where the furrow will ultimately form. There is a considerably body of experimental observation that has been interpreted to suggest that the stimulus emanates from the equatorial regions between two apposed asters (eg Rappaport & Rappaport, 1984; Rappaport, 1971). It is difficult to see how such a pattern of stimulation could arise in this scheme, but of course this is not a valid reason for rejecting it. However, several of the observations that have been taken to be suggestive of the presence of an equatorial stimulation can be also be explained with direct astral stimulation schemes (White & Borisy, 1983; White, 1985). Rather than enter into a lengthy debate about the relative merits of the two proposals, we would like to explore the consequences of the astral relaxation theory and to discuss observations on some unusual cleavage configurations that occur in the early development of C. elegans in the context of equatorial and astral stimulation.

THE ASTRAL RELAXATION THEORY OF FURROWING

The most economical explanation of furrowing, given that there is a generalised activation of cortical contraction prior to furrowing, is that a stimulus emanating from the asters acts to lower the cortical tension in their immediate vicinity. There will consequently be a lower tension in the equatorial regions than the adjacent regions in the cortex, thereby producing a localised equatorial constriction. This theory was originally developed by Wolpert (1960). Although this theory is very attractive it is not in itself a sufficient explanation of furrowing. Indeed any theory that suggests that furrowing occurs because of a difference in isotropic cortical tension between the equatorial regions and adjacent regions cannot work as it stands because of the following simple physical argument.

If the cortical region is reasonably thin, then any activation of randomly oriented tension elements within it can be considered to develop

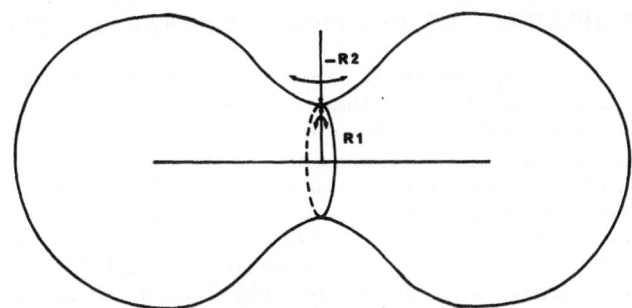

FIGURE 3.   A differential in isotropic surface tension between between
the equator and the adjacent regions cannot by itself produce
cleavage. A cleavage furrow is a saddle surface; the two principal
radii of curvature of the surface, R1 and R2, will therefore be of
opposite signs. The inward directed pressure due to surface
tension forces (T) is T/R1 + T/R2. In the configuration drawn, R1
is the same size as R2, but is of opposite sign. There cannot
therefore be any inward directed force from surface tension in this
situation. From White, 1985.

a surface tension. The pressure exerted by a curved surface which has
an isotropic surface tension T is :

$$T/R1 + T/R2$$

where R1 and R2 are the principal radii of curvature of the surface. A
cleavage furrow is a saddle surface (Fig. 3), where the two radii of
curvature have opposite signs. When these two radii have similar
absolute values, which would typically occur at some intermediate point
in the development of a furrow, the two halves of the expression cancel
out. Surface tension forces can therefore no longer generate any
inward-directed pressure at the equatorial regions in this situation, so a
furrow could not progress to completion.
    The astral relaxation model as it stands assumes an isotropic
surface tension. However, the presence of a contractile ring at the
cleavage furrow, with its circumferentially oriented contractile elements,
indicates that the tension exerted by the contractile cortex at the
equator is likely to be highly anisotropic. The pressure generated by a
curved surface in the case of anisotropic surface tension is:

$$T1/R1 + T2/R2$$

where T1 and T2 are the components of the surface tension along the
principal axes of curvature. If the tension in the equatorial orientation
(T1) is very much greater than the tension in the polar orientation (T2)
then furrowing can progress to completion, in spite of the negative
influence of the T2/R2 term.
    These arguments emphasise the importance of the orientation of the
contractile elements within the contractile ring and beg, yet again, the
question of how this organelle is formed by the interaction of the MA
with the cortex.

We have implicitly assumed up to now that tension elements are fixed in the plane of the cortex. This is unlikely to be the case however. The components of the cortex probably maintain their cortical location by means of some attachments to the plasma membrane. Such attachments will not constrain them locally because a membrane is essentially a two-dimensional fluid and could offer only viscous resistance to lateral movements. There is therefore every reason to suppose that cortical elements have a high degree of lateral mobility in the plane of the membrane. This will have the significant consequence that, if gradients of tension exist in the cortex, then contractile elements would pull themselves into the regions of highest tension (White & Borisy, 1983). Thus gradients of tension would result in the redistribution of cortical elements. A uniformly activated cortex would therefore be in a metastable state, because, if there is a local weakening of the cortex, the adjacent contractile elements will move away from this region up the tension gradient. This will continue until all the elements have aggregated on the far side of the cell, a situation that possibly arises during capping.

During cytokinesis we have argued that the cortical tension will be greatest in the equatorial regions. In the case of the astral relaxation theory it has been proposed that this is caused by the reduction of a previously uniform tension in the vicinity of the asters. Equatorial stimulation theories propose that the equatorial regions are directly activated in some way by the apposed asters. Either way the resulting gradient of tension towards the equator will cause mobile tension elements to aggregate in this region. Because of the unidirectional compression of cortical elements that occurs when this happens, these elements will also become partially circumferentially oriented on reaching the equator (White & Borisy, 1983). Thus by the simple expedient of allowing tension elements to be laterally mobile it is possible to explain the location and the form of the contractile ring.

Contractile rings are generally rather narrow being situated at the apex of the cleavage furrow and the constituent filaments are aligned, being essentially parallel. In the scheme that we have outlined above the contractile ring is initially rather broad and the filaments only partially aligned. However as the equatorial regions flatten into a cylindrical conformation, tension-producing elements will be further aligned circumferentially as components of their internally generated tension will drive them into this orientation (White & Borisy, 1983). When furrowing commences, elements will similarly drive themselves into the base of the furrow, like an elastic band slipping up the neck of a greasy wine bottle (White & Borisy, 1983). The contractile ring is therefore regenerative, becoming sharper and more aligned as it progresses. Computer simulations of this model can produce realistic cleavage configurations (Fig. 4) and have demonstrated that, in principle, the model is capable of accommodating many of the naturally occurring variations of cleavage configuration that occur (White & Borisy, 1983). If the lateral mobility of tension elements is allowed, then both the astral relaxation and equatorial stimulation schemes will produce cleavage figures in the computer simulations, but those using equatorial stimulation were found to give unnaturally sharp furrows, producing two nearly hemispherical daughter cells (J.G. White, unpublished observations).

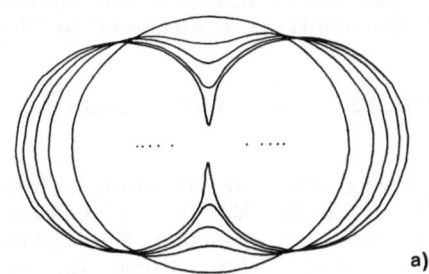 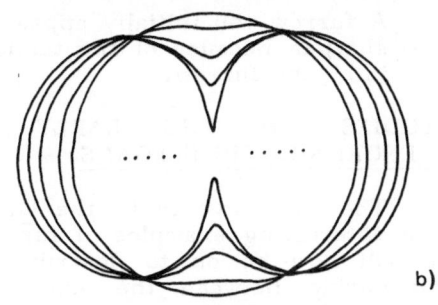

FIGURE 4.   Comparison of a sequence of computer simulations of the astral relaxation model with laterally mobile tension elements (a) with measurments made on a cleaving sea urchin egg (b) from data described by Hiramoto, 1958. From White & Borisy, 1983.

SUMMARY OF ASTRAL RELAXATION MODEL WITH MOBILE TENSION ELEMENTS

The basic tenets of the model that has been developed are:

A)   That cleavage occurs as a consequence of differential cortical tension, the highest tension being present in the equatorial regions.

B)   Cortical tension is generated by linear tension producing elements which are free to move in the plane of the cortex and which are initially evenly distributed and randomly oriented.

C)   These elements are brought into an initiall state of uniform activation.

D)   The presence of the asters of the MA modulate this activation by reducing the tension of elements in their immediate vicinity.

E)   Elements are free to move in the plane of the cortex.

The inevitable consequences of these tenets are:

A)   The relaxing influence of the asters will result in the establishment of gradients of cortical tension, the equatorial regions having the highest tension.

B)   Tension elements will move up the gradient of tension towards the equator.

C)   In so doing they will become preferentially oriented circumferentially.

D)   The cell will start to change shape as a consequence of the differential cortical tension, initially flattening in the equatorial regions. This will result in a further circumferential orientation of the contractile elements at the equator.

E) A furrow will initially appear as a broad band, but this will rapidly sharpen because of the tendency of the contractile elements to "fall into" the furrow.

STUDIES ON THE EARLY CLEAVAGES IN THE EGGS OF CAENORHABDITIS ELEGANS

The early events in the C. elegans egg after fertilisation exhibit some interesting examples of furrowing in which the MA is not bisected. We will now go on to describe these events and discuss the spatial relationship between the asters and the cortex when these furrows appear. We will then attempt to interpret these observation in the light of current ideas on the mechanisms of cytokinesis.

The early development of the C. elegans egg has been studied by microcinematography (Nigon et al., 1960), and more recently by immunofluorescence studies visualising tubulin (Albertson, 1984) and actin (Strome, 1986). These studies, together with our own on drug treated embryos, have revealed some unusual configurations of the cleavage furrow relative to the MA. Mutants of C. elegans have been isolated which exhibit abnormalities in these early events (Wood et al., 1980; Schierenberg et al., 1980), opening up the possibility of a genetic analysis of cytokinesis in this organism

PSEUDOCLEAVGE IN THE C. ELEGANS EGG

The oocyte, which is arrested in meiotic prophase prior to fertilisation, resumes meiosis upon sperm entry, undergoing two meiotic divisions and excluding two polar bodies. The oocyte pronucleus has no associated centriole (Albertson, unpublished data) and the meiotic spindle is barrel-shaped (Albertson, 1984). The entering sperm relieves the meiotic block and introduces a centrosome with an associated centriole along with the paternal DNA. The centrosome soon splits as the male and female pronuclei appear. Asters start to develop around the centrosomes sending microtubules out to the cortex (Fig 5b). Concomitantly with these events the whole of the anterior surface of the egg starts to ruffle and eventually a pseudoceavage furrow forms around the equator of the cell (Fig. 5a). Cortically located actin becomes depleted from the posterior regions of the cell at this time (Strome, 1986; Fig. 6b), perhaps explaining why the motile activity is confined to the anterior regions. The female pronucleus migrates towards the male pronucleus, slowly at first, but speeding up as they get closer together (Albertson, 1984). The pseudocleavage furrow relaxes when the pronuclei meet (Fig. 5c & d; 6d), at which time the asters are considerably enlarged (Fig 5d). The posterior surface of the egg is smooth and appears relaxed throughout these events (Fig. 5a & c).

The pseudocleavage furrow is unusual in that it does not bisect a MA. Although there is no MA at this time there are two apposed asters present during the period of pseudocleavage (Fig. 5b & d). One would therefore expect that the cleavage furrow would pass between them along the length of the cell instead of constricting at the equator, leaving the two asters to one side of the furrow.

THE FIRST CLEAVAGE IN THE C. ELEGANS EGG

After the two pronuclei have come together the whole pronuclear/aster complex rotates so that the asters come to lie along the longitudinal axis of the egg. The asters continue enlarging as the pronuclear membranes break down and a central spindle is formed (fig.

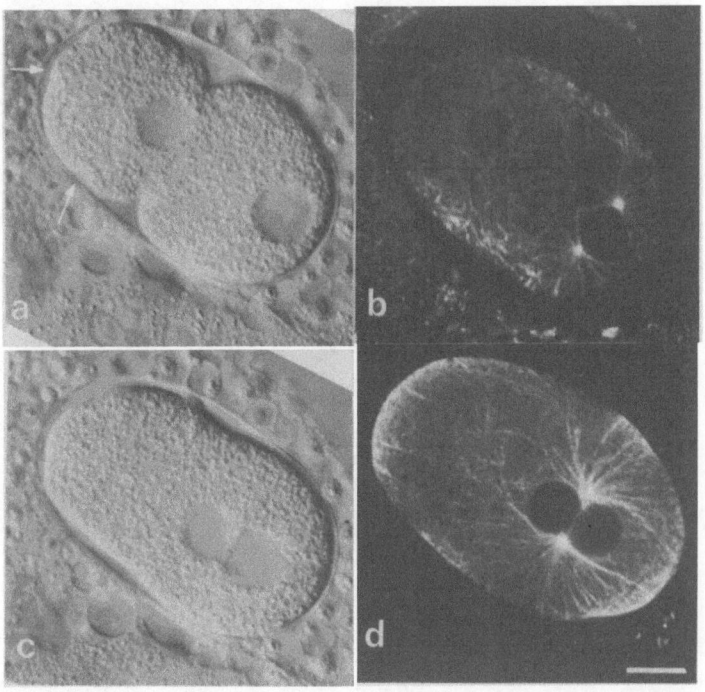

FIGURE 5.   Nomarski images of live eggs during pronuclear migration compared with antitubulin stained eggs at approximately the same stage. During the early stages of pronuclear migration there is considerable ruffling activity in the anterior end of the cell (white arrows in a) and a pseudocleavage furrow develops.   At this time two asters are enlarging on the newly separated centrosomes (b). As the pronuclei come together (c) the asters enlarge (d) and the pseudocleavage furrow regresses.   Cells were permeabilised by freeze cracking and were then fixed in methanol for 4 min. followed by acetone 4 min.   They were stained for tubulin using the YL1-2 antibody (Kilmartin et al., 1982).   Scale bar 10μ, posterior bottom-right.

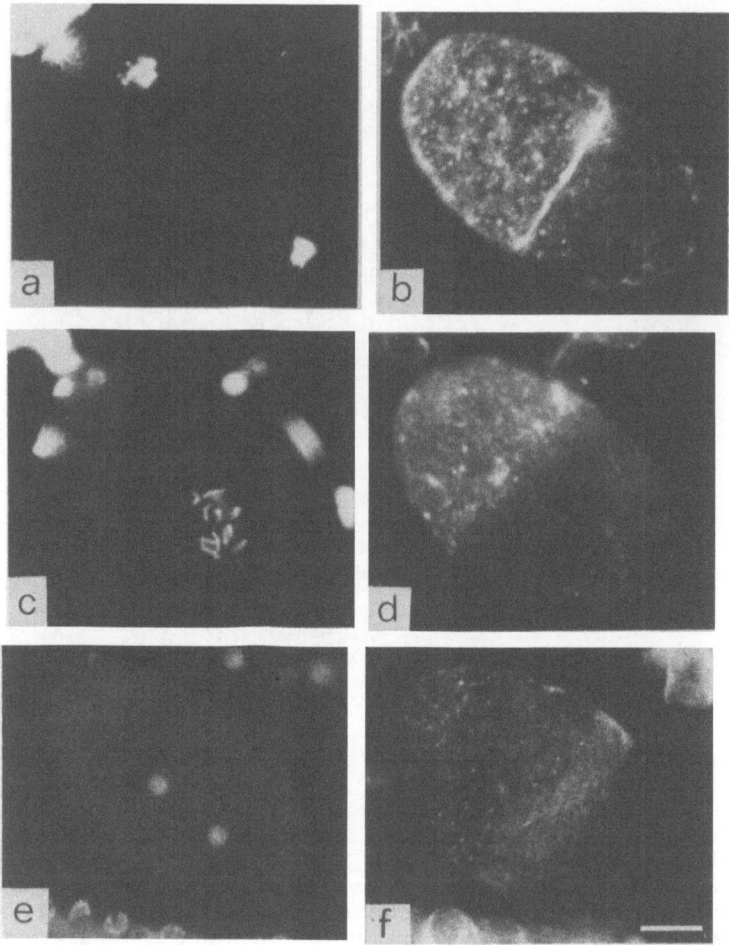

FIGURE 6.   Sequence of eggs at different stages before first cleavage
double stained with DAPI to visualse DNA (left hand panels), and
phalloidin to visualise F-actin (right hand panels).   The actin
pictures were taken at the top focal plane of the cell in order to
visualise cortical actin.   At pseudocleavage the pronuclei are still
well separated (a) and actin staining is localised to the anterior
cortex (b). There is a thin bright line of actin staining in the
contractile ring of the pseudocleavage furrow (b).   When the
pronuclei meet and the chromatin starts to condense (c), the
pseudocleave furrow regresses but the anterior localisation of
cortical actin persists for a time (d).   At telophase of the first
division the chromosomes have separated and de-condensed (e) and
a broad band of actin accumulates in the equatorial regions of the
cell prior to cleavage (f).   Scale bar 10 μ, posterior bottom-right.
From Strome, 1986.

7a & b). The cell then enters metaphase. During anaphase-B the mitotic spindle elongates and the posterior aster moves backwards with violent, rapid rocking movements and becomes attenuated in size (Albertson, 1984). The anterior centrosome stays more or less in the same place as this is happening and remains the same size. The centre of the spindle is therefore displaced towards the posterior end of the cell. The cleavage furrow forms at telophase (Fig. 7c), bisecting the MA to produce a large anterior cell and a smaller posterior cell. Thus, in contrast to pseudocleavge, the first cleavage in the C. elegans egg follows the classical patern of cytokinesis.

A broad belt of actin filaments appears in the equatorial regions just prior to the appearance of the cleavage furrow (Fig. 6e & f). This contrasts with the narrow dense band of filaments seen in the pseudocleavage furrow (Fig. 6b) and perhaps is suggestive that the contractile ring might arise by the migration of contractile elements towards the equator.

NOCODOZOLE BLOCKED FIRST CLEAVAGE

If eggs are treated with nocodozole (10 µg/ml) an attenuated spindle will form and there is often an abortive attempt at cleavage. Under these conditions the spindle ends up in a transverse orientation at the posterior end of the cell (Fig. 7e & f). When this spindle enters telophase, a transverse furrow is seen in the cell (Fig 7e). Thus, in an analagous situation to pseudocleavage, the cleavage furrow does not bisect the MA but forms in an orthogonal orientation to the mid-plane of the MA. At slightly lower doses of the drug a larger spindle develops, but still oriented transversely. An equatorial furrow still forms but, interestingly, an additional furrow often cuts into the posterior end of the cell to bisect the spindle (Strome and Wood, 1983). This furrow meets up with the equatorial furrow to produce a trefoil configuration (data not shown). This is very reminiscent of cleavages where a polar lobe is produced (Conrad & Williams, 1974), in this case the polar lobe is the cytoplast at the end of the cell. In common with pseudocleavage and also polar lobe formation, the equatorial furrow in drug-treated eggs with transverse MAs does not progress to completion. When a furrow forms to bisect the transverse spindle, this will often go to completion giving rise to a small cell and a larger cell that inherits the contents of the anterior "polar lobe".

MITOTIC BEHAVIOUR OF THE EMBRYONIC MUTANT ZYG-9

Embryonic mutants of C. elegans have been isolated with a phenotype that closely resembles nocodozol treated embryos (Hirsh et al., 1976; Schierenberg et al., 1980; Wood et al. 1980; Albertson, 1984). Temperature sensitive mutants of zyg-9 fail to develop at the non-permissive temperature. Analysis of video recordings has shown that the female pronucleus fails to migrate towards the sperm pronucleus (Albertson, 1984). A small first cleavage MA is made around the stationary male pronucleus. Rotation does not occur and the MA remains in a transverse orientation at the posterior end of the cell. The asters of the MA seem considerably smaller than those seen in a wildtype MA. An equatorial cleavege furrow forms, as in the case of pseudocleavage and the nocodozol-treated embryos. A furrow also bisects the MA and joins up with the equatorial furrow giving a trefoil configuration. Thus the configuration of the cleavage furrows relative to the position of the MA is almost identical to the nocodozole treated embryos.

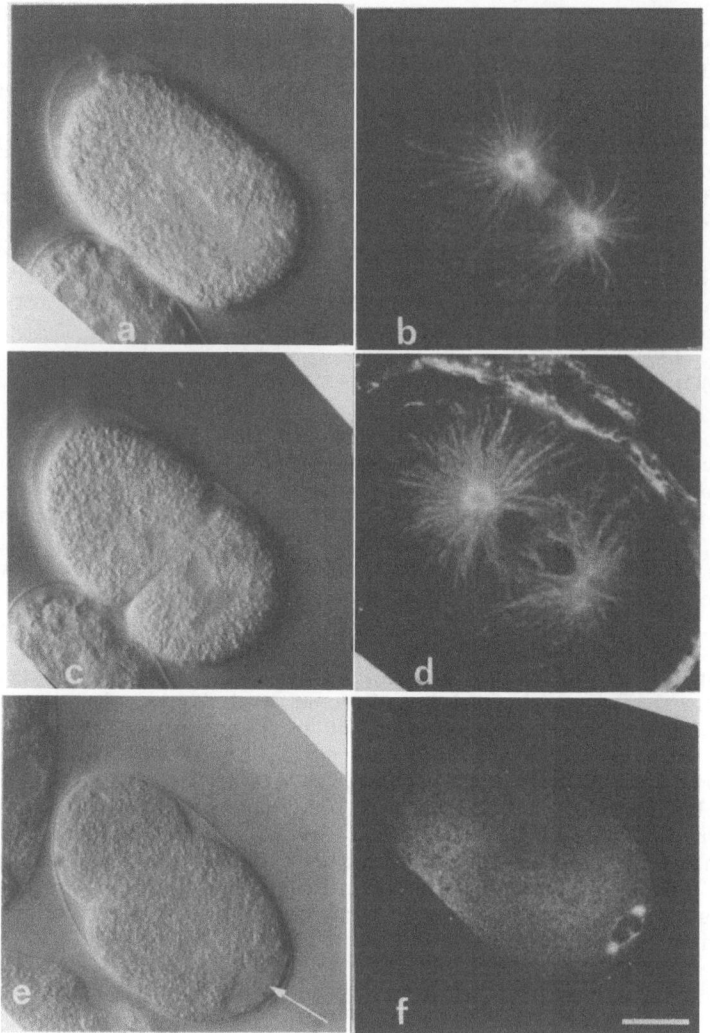

FIGURE 7.   Sequence of Nomarski pictures of live eggs (left hand
    panels) together with antitubulin stained eggs at approximately the
    same stages of development (right hand panels).   At metaphase a
    mitotic spindle forms (a) and the chromosomes congress onto the
    mitotic plate (dark line on spindle in (b)).   Cytokinesis occurs at
    telophase (c) as the daughter nuclei are reforming (d).   The
    cleavage is asymmetric as the posterior aster has become attenuated
    and the MA displaced towards the posterior end of the cell (d).
    When eggs are treated with 10 μg/ml nocodozole a very small MA is
    formed.   The MA forms transversely and migrates back to the
    posterior end of the cell, staying in a transverse orientation (e,f).
    An equatorial furrow forms when the cell enters telophase (e,f).
    Scale bar 10μ, posterior bottom-right.

POSSIBLE INTERPRETATION OF UNUSUAL CLEAVAGE FIGURES

We have described three cases of furrowing in the early egg of C. elegans, in which the position cleavage furrow breaks the usual rule of bisecting the MA. In all these cases a furrow forms around the equator of the cell whilst the transversely oriented MA is situated at the posterior end. In the case of pseudocleavage, cortical actin is depleted from the posterior end of the cell in the vicinity of the asters (Strome, 1986; Fig. 6b & d). The distribution of cortical actin in the other two cases is not known. It seems very hard to explain the position of the cleavage furrow in these circumstances by an equatorial stimulation scheme. The astral relaxation scheme with mobile tension elements seems at first sight to be equally in trouble, but on closer examination can provide an explanation. In all the cases described the asters are situated rather close together at the posterior end of the cell (in the case of zyg-9 and some of the low-dose nocodozol embryos, the asters are further apart and an additional furrow forms to bisect the MA). The two closely apposed asters would be expected to lower the cortical tension in their vicinity and therefore cause the the cortical contractile elements to migrate to the region of highest tension, which in this case is in the anterior regions of the cell away from the MA. This may explain the caps of actin filaments seen at the anterior end during pseudocleavage. Computer simulations have shown that asters that are a long way apart in an elongated cell can act independently, each elliciting a cleavage furrow (Fig. 8, White & Borisy, 1983). The C. elegans egg is elongated because of the constraints of the outer egg shell. So it seems as if the two closely apposed asters in the situations that have been described, could effectively act as a single aster which interact with the posterior cortex of the cell. As has been shown with the simulations with elongated cells, the anterior migration of the cortical contractile elements is so rapid in these situations that they pile up in a wave front which can contract and regenerate to form a contractile ring. Thus it appears as if these unusual cleavage configurations can be readily explained by the astrally mediated relaxation of laterally mobile cortical elements.

One other interesting point to emerge from these observations is that none of the cleavage furrows that do not bisect a MA progresses to completion. They all eventually regress. This may suggest that some component of the central spindle (possibly the residual mid-body) is necessary for the completion of cytokinesis and the final pinching off and resealing of the plasma membranes.

CONCLUSIONS

We have argued that cortically located contractile elements can act to produce tension in the cortex of a cell, yet are laterally mobile and are free to move in the plane of the cortex. Gradients of tension in the cortex will cause the elements to migrate to the regions of highest tension. Cytokinesis is considered to occur in four steps: 1) An initial MA independant global activation of cortical tension elements. 2) A local reduction in this cortical tension in the vicinity of the asters. 3) A consequential redistribution and circumferential alignment of the cortical tension elements into an equatorial band. 4) A regenerative sharpening of the equatorial band of elements into a contractile ring as furrowing commences.

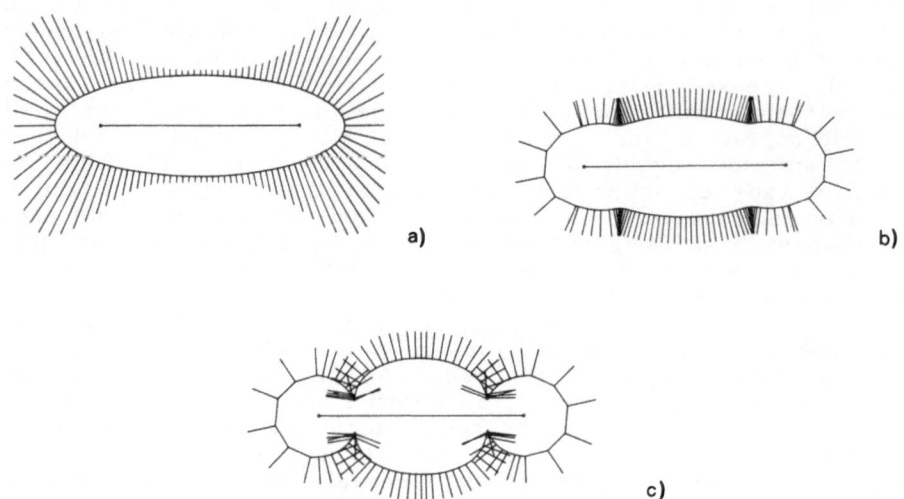

FIGURE 8.   Computer simulations of widely separated asters in an
elongated cell, using the astral relaxation model with mobile tension
elements.   The length of the spikes emanating from the cells
represents the magnitude of the stimulus from the asters at that
point, and the density of the spikes represents the density of the
tension elements.   In the situation depicted the asters are too far
apart to directly interact with each other and so, in effect are
acting independently.   Because the cells are elongated and the
asters close to the cortex, the cortical tension elements rapidly
migrate away from the astral region accumulating into a wavefront
which develops into a contractile ring.   The computer simulations
therefore predict that in a configuration such as this it is possible
to get mono-astral furrowing.

References

Albertson, D.G. (1984). Formation of the First Cleavage Spindle in Nematode Embryos. Develop. Biol. 101, 61-72.

Asnes, C.F. & Schroeder, T.E. (1979). Cell cleavage. Ultrastructural evidence against equatorial stimulation by aster microtubules. Exp. Cell Res. 122, 327-338.

Berlin, R.D. & Oliver, J.M. (1978). Analogous ultrastructure and surface properties during capping and phagocytosis in leukocytes. J. Cell Biol. 77, 789-804.

Bluemink, J.G. (1972). Cortical wound healing in amphibian eggs: an electron microscopal study. J. Ultr. Res. 41, 94-114.

Conklin, E.G. (1917). Effects of centrifugal force on the structure and development of the eggs of Crepidula. J. Exp. Zool. 22, 311-419.

Conrad, G.W. & Rappaport, R. (1981). Mechanisms of cytokinesis in animal cells. In Mitosis/Cytokinesis (ed. A.M. Zimmerman & A. Forer). Academic Press, New York.

Conrad, G.W.& Williams, D.C. (1974) Polar lobe formation and cytokinesis in fertilised eggs of Ilyanassa obsoleta. Develop. Biol. 36. 363-378.

DeBrabander, M. & DeMay, S. (Eds.)(1980) Microtubules and microtubule inhibitors. Published by Elsevier/North Holland, Amsterdam.

Euteneur, U. & McIntosh, J.R.(1981). Structural polarity of kinetochore microtubules in PtK2 cells. J. Cell Biol. 89, 338-345.

Gingell, D.G. (1970). Contractile responses at the surface of an amphibian egg. J. Embryol. Exp. Morph. 23, 583-609.

Hara, K., Tydeman, P. & Kirschner, M. (1980). A cytoplasmic clock with the same division period as the division cycle in Xenopus eggs. Proc. Natl. Acad. Sci. USA 77, 462-466.

Hiramoto, Y. (1956). Cell division without a mitotic apparatus in sea urghin eggs. Exp. Cell Res. 11, 630-636.

Hiramoto, Y. (1963). Mechanical properties of sea urchin eggs. Exp. Cell Res. 32, 59-75.

Hiramoto, Y. (1971). Analysis of cleavage stimulus by means of micromanipulation of sea urghin eggs. Exp. Cell Res. 68, 291-298.

Kappel, D.E., Oliver, J.M. & Berlin, R.D. (1982). Surface functions during mitosis. III. Quantitative analysis of ligand-receptor movement into the cleavage furrow: diffusion vs. flow. J. Cell Biol. 93, 950-960.

Kilmartin, J.V., Wright, B. & Milstein, C.(1982) Rat monoclonal antitubulin antibodies derived by using a new non-secreting rat cell line. J. Cell. Biol. 93, 576-582.

Landau, J.V., Marsland, D. & Zimmerman, A.M. (1955). The energetics of cell division: effects of adenosine triphosphate and related substances on the furrowing capacity of marine eggs (Arbacia and Chaetopterus). J. Cell Comp. Physiol. 45, 309-329.

Mabuchi, I. & Okuno, K. (1977). The effect of myosin antibody on the division of starfish blastomeres. J. Cell Biol. 74, 251-263.

McIntosh, J.R., Kent, L.M., Edwards, M.K. & Ross, B.M. (1979). Three dimensional structure of the central mitotic spindle of Diatoma vulgore. J. Cell Biol. 83, 428-442.

Mitchison, J.M. & Swann, M.M. (1955). The mechanical properties of the cell surface. J. Exp. Biol. 32, 734-750.

Odor, D.L.& Renninger, D.F., (1960) Polar body formation in the rat oocyte as observed in the electron microscope. Anat. Rec. 137, 13-23.

Rappaport, R. (1960). Cleavage of sand dollar eggs under constant tensile stress. J. Exp. Zool. 144, 225-231.

A LIFE SCIENTIST'S VIEW OF THE "MECHANICS OF CELL DIVISION"

or: UNKNOWN, FORGOTTEN AND NEGLECTED MECHANISMS OF MITOSIS

Neidhard Paweletz

Institute of Cell and Tumor Biology
German Cancer Research Center
D-6900 Heidelberg, F.R.G.

INTRODUCTION I or: THE DEFINITION OF MECHANICS

The ancient Greeks defined mechanics as the art of constructing machines. Nowadays, mechanics is primarily understood as one of the main disciplines of physics. It is a term applied to the theories of force such as motion and stress, not necessarily taking into account the matter acted upon or acting. Mechanics is subdivided into three parts: kinematics, dynamics and statics. Kinematics is concerned with pure movements without considering their origin. Dynamics deals with movements in relation to those forces which cause them. Statics is the theory of the composition of forces and their equivalence (from an encyclopedia). According to these definitions, mechanics entails an ordered cooperation of forces or stress and their respective substrates. There is no space for non-physical dimensions and principles in these definitions which reinforce the impression of the unchangeable on the one hand and the unbroken, concerted combination of parts on the other hand, which does not leave anything to chance.

The expressions "mechanics" and "mechanical" are, however, also used in our colloquial language. They are endowed with an obvious meaning of rigidity and machine-like behavior. Good examples of this meaning are the mechanical dolls which can move like human beings (though a bit mechanical, see also break dance) in which the ensemble of all parts cooperates in an orderly way. These dolls, however, are not equipped with their own, internal energy which must be introduced from outside.

The understanding of "mechanics" by the layman which surmounts physical dimensions makes it extremely difficult to consider and to designate processes of the living cell as mechanical. What is the reason for terming biological events "mechanical" and relating cell division (which is one of the most basic processes of life) to "mechanics"?

INTRODUCTION II or: THE FORMER UNDERSTANDING OF "MECHANICS OF CELL DIVISION"

To explain and interpret biological events in the last few centuries up to the end of the 19th century essentially two contradictory concepts were used. While the vitalists believed that in addition to the real existence of the organisms and all processes which take place in them a special vital energy (vis vitalis) must be active during evolution and development of living organisms and for the maintenance of life, the mechanists assumed that they could explain all phenomena of life on the basis of physical and/ or physicochemical criteria. A series of important discoveries in the last century, e.g. the synthesis of urea, the production of an organic (organismic) compound out of anorganic ingredients by Woehler, led to an increasing weakening of vitalism, but it did not disappear completely. It was revived for a short time in this century by the biologist and philosopher Hans Driesch. The decline of vitalism, however, was paralleled by an additional increase in strength of the mechanistic way of explanation. By means of this idea, it should be possible to comprehend all biological events including cell division as mechanical processes. The living cell must be considered as a small but highly efficient machine, movements of cells or parts within the cell as mechanics.

The mechanistic view of mitosis recognizes that during the process of cell division the chromosomes are moved, during cytokinesis the whole cytoplasm is divided by mechanical processes to give rise to two daughter cells. These processes, in particular, are named the "mechanics of mitosis". It comprises not only the movements themselves but also the forces which enable the cell to perform them. Defined mechanisms (cooperation of the parts of the machinery) lead to an orderly arrangement and distribution of chromosomes or to the division of the cytoplasm.

In 1929, Wassermann published a detailed review on the events of cell division and a summary of the relevant literature on mitosis known so far. In this epoch-making book he described four defined mitotic events as parts of "the mechanics of mitosis". These were 1. the rearrangement of chromosomes during the incorporation into the spindle as soon as the chromosomes leave the nuclear envelope (during its breakdown) (prophase-prometaphase); 2. the arrangement of the chromosomes into the equatorial plane (prometaphase-metaphase); 3. the movement of chromosomes to the poles, which occurs during anaphase; 4. and finally the cleavage of the cytoplasm to develop two daughter cells independent of each other. Wassermann defined and restricted the "mechanics of cell division" as the essential movements of the chromosomes and cytokinesis. This definition implied that all other movements of chromosomes or of cytoplasmic organelles which occur during mitosis were excluded. Nowadays, Wassermann's interpretation of mitosis must be classified as mechanistic.

During the first half of our century inspite of its revival by Hans Driesch vitalism has nearly vanished completely. During this period, the view of biologists was strongly mechanistically influenced and formed. A large number of excellent biological publications were published in the "Archiv für Entwicklungsmechanik der Organismen". With increasing knowledge of the mitotic events, the view of biologists became more differentiated and many authors were more cautious in using the term "mechanics of mitosis". In 1953 Schrader published his book on mitosis with the subtitle "the movements of chromosomes in cell division". It mainly deals directly with those mitotic events which in former times were termed "mechanics of cell division", but in this book the expression "mechanics" can only be found once in the table of contents as "metaphase mechanics" in relation to the arrangements of chromosomes into the metaphase plate (this means prometaphase movements). The anaphase movements and cytokinesis are no longer termed as "mechanics".

In other reviews on mitosis (Hughes, 1952; Mazia, 1961; Levine, 1963; Bajer and Molè-Bajer, 1972) the term "mechanics of cell division" no longer exists in the strict sense. Mazia (1961) in his extensive and detailed description of problems of mitosis only used the term "mechanisms". This expression must, of course, be seen in relation to the term "mechanics". It is, however, not burdened with all the emotional content of earlier periods, which has led to the controversy of vitalistic versus mechanistic ideas. In this context, Mazia inquired after the probable mechanism of metakinesis (arrangement of the chromosomes into the metaphase plate) and he discussed the mechanisms of chromosome movements during anaphase. There is no additional mechanism which refers to other chromosome movements but Mazia described another mechanism, that of cytokinesis, the division of the cytoplasm. There is a distinct alteration in the way of description and interpretation which in turn reflects new ideas and views of the respective generation of scientists. Nowadays, the living cell is no longer a well developed machine but a structure with its own laws which are different from the physical and chemical ones. The cell is considered as a living being in which processes occur in many different ways and in which a system of interdigitating events exists, which, however, is not set into motion by a mysterious vital energy (vis vitalis) on the one hand but on the other hand the cell is also not only a combination of levers, cog-wheels, pistons or cylinders. About one decade after Mazia's review Bajer and Molê-Bajer (1972) emphasized in their book the dynamic and highly flexible aspects of mitotic events and they spoke of dynamics of the spindle which was distinguished from chromosome movements. They called mutual influence of the chromosomes during the mitotic stages "mechanical interactions".

Using the key word "mechanics of cell divisions" to search for the relevant literature of approximately the last ten years the output is extremely low. The movements of chromosomes are no longer referred to or included in the "mechanics of mitosis" and even for cytokinesis authors mostly speak of "mechanisms" rather than "mechanics". In only very few cases, cytokinesis is still termed "biomechanics of cytokinesis" (see e.g. Akkas, 1980). The journal "Archiv für Entwicklungsmechanik der Organismen" has recently changed its title to adapt to the new ways of thinking. It can only be a matter of time until the "Journal of Biomechanics" will also alter its name.

MATERIALS AND METHODS or: WHAT MECHANISMS OF MITOSIS MUST BE TAKEN INTO CONSIDERATION?

Examining the respective relevant literature, we can state that the term "mechanics of cell division" is only used for two main processes at all, namely the movement of chromosomes (during prometa- and anaphase) and the cleavage of the cytoplasm (cytokinesis). The expression "mechanism", too, is only related to these events. If "mechanics" and "mechanism" really are only seen in relation to movements of cellular structures, the question arises as to whether only the mentioned few chromosome movements and cytokinesis exist during the course of mitosis or whether there are further processes and movements which also should have their "mechanisms". Nowadays, there is no reason why the term "mechanics" or "biomechanics" should be used only for cytokinesis since there is no more movement than during chromosome translocation.

We have to check all mitotic events to find out whether, in addition, similar and other mechanisms become active, which have nothing to do with chromosome movements or cytokinesis.

For this purpose, we have to describe at least briefly the phenomenology of mitosis and try to list all events which occur during the division process. Cells which go into mitosis can often be distinguished from their neighbors by an enlarged nucleus. In some cases, rotation of the latter can be observed in living cells. The chromatin meshwork within the prophase nucleus gets coarser until individual chromosomes can be identified. In the meantime the cell has partly separated from its neighbors and starts to round up. The nucleolus becomes disintegrated. In the cytoplasm in the

direct vicinity of the nucleus a small spindle is forming, and the centrosomes become separated from each other. The nuclear envelope is decomposed
into small patches, the chromosomes are incorporated into the developing
spindle. The cell nearly becomes a sphere (prophase). The centrosomes
migrate to the prospective poles of the spindle. The condensation of the
chromosomes occurs at its culmination point. They are arranged into the
metaphase plate (prometaphase, metaphase). The bipolar spindle is completely
formed, the chromosomes are transported to the respective poles. The spherical cell becomes an ellipsoid (anaphase) and starts to cleave until cleavage
is complete. The spindle is disappearing and around the chromosomes the
cell begins to form a new nuclear envelope, the chromosomes decondense, the
nucleolus reforms (telophase). The ellipsoid form of the cell in anaphase-
telophase changes into two nearly spherical daughter cells which at the
beginning remain connected by a thin cytoplasmic bridge which carries the
Flemming body. Later on they are separated from each other definitely. The
spherical daughter cells spread and flatten more and more and try to re-
establish themselves within the neighboring cells (reconstitution phase,
early G_1 phase).

This brief review has shown clearly that the mitotic processes can
be grouped into three series of events (Lettré, 1961): 1. Within the nucleus and after the breakdown of the nuclear envelope in the chromatin,
many changes occur. They must be distinguished from 2. the alterations
within the mitotic apparatus (spindle) which in turn 3. are accompanied by
processes in the rest of the cytoplasm. By appropriate experimental opera-
tions, it is possible to uncouple these chains of events from each other or
to eliminate one or the other of them. In addition, the same can happen to
the various different steps of these series of events. The dissociability
of the different mitotic stages demonstrates that on the one hand mitosis
cannot be compared to the work of a machine, but on the other hand it also
shows that much more than only a maximum of four mechanisms must be active
to guarantee an ordered cell division.

Which Mechanisms must Operate for a Correct Mitosis?

Considering the different mitotic stages in detail, the following
processes occur of which the mechanisms are mostly unknown and inexplicable
up to now. In prophase there must be 1. a rearrangement of the cytoskeleton in interphase to enable the cell to round up to form the typical mitotic
cell and to detach and separate from its neighbors partly or completely.
2. The condensation of the chromatin which has already started in the preceding S-phase (Mazia, 1978) continues until individual chromosomes become
visible (Comings and Okada, 1970). 3. With proceeding prophase the condensed chromosomes are translocated within the still complete nucleus to
the side where the centrosomes are located in the cytoplasm (Schrader,
1941; Paweletz, 1974). 4. The nuclear envelope forms a depression and later
on folds and indentations (Roos, 1973). 5. The two centrosomes (each with
one pair of centrioles) separate from each other and begin their migration
to opposite regions of the cell, the prospective poles of the bipolar
spindle (Rattner and Berns, 1976). This bipolarization process represents
one of the most important steps of mitosis (Mazia, 1978). 6. A rearrangement, a reorientation and a disintegration occur in the membranous system
of the cell especially of the Golgi apparatus within the cell center
(Paweletz and Fehst, 1984a). 7. The nucleoli as accumulations of rRNP around
the nucleolus organizing regions (NOR) of the respective chromosomes start
to disintegrate and segregate by distributing the nucleolar material within the nuclear area which then precipitates onto the condensed chromosomes
(Paweletz and Risueño, 1982). 8. Around the centrosomes microtubules (MTs)
form asters, the formation of the bipolar spindle starts (Roos, 1973).

According to earlier nomenclature and interpretation (e.g. Wassermann, 1929; Mazia, 1961) all these events are neither parts of the "mechanics of cell division" nor are they ascribed to the mechanisms of mitosis though at least some movements of chromosomes and of parts of the spindle (centrosomes) can be observed.

In prometaphase the list of mechanisms which are mostly neglected or forgotten and cannot be explained can be continued: 1. The chromosomes are taken up into the spindle as soon as 2. the nuclear envelope has become disintegrated into small pieces (see Rickards, 1981). Thereafter 3. the chromosomes become arranged in the metaphase plate. During this process 4. the two poles separate more from each other so that the spindle becomes erect (stretch of the spindle; see Schrader, 1953). 5. Parts of the disintegrated nuclear envelope form paired cisternae (Taura, 1978) and become 6. translocated into the mitotic apparatus together with elements of the likewise disintegrated Golgi apparatus and the endoplasmic reticulum (ER) (Paweletz and Fehst, 1984b). 7. The condensation of the chromatin continues. 8. The rounding up of the cytoplasm is now nearly complete.

During metaphase which must be considered as the shortest of all mitotic stages, the chromosomes are in an orderly way arranged in the metaphase plate but they 1. move slowly up and down thus oscillating around the equatorial plane (see Bajer and Molè-Bajer, 1972). 2. The condensation of the chromatin has reached its summit (Mazia, 1978). 3. The membranous structures of the mitotic apparatus have assumed their metaphase-specific arrangement (Moll and Paweletz, 1980). There are no other essential movements during this part of mitosis but some mechanisms last from the previous stages.

In the subsequent anaphase, a series of new mechanisms become effective to continue cell division according to order. 1. The chromosomes start their decondensation (Mazia, 1978). 2. The chromatids or chromosome halves separate from each other. This process is completely independent from MTs (mitotic poisons have no effect), it is still poorly understood (see Bajer and Molè-Bajer, 1972). The most striking event, however, is 3. the movement of the chromosome halves within the spindle towards the poles. These movements have ever since been in the focus of attention and interest in investigations on problems of mitosis, since many scientists supposed that the respective mechanisms would be easy to study and to explain. Until now there is no universal explanation for chromosome movement during anaphase (see Fuge et al., 1985). The concentration on this specific mechanism has probably led to the neglect of other mechanisms of mitosis, which have been forgotten or not studied at all. The question arises, as to whether chromosome movement in anaphase is the most important event which leads to an orderly division of the cell or whether other processes are as important or even more important. If we concentrate our attention in this respect only on the movements of chromosomes in the strict sense the movements to arrange the chromosomes into the metaphase plate (prometaphase) are atleast as important as the anaphase movements since no orderly distribution of chromosomes during anaphase can take place without an orderly metaphase plate.

During anaphase, further mechanisms are effective, namely 4. a new part of the mitotic apparatus, the midbody, is formed while simultaneously drastic alterations in the two half-spindles between pole and chromosomes occur (see Fuge, 1977). 5. In the midbody the osmiophilic streak develops, the region in which the MTs of the two spindle halves overlap (Paweletz, 1967). Its formation is mostly unknown. 6. The membranous system of the mitotic apparatus undergoes further changes (Schroeter et al., 1985). 7. During anaphase a new process, which in the ideas of some scientists is identical with the "mechanics of cell division" (but which is not fair), cytokinesis, the cleavage of the cytoplasm starts. 8. In some special cell types, e.g. the fertilized sea urchin eggs, the culminating point of bipolarization which has already started in metaphase can be observed. By this process, the two poles for the next mitosis are formed. The mechanisms which lead to bipolarization could be studied extensively in the last few years and could be elucidated (Paweletz et al., 1984; Mazia, 1984). This process which takes place in fertilized sea urchin eggs during meta-, ana-, and telophase is going on in mammalian cells as the bipolarization during prophase for the just occurring mitosis (Rattner and Berns, 1976).

During telophase, too, a series of events can be observed of which the underlying mechanisms have attracted only little interest until now and are therefore almost completely unexplained. 1. After the chromosomes have reached their destination they initiate the formation of a new nuclear envelope (Roos, 1973). Depending upon the distance between individual chromosomes at the poles, either karyomeres (each chromosome or small groups of them have their own envelope) or one new nucleus is formed. 2. If the telophase results in karyomeres, they have to move towards each other (lateral interaction?) until they can fuse (how do they fuse?) to form one large nucleus. This process can best be studied in telophase of the fertilized sea urchin egg in which the chromosomes mostly move to a plate-like pole (flat, stretched centrosome) instead of to a dot-like structure (pole in mammalian cells) and therefore are far enough apart not to form one nucleus at the beginning (Paweletz et al., 1984). During the formation of the new nuclear envelope, small cisternae of the old ER-NE (nuclear envelope) complex attach to the chromosomes and fuse to form one envelope. This reconstruction of the nuclear envelope goes hand in hand with the reformation of the nuclear lamina (for recent literature, see Schatten et al., 1985). 3. While the half spindles are back-formed more and more until they disappear completely, the midbody is lagging but also decreases in size. In the middle of the cytoplasmic bridge which has formed out of the cleavage furrow and which still connects the two daughter cells, the Flemming body emerges, the remnants of the midbody and the former osmiophilic streak (Paweletz, 1967). 4. The cleavage continues until finally the daughter cells are separated from each other completely. The sealing of the cleavage furrow and the detachment of the Flemming body requires special precautions of the cell (Sanger et al., 1985). 5. The chromatin decondenses more and more until the typical interphase meshwork is formed. 6. Around the NORs the nucleolar material accumulates to give rise to pronucleoli which in turn transform into true nucleoli. Part of the old material which has been precipitated onto the chromosomes now is detached and is reused for the new nucleolus (Paweletz and Risueño, 1982). 7. The membranous elements of the Golgi apparatus which have survived the disintegration during prophase and the distribution within the mitotic apparatus during the subsequent stages start to gather in the vicinity of the new nucleus and fuse to form new dictyosomes (Paweletz and Fehst, 1984a). 8. The daughter cells lose their spherical shape, spread and flatten and re-establish and settle in the surrounding tissue (Paweletz and Schroeter, 1974). 9. In the fertilized sea urchin eggs the poles for the next mitosis are ready to form the new spindle (Boveri, 1900; Paweletz and Mazia, 1984).

This enumeration of various different mechanisms in the different mitotic stages cannot be considered to be complete. However, it enumerates most of the essential processes which together constitute mitosis. While some of the events go on over long periods comprising several mitotic stages, some others are stage-specific and restricted to specific states. Some of these processes are unseparably interdigitated with others and an experimental separation would lead to abortive mitoses. Other mechanisms are independent of each other, some can even lack with only minor disturbances of the course of mitosis (see Lettré, 1961).

By all means this even incomplete summary of mechanisms during cell division clearly demonstrates the extraordinary complexity of mitotic events and the intense cooperation of individual processes. In addition, it becomes clear that a rather large number of mechanisms are active in guaranteeing an exact partition of cellular structures and compounds. Therefore, it appears extremely difficult to comprehend why the number of mechanisms is restricted to three in the earlier literature (Mazia, 1961: prometaphase, anaphase, cytokinesis) or four parts of "mechanics of cell division" (Wassermann, 1929; early prometaphase (prophase?), late prometaphase, anaphase and cytokinesis). Either "mechanics of mitosis" is used for all movements, translocations and distributing events during mitosis or for none. The same is true for the number of effective mechanisms (see Mazia, 1961).

Each listed mechanism has its own meaning for the continuation of mitosis. Therefore, it should go without saying that all of them should be studied intensively. It can be predicted that the understanding of mitosis will only be possible if all events and the driving mechanisms involved can be explained.

In the context of the present publication, it is not possible to describe and explain all processes in detail. In addition, the movements of chromosomes and their underlying mechanisms as well as the mechanisms of cytokinesis have been described and discussed so often and extensively, and so many hypotheses of chromosome movement have been proposed (Wassermann, 1929; Schrader, 1953; Bajer and Molè-Bajer, 1972; Fuge et al., 1985) that I do not wish to lengthen this list. However, as we have seen above, there are several other mechanisms which have not yet been seen at all, which have been mentioned but are forgotten now or which have been neglected since their first description. From this list of the "unknown" mechanisms, I shall select two of them which must be seen in relation to movements of cellular structures and therefore should be named mechanisms of cell division or even more precisely dynamics of mitosis.

RESULTS I or: THE PHENOMENOLOGY OF TRANSLOCATION OF CHROMOSOMES WITHIN THE PROPHASE NUCLEUS

It can often be observed, in particular in large polyploid nuclei, that during prophase, during which the chromosomes still remain within the nuclear envelope, chromosomes are translocated from the interior of the nucleus to its periphery and, in addition, from all over the nucleus to accumulate at that side where the cell center can be found (see Paweletz, 1974; Rickards, 1975, 1981 and literature therein). In the following, this process will be described in more detail and will be carefully analysed. I shall try to make this process understandable and to establish the mechanisms which lead to this translocation.

Let me first describe the phenomenology of early mitotic stages. It is astonishing that most of the mitotic stages are not terminated by defined phenomena but that there is mostly a continuous transition from one stage to the next. The beginning of prophase, the transition from interphase to mitosis, cannot be determined exactly since the condensation of chromatin, the important sign for the identification of mitotic figures, already starts during the preceding S phase (Mazia, 1978) and early prophase chromosomes are not easily distinguishable from a coarse chromatin network. Prophase, however, is clearly terminated by the breakdown (disintegration) of the nuclear envelope (for literature, see Mazia, 1961). This time-point can more or less be shifted into one or the other direction depending on whether determined by light or electron microscopy. It is significant to define the

end of prophase (see Rickards, 1981 for discussion) by the first signs of
opening of the nuclear envelope which can only be appraised by means of electron microscopy. By these criteria it is not difficult to define the transition from pro- to prometaphase in cells with an open spindle, these are
cells with breakdown of the nuclear envelope. In cells with an intranuclear
(closed) spindle, the definition of prophase in the above mentioned sense
is much more difficult (for literature and discussion, see Fuge, 1977); I
shall describe here only phenomena which occur in cells of higher animals;
to be concrete, cells of the tumor strain HeLa, a human cervix uteri carci-
noma. The phenomena, however, can also be seen in other cell types.

In cells with relatively small chromosomes, the central part of the
nucleus in prophase increasingly becomes devoid of chromatin, and the
chromosomes develop into shorter bodies. These are attached to the nuclear
envelope either directly or by means of bridging structures or compounds
(Comings, 1968). Already in the very early stages before the chromosomes are
sufficiently condensed, a shallow concavity in the prophase nucleus becomes
visible to give rise to a kidney-shaped form which is characteristic for
early prophase nuclei. Within the cytoplasm of this area, the centrosome
(which at this time contains four centrioles) and the Golgi apparatus, which
often surrounds the cell center in a shell-like manner are found. The centrosome is the origin of MTs which form an aster-like configuration. Towards
the depression of the nucleus MTs increase in length and number, some of
them being in direct contact with the outer nuclear membrane (Bajer and
Molè-Bajer, 1969). Some of these MTs are touching the nuclear surface tan-
gentially. Sometimes they can be connected to the nuclear envelope by means
of cross-bridges. Only a few MTs are present within the cytoplasm opposite
to the cell center.

At the beginning of the translocation process, the chromosomes attached
to the nuclear envelope appear to be homogeneously distributed over the
inner surface of the nucleus. They occupy most of the nuclear periphery. The
funnel-shaped depression of the nuclear surface becomes increasingly deeper.
With proceeding prophase the originally simple indentation becomes more
complicated owing to secondary folds which later on in turn form tertiary
indentations and so forth (Figs. 1-4). In more advanced stages of prophase
the majority of chromosomes can be found accumulated within the nucleus at
the side towards the cell center while in the opposite part large regions
are free of chromosomes (Figs. 1,3). On the cytoplasmic side, the indentations and folds of the nuclear envelope contain small bundles of MTs which
mostly end at the outer nuclear membrane. Some MTs appear slightly bent and
often transversely cut MTs can be observed near the bottom of the folds
(Fig. 4). This indicates that the bundles of MTs originating at or near the
centrosomes partly bend at the base of the folds. This in turn must be considered as an indication that the MTs grow towards a resisting structure,
probably the nuclear envelope. The region of the deep indentations and folds
of the nuclear envelope broadens until two such areas are found which are
separated by a bulging part of the nucleus (Figs. 1,3). Within each of
these two depressions, one centrosome (one pair of centrioles surrounded
by the osmiophilic cloud) is found (Fig. 1). This indicates that the bipolarization process, the formation of the two mitotic poles out of one cell
center has started (see also Rattner and Berns, 1976). The chromosomes
gather in two groups around the respective regions of the indentations
(Fig. 3). Simultaneously, the opposite part of the nucleus becomes completely
devoid of chromosomes. At the smooth part of the nuclear envelope at which
chromosomes are no longer present (Fig. 3), structural disturbances in the
course of the two nuclear membranes become obvious. Restricted to small
parts of the nuclear surface, the perinuclear cisterna appears inflated
(Fig. 4). Small groups of membrane-bound vacuoles and vesicles accumulate
in the cytoplasm in the neighborhood of such disturbances. Every now and
then, one of these structures is attached to the nuclear surface. While

such disturbances of the nuclear envelope can be seen distributed all over the nuclear surface at the beginning of prophase, they become restricted to the smooth part of the nuclear envelope just opposite to the identations and folds. In some cases the outer nuclear membrane is discontinuous and thus the membranes of the attached vesicles or vacuoles try to bridge over this gap. In some of such disturbances, openings of the nuclear envelope (outer and inner membrane) can be seen to which short pieces of cisternae of the ER are attached (Fig. 5). Such cisternae become incorporated into the nuclear envelope to guarantee a continuity of the nuclear cover. At first sight, these areas might be considered as first indications for the breakdown of the nuclear envelope. However, they are already visible at the beginning of prophase far away from the start of breakdown. Careful ultrastructural analyses of the whole nuclear envelope demonstrate clearly that apart from these disturbances the nuclear envelope remains continuous and intact at least until the chromosomes have all gathered within the region of the folds and indentations. The translocation of the chromosomes within the prophase nucleus is nearly completed. Until this time, no MT is visible within the nucleus in spite of careful and intense studies.

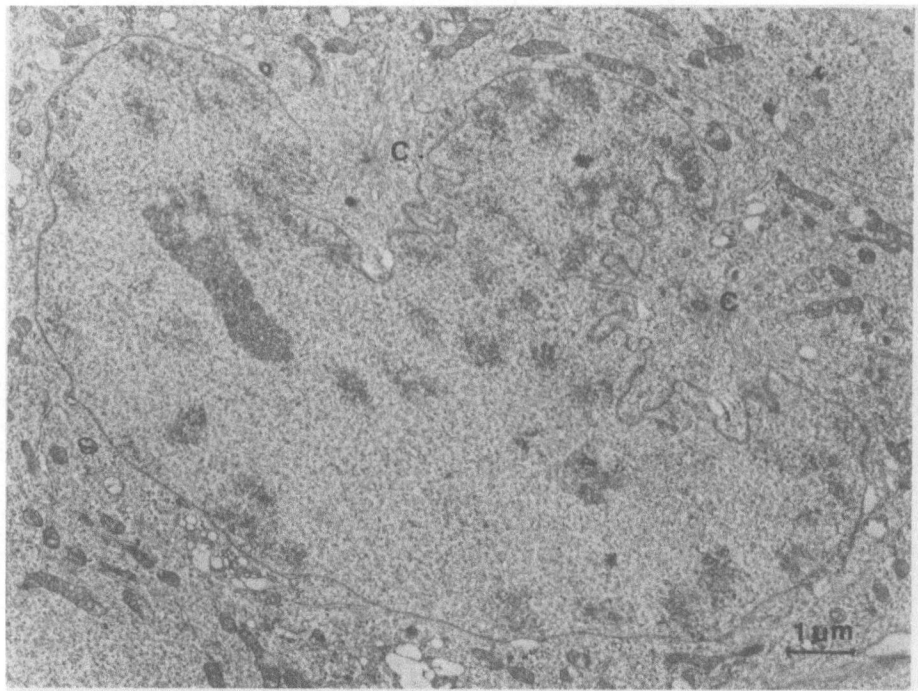

Fig. 1. Cell in prophase. Two depressions which both contain a centrosome (C) have formed. The chromosomes are located in the periphery of the nucleus. Microtubules radiate out from the centrosomes (from Paweletz, 1974).

In his detailed description and critical appreciation of the results on chromosome movements in early mitotic stages, Rickards (1981) critisizes that in HeLa cells (Paweletz, 1974) the polarization i.e. the gathering of chromosomes on one side of the prophase nucleus (Schrader, 1953) can only be seen when the nuclear envelope has already opened at least partly which would mean according to the criteria of prophase that the cell is already in early prometaphase. This, however, is not correct for our own results. The polarization process can be unambiguously observed before even the beginning of the breakdown of the nuclear envelope. This can be proven by light and electron microscopic data (see also Paweletz, 1974). Rickards (1981) uses Fig. 5 of the paper by Paweletz (1974) to defend his hypothesis that translocations of chromosomes can only be seen after the opening of the nuclear envelope. This figure represents indeed an early prometaphase nucleus but only when the polarization is already completed and the chromosomes have finished to accumulate inside the nucleus at the side of the centrosomes. Fig. 4, which he also criticizes, shows a prophase nucleus in which (contrary to his interpretation) an as yet incomplete accumulation of chromosomes can be found on one side of the nucleus. According to generally accepted characteristics of prophase, the translocations of chromosomes in this case must unequivocally be called movement of chromosomes during prophase. They are clearly distinguishable from prometaphase movements which only start later.

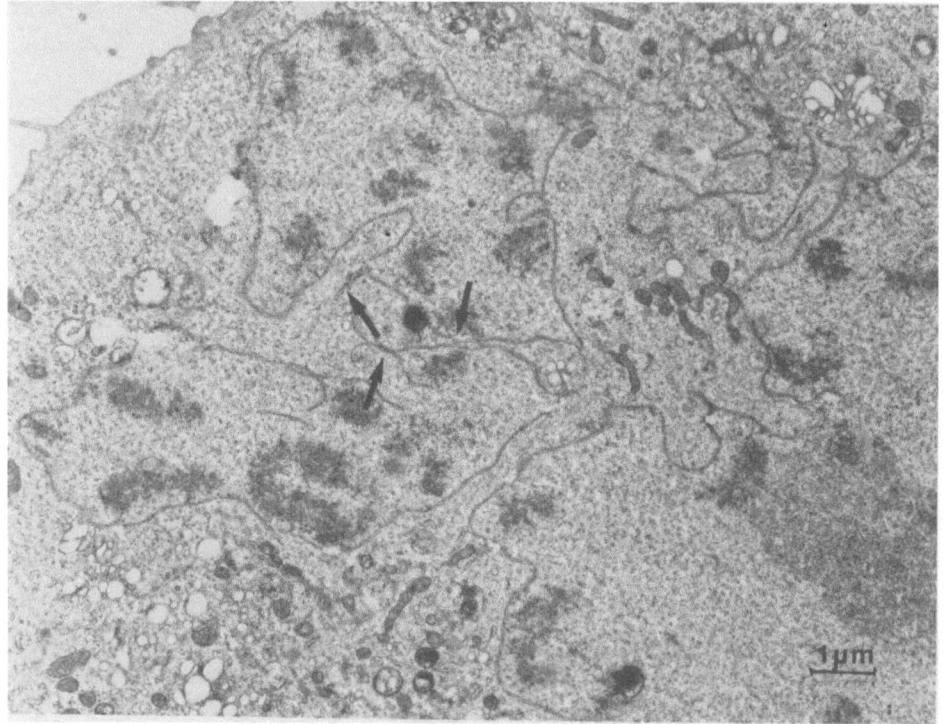

Fig. 2. Cell in prophase. View on top of the region of folds and indentations. Chromosomes are abundant in this part of the nucleus. Parts of the nuclear envelope come very near to each other to form paired cisternae (arrows).

Within the cytoplasmic side of the folds of the nuclear envelope, as was mentioned already earlier in this text, small bundles of MTs can be found which mostly end at the outer nuclear membrane. At a definite time point, it can be assumed that this coincides with the end of the translocation movements of the chromosomes, the nuclear envelope starts to open at a few of the bases of the respective crypts (Fig. 4). Up to now it cannot unambiguously be decided in this particular case whether the MTs are directly involved in the disintegration of the nuclear envelope at this site as the pressure of their growth initiates the process or whether the breakdown also occurs without any MT. It can, however, be stated that MTs are always present at sites of the opening of the nuclear envelope and that they penetrate this orifice immediately after the beginning of the disintegration. From now on, MTs can be found within the nuclear area near the accumulated chromosomes. Experiments with microtubular poisons reveal a breakdown of the nuclear envelope all over the nuclear surface, but folds and indentations are lacking. An exact analysis of the course of these MTs within the nucleus reveals small bundles and single MTs which often cross each other, indicating that they seem to originate from the two separate centrosomes.

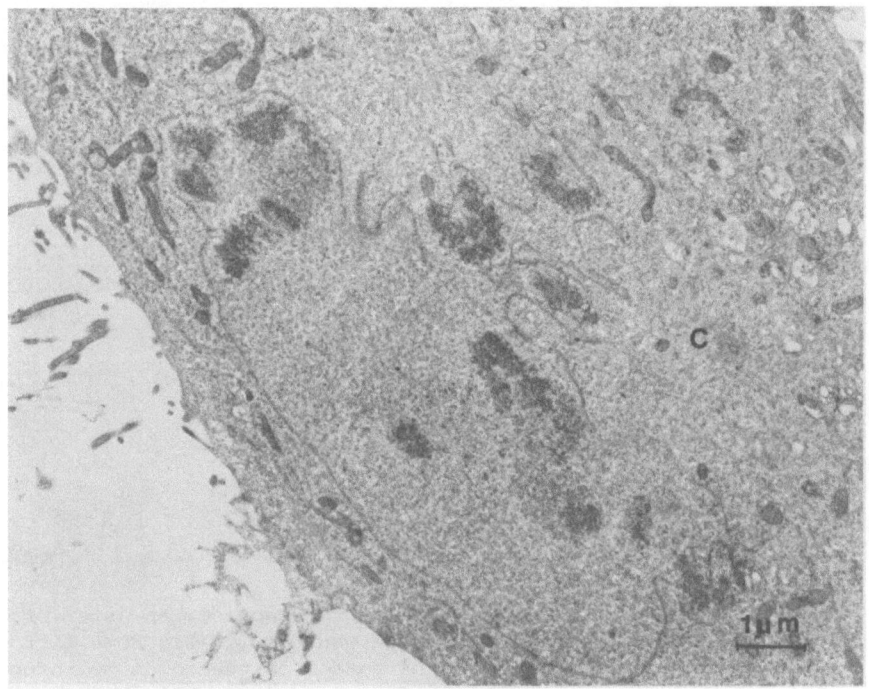

Fig. 3. Cell in early prometaphase. While the chromosomes have gathered in the regions of folds of the nuclear envelope the opposite part of the envelope is free of chromosomes (from Paweletz, 1974).

In these stages, kinetochores of the chromosomes still inside the nucleus cannot yet clearly be identified, but MTs get into direct contact with parts of the chromosomes which might be kinetochores. The two regions of the nucleus with the deep indentations and folds separate more and more, so that in some cases they can already be found at opposite ends of the nucleus before the end of the breakdown of its envelope. In general, however, the disintegration starts and ends earlier and the centrosomes in the cyto-plasm which together with the regions of indentations of the nuclear enve-lope have separated, have not yet reached the polar positions. The separa-tion of the centrosomes must then continue without a nuclear envelope. In most cases, a bulge of the nucleus has formed between the two separating centrosomal foci. This nuclear protrusion inhibits a direct microtubular connection between the migrating centrosomes. In addition, no MTs are seen near the surface of this nuclear bulge parallel to the nuclear envelope. The course of the nuclear membranes appears intact for a rather long time in the region which is now devoid of chromosomes (Fig. 3). This gives some indications that at least at the beginning MTs are somehow involved in the disintegration of the nuclear envelope.

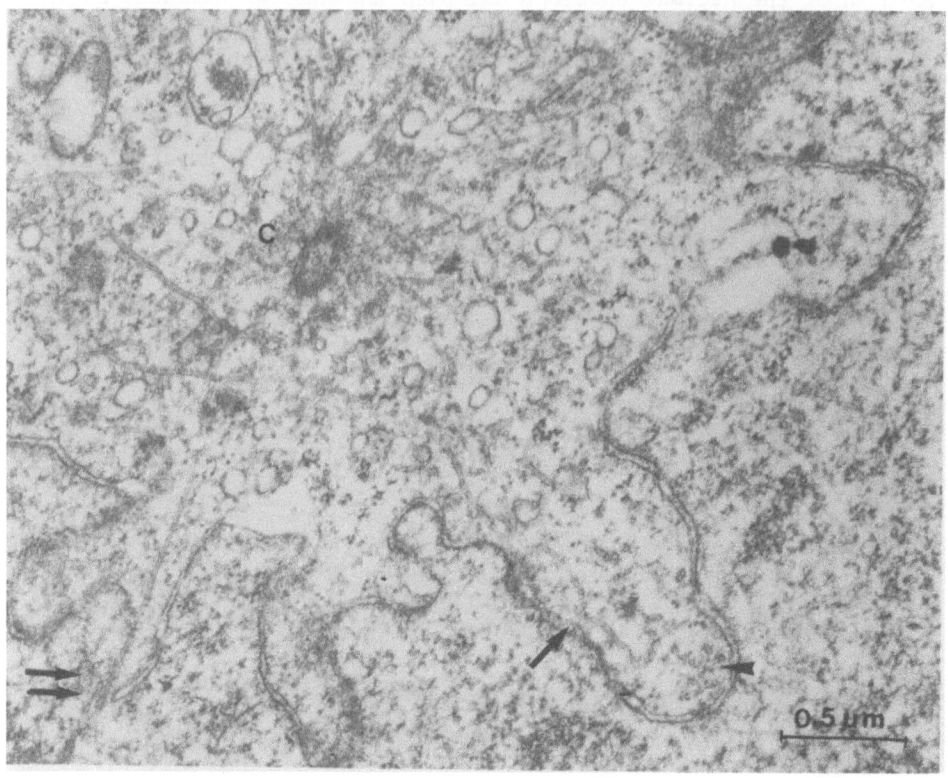

Fig. 4. Part of a cell in prophase - early prometaphase. An aster of microtubules has formed around the centrosome (C). Some microtubules are bent (arrow), at the bottom of the folds microtubules are cut transversely (arrow heads). One microtubule has penetrated the nuclear envelope (double arrow) (from Paweletz, 1974).

The indentations and folds can be so deep that in some areas two neighboring parts of the nuclear envelope come so close together that they attach to each other to form paired cisternae (Fig. 2) (for definition and literature, see Taura, 1978). These specific parts of the nuclear envelope remain intact for a long time after complete disintegration and can be observed in the neighborhood of the chromosomes during prometaphase, in some cases even during later stages.

DISCUSSION I or: THE MECHANISMS OF TRANSLOCATION OF CHROMOSOMES WITHIN THE PROPHASE NUCLEUS

The brief description of the phenomenology of prophase up to the beginning of prometaphase (see also Paweletz, 1974) has clearly indicated that more than only the prometaphase and anaphase movements of chromosomes are realized during mitosis. These "new" movements require other mechanisms than those of prometaphase and anaphase which enable the cell to translocate the prophase chromosomes. These movements have mostly been overlooked and neglected in the last decades. It is therefore necessary to ask now how these mechanisms function and how the translocations can be explained.

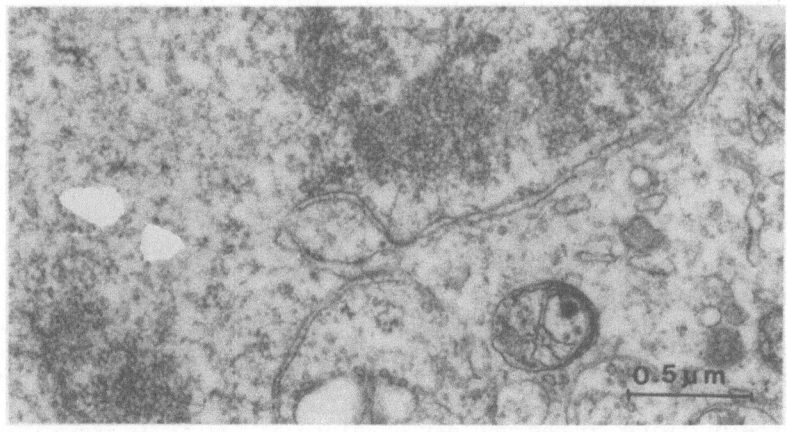

Fig. 5.   Short cisternae of the ER are integrated into the nuclear enve-
          lope.

The chromosome movements briefly described here have been known for a long time (for literature on this special problem, see Rickards, 1981). At the end of the last century, Rabl (1885) described a grouping of chromosomes on one side of the prophase nucleus where the cell center is located (Rabl orientation of chromosomes). This Rabl orientation can be observed in different plant and animal cells. In the early stages of the meiotic prophase of various different types of organisms an arrangement of chromosomes resembling a Rabl orientation is often found (see e.g. Rhoades, 1961; Church, 1981), which in this case is similar to the bouquet stage. During the division of the spermatocytes of the earwig Anisolabis, Schrader (1941) could demonstrate an extreme polarization of the prophase chromosomes which leads to an anaphase-like configuration. Rickards (1981) describes several different processes and compares them with the saltatory movements of vesicles and granules within the aster of the mitotic or meiotic spindle. Unfortunately he does not decide definitely which mechanisms could be active. He also believes, however, that real prophase movements are different from the other chromosome translocations. He assumes that MTs are only transmitters of forces while the forces themselves are produced by contractile proteins. In a series of studies (for literature, see Rickards, 1981), contractile proteins could be demonstrated within the nucleus. Rickards (1975, 1981) proposes a model in which MTs outside the nucleus cooperate with the (saltatory) movements of the chromosomes within the nucleus provoked by the action of contractile elements within the nucleus.

The movements of chromosomes within the prophase nucleus described here (see also Paweletz, 1974) must be clearly distinguished from those described by Rickards (1975) since firstly from the beginning they are not saltatory but exactly orientated to a destination in the region of folds and indentations in the neighborhood of the centrosomes within the cytoplasm. Secondly they occur only shortly before the breakdown of the nuclear envelope and can go over directly into prometaphase movements which occur without the nuclear envelope, whereas Rickards (1975) reports chromosome movements about one hour before the disintegration of the nuclear envelope. Thus the mechanism proposed by Rickards (1975, 1981) can be ruled out.

It is now well known that the interphase chromosomes are not arranged at random within the nucleus and therefore cannot move and orientate at random but are included in an ordered system. This order is guaranteed by the attachment of definite parts of the chromosomes to the inner nuclear membrane (Comings, 1968; Comings and Okada, 1970) either directly or by means of bridging proteins (lamins, perichromin). They remain connected until disintegration of the nuclear envelope. The chromosomes start their condensation in the periphery (probably at the attachment site) and this continues toward the central regions of the nucleus. The first parts of condensed chromosomes are therefore found in the periphery of the nucleus and sometimes also at the nucleolus. With proceeding condensation, the long chromosomal threads become shorter, which results in an emptying of the nuclear central parts (see also Comings and Okada, 1970). It has not yet been unequivocally proven whether the kinetochores themselves are the attachment sites at the nuclear envelope or whether other parts of the chromosomes are responsible for this attachment. Taking the results from meiotic prophase into account (see e.g. Church, 1981) during which in the bouquet stage the chromosomes are attached with their ends rather then with the kinetochores, one can assume that it might be very similar during the mitotic prophase (remember the Rabl orientation). If, however, the chromosomes are attached once or twice to the nuclear envelope at all it is conceivable to conclude that the location of the kinetochores is also more or less fixed during interphase and prophase and with increasing condensation the kinetochores also approach the nuclear envelope from inside (see also Brenner et al., 1981).

The mechanism of condensation is rather obscure. It can, however, be assumed that changes in the internal ionic milieu play an important role. It has been demonstrated since long ago (see Wassermann, 1929) that the prophase nucleus is swelling instead of shrinking. This results in an influx of water into the nucleus which in turn alters the ionic environment. It must, however, be doubted whether contractile proteins, which could be found in some types of nuclei, are involved in the condensation process as was proposed by Rickards (1981).

In early prophase stages, the number of MTs around the centrosomes increases, the majority of them is growing towards the nucleus. These MTs are obviously responsible for the formation of folds and indentations by their outgrowth. This becomes evident by bundles of or single MTs within the indentations which touch the nuclear envelope directly and those which are bent at their ends at the bottom of the folds. The indentations are not formed by a shrinkage of the volume of the nucleus since they can only be seen on one side of the nucleus just in the neighborhood of the centrosomes; they are not distributed at random around the whole circumference of the nucleus. In addition, a swelling of the nucleus can be observed rather than a decrease in volume. The nucleus apart from the region with the folds appears tight, which also contradicts shrinkage. It can, however, not be ruled out that the interior of the nucleus is also involved in the development of the folds as the condensing chromosomes might pull from the inside at special areas of the nuclear envelope. A cooperation between MTs on the cytoplasmic side of the nucleus and the chromosomes inside could thus also result in folds and indentations.

This deforming activity of MTs and the nuclear envelope cannot only be shown in nuclei of mammalian cells but it seems to be a more common phenomenon e.g. in all cells with an intranuclear spindle (see e.g. Kubai and Ris, 1969). In the nuclei of fertilized sea urchin eggs deep indentations can also be demonstrated in the nuclear surface during the first bipolarization process before and during prophase. Here, too, MTs must be involved since they are found within the folds often ending at the outer nuclear membrane (Paweletz and Mazia, 1987). Thus, it can be concluded that in general MTs growing towards the nucleus from outside are responsible for the deformation of the nucleus.

Two phenomena at the nuclear envelope are striking during prophase. These are firstly the disturbances in the course of the nuclear membranes in particular on the side opposite to the indentations and secondly the folds and indentations themselves. Since the nucleus does not become smaller during the formation of the folds but rather larger, new parts of the nuclear envelope must be formed during prophase. Judging from the structural evidence it must be concluded that the growth of the nuclear membranes occurs in a polar manner. While the nuclear surface increases in size on one side (the tight), this is nearly compensated by the formation of folds which leaves the surface of the ellipsoid part of the prophase nucleus nearly constant. This results in a shift of special regions of the nuclear surface from one side (the tight one) of the nucleus to the other (the convoluted one). Together with this shift of membranes, the chromosomes which are attached to these nuclear membranes are also shifted towards the regions of the folds in the neighborhood of the centrosomes. During a careful observation of the behavior of chromosomes in the living cell of this stage, one can recognize that some chromosomes do not move along the shortest way just across the nucleus but along an arc which corresponds to the curvature of the nuclear envelope. As signs for the polar growth of the nuclear envelope, the disturbances in the course of the membranes and the incorporations of short cisternae of the ER into the nuclear envelope must be assumed.

Thus, a movement of chromosomes is realized here in which neither microfilaments nor MTs are directly involved. The cooperation of polarized growth of the nuclear envelope together with the increase in size of MTs originating at or near the centrosome renders the translocation of the prophase chromosomes possible. Similar mechanisms of cooperation between MTs in the cytoplasm and the chromosomes attached to the nuclear envelope inside the nucleus can be found in some protozoa (e.g. Kubai and Ris, 1969).

It is not possible to determine the exact size of the nuclear surface before and after the formation of the indentations. A prudent estimation of the length of the nuclear envelope in a few representative ultrathin sections reveals that the involuted part of the envelope has nearly double the length of the smooth part. This could mean that more than one third of the surface is newly formed during prophase. This would be approximately sufficient to translocate many chromosomes from one side of the nucleus to the other.

During prophase, we could observe that out of one deep funnel-shaped indentation with secondary and tertiary folds, two depressions emerge. In each of these two cytoplasmic regions, one centrosome with the adjacent aster of MTs is found. We can therefore conclude that by the separation of the centrosomes the depressions are also shifted until they have reached the polar positions. The shift of the indentations depends on the velocity of the migrating centrosomes. This enables us to find stages in which the nuclear envelope starts to disintegrate before two depressions become visible, while on the other hand the centrosomes can have arrived at their polar positions without the disintegration of the nuclear envelope.

The question now arises as to by which mechanisms the centrosomes are separated from each other, i.e. the bipolarization process takes place. Rattner and Berns (1976) have carefully analysed the process in PtK cells by means of electron microscopy and revealed loose bundles of MTs between the separating centrosomes (centrioles). They conclude from their results that the centrosomes are pushed apart by bundles of MTs increasing in length. This formation of a small central spindle between the two centrosomes has been known for a long time as the centrodesmose (see Wassermann, 1929).

In the fertilized sea urchin egg bipolarization takes place during the alterations of the shape of the centrosomes in the course of mitosis. Here the development of the two poles for the next mitosis occurs during the preceding mitosis. MTs are also involved here in the changes of the shape of the centrosome (Paweletz et al., 1984; Mazia, 1984).

For HeLa cells, the interpretation of the separating mechanism is more difficult. It is clear that between the two separating indentations with their centrosomes, a strong bulge of part of the prophase nucleus is present which prevents a direct connection of the two centrosomes by means of elongating MTs. We can, however, suppose that exactly this nuclear protrusion serves as a strong counterfort at which the microtubules of the respective asters can act. The asters increasing in size seem to push apart the centrosomes together with the depression of the nuclear surface.

As was described earlier in this paper, bundles of MTs grow into the nucleus as soon as the nuclear envelope starts to open. The MTs originating from the two centrosomes come into contact with the chromosomes through the holes in the nuclear envelope. The chromosomes are still within the cage of the nuclear membranes. It cannot be determined unequivocally if these MTs come in contact directly with the kinetochores or if other parts of the chromosomes act as anchoring points for the MTs at the beginning of the incorporation of the chromosomes into the bipolar spindle. Recently, several studies (see Paweletz and Mazia, 1979; Euteneuer and McIntosh, 1981; Mitchison and Kirschner, 1985) have revealed that kinetochores can capture MTs and cap them then rather than acting as an MTOC (microtubule organizing center). The most characteristic components of the kinetochores seem to remain present during interphase and of course prophase. This could be demonstrated by Brenner et al. (1981) by means of autoimmune antibodies which react with kinetochore proteins. For our problem, we can assume that the MTs which have penetrated the nuclear envelope and entered the nuclear area are captured by the (perhaps immature) kinetochores. This means that the chromosomes are in direct connection with the two centrosomes, the prospective poles of the bipolar spindle, before the nuclear envelope has disappeared completely. Thus, the chromosomes are incorporated into a new cage (the mitotic apparatus) before they have left the old one (nuclear envelope). This mechanism offers a high degree of safety against the loss of chromosomes during the arrangement into the spindle. Simultaneously, this principle guarantees the preservation of the orderly arrangement of the chromosomes in the interphase nucleus through mitosis into the next interphase chromatin meshwork. Lettré and Lettré (1959) had postulated the persistence of spindle fibers during mitosis and interphase to explain special phenomena (e.g. the Rabl orientation) of mitosis. In the light of our ideas just presented above, the assumption of the persistence of spindle fibers is no longer necessary. Nevertheless, it is possible to maintain the order of the chromosomes by transferring it like the relay in a relay-race from the nuclear envelope to the spindle and from the mitotic apparatus to the next nuclear envelope without persisting spindle fibers.

After the engagement of all chromosomes with the bipolar spindle and the complete disintegration of the nuclear envelope the prometaphase movements of the chromosomes can start.

In the last pages of this part of the article we could see that besides the "classical" chromosome movements during prometaphase and anaphase which are driven by typical mechanisms and which have formerly been termed "mechanics of cell division" additional chromosome translocations occur during mitosis. This fact renders a restriction to only three or four parts of the "mechanics of mitosis" or mechanisms impossible. This prophase movement must thus also be ascribed to the mechanisms, the dynamics of mitosis.

RESULTS II or: THE DEVELOPMENT OF MEMBRANOUS COMPARTMENTS WITHIN THE MITOTIC APPARATUS

During the enumeration of processes in the different mitotic stages the existence and changes of membranous structures in and around the mitotic apparatus was mentioned frequently. I have selected this process as another example for the existence of more mechanisms than noted before. They have hardly been recognized before (see Hepler and Wolniak, 1984; Paweletz and Fehst, 1984b). In the following, I shall briefly describe the phenomenology of this process and then discuss the mechanisms of translocation of these membranous structures.

Since the beginning, the attention of the students of problems of mitosis was focused on the chromosomes and the spindle fibers. After the discovery of MTs as the elements of the spindle fibers by de Harven and Bernhard (1956) they have fascinated or even mesmerized the researchers into neglecting all other components of the spindle. One other of the most important components of the spindle are its membranes. At the beginning of the 1960's Porter and Machado (1960) described a system of membranes in the mitotic apparatus of cells of onion roots but this paper was mostly forgotten. In 1975, Harris reported on large quantities of membranous vesicles and cisternae in the mitotic apparatus of fertilized sea urchin eggs. Little was known about their function. At the beginning of the 1980's two groups of scientists started independently of each other to study the appearance and role of membranes of the mitotic apparatus (Hepler, 1980; Moll and Paweletz, 1980; Paweletz and Finze, 1981). Today there are already a few reviews on the membranes of the spindle (Paweletz, 1981; Hepler and Wolniak, 1984; Paweletz and Schroeter, 1986) and a series of original papers on this topic (for literature, see Paweletz and Schroeter, 1986).

During prometaphase and metaphase, the mitotic apparatus of HeLa cells is enwrapped in a sheath of short pieces of the ER-NE complex (endoplasmic reticulum, nuclear envelope; Figs. 6,7). In addition, cisternae can also be seen around the (virtual) long axis of the spindle from the polar region to the chromosomes (Fig. 6). Whereas the sheath around the mitotic apparatus is nearly continuous by overlapping of short pieces of cisternae (Fig. 7), it opens in the polar region (Fig. 6). Here an aster composed of MTs and cisternae of the ER-NE complex has formed. Though it cannot be proven unequivocally whether the membranes are orientated by MTs or the MTs aligned according to the course of the membranes, it can be assumed that they support each other and develop a solid buttress which is essential for the function of the spindle.

In addition to the cisternae, large numbers of vesicles of different caliber are also present in the mitotic apparatus (Figs. 6,7). While the origin of the cisternae can be derived easily from their structural appearance, it is more complicated for the vesicles to define their descent. The cisternae belong to the compartment which is formed by the ER and the disintegrated nuclear envelope (NE). Enzyme histochemical and immunocytochemical investigations (Schroeter et al., 1985; Paweletz and Schroeter, 1986) with antibodies against proteins of the ER and of the Golgi apparatus respectively have demonstrated that at least two populations of vesicles exist within the spindle. There are vesicles derived from the ER and those from the Golgi apparatus (see also Paweletz and Fehst, 1984a).

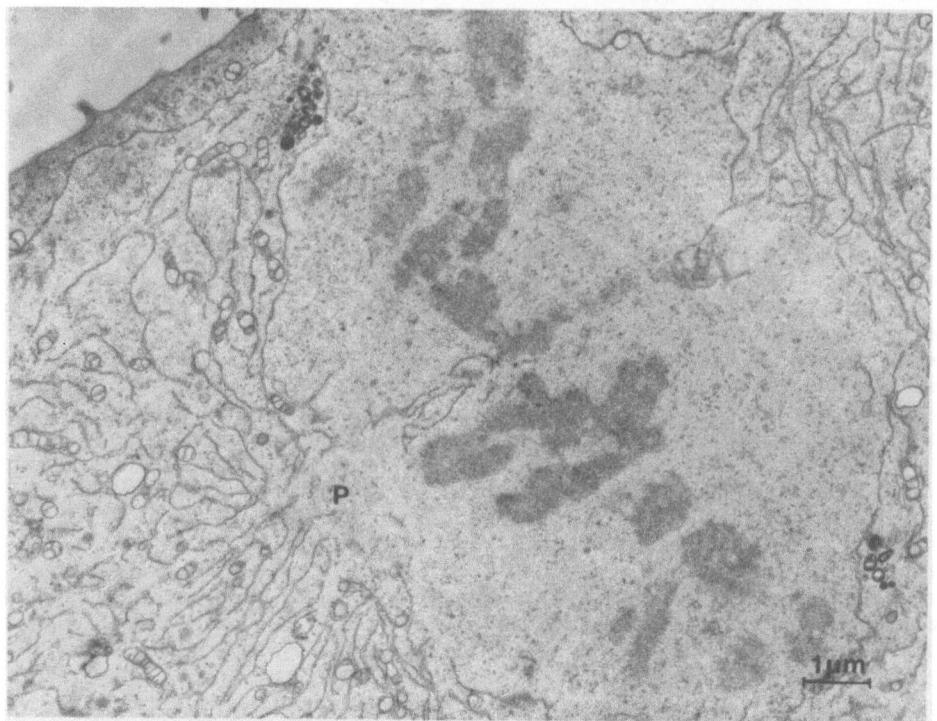

Fig. 6.   Cell in metaphase. Membranes are surrounding the mitotic apparatus, an aster-like arrangement of membranes is formed at one
of the two poles (P). Cisternae are also visible between poles
and chromosomes. Vesicles are spread all over the spindle region
(Courtesy, Dr. Moll).

During prophase, the cisternae of the dictyosomes disintegrate into
vesicles which accumulate in the cytoplasm around the centrosomes. As
soon as the nuclear envelope has opened to let small bundles of MTs make
contact with the chromosomes, vesicles can be seen within the nuclear domain. They must have been transported there. During prometaphase, they
become distributed all over the spindle area. In all mitotic stages, a
slight accumulation of them is found in the polar region and near the
kinetochores. During metaphase, vesicles in direct connection with cisternae within the mitotic apparatus stand out clearly. Though it cannot
be proven definitely whether the vesicles fuse with the cisternae or whether
they are pinched off, a direct structural interaction between the two membranous structures can clearly be demonstrated. This enables a transition,
a mixture and an exchange of the respective contents. This phenomenon seems
to be of great importance for the overall membranous compartment of the
mitotic apparatus.

As soon as the midbody of the spindle has formed during anaphase,
vesicles can also be seen in this part, in some cases even within the
bundles of MTs. During the formation of the osmiophilic streak, the zone
of overlapping microtubules, the midbody is rather densely filled with
vesicles. When the nuclei of the daughter cells are reformed small accumulations of vesicles can be found near the nuclei on the side of the

cleavage furrow. Many of these vesicles group together first and later on arrange themselves in short rows. The vesicles in these rows fuse to establish the first cisternae of dictyosomes of the new Golgi apparatus. The number of cisternae per dictyosome increases with proceeding reconstruction of the interphase configuration of the daughter cells. During these events direct connections between vesicles and cisternae of the dictyosome become visible. In this case, we can assume that they represent fusion products rather than showing a pinching off process. These results show clearly that elements of the disintegrated Golgi apparatus are preserved during mitosis in the form of vesicles. These vesicles serve as part of the material which reconstructs the new Golgi apparatus (see also Paweletz and Fehst, 1984a).

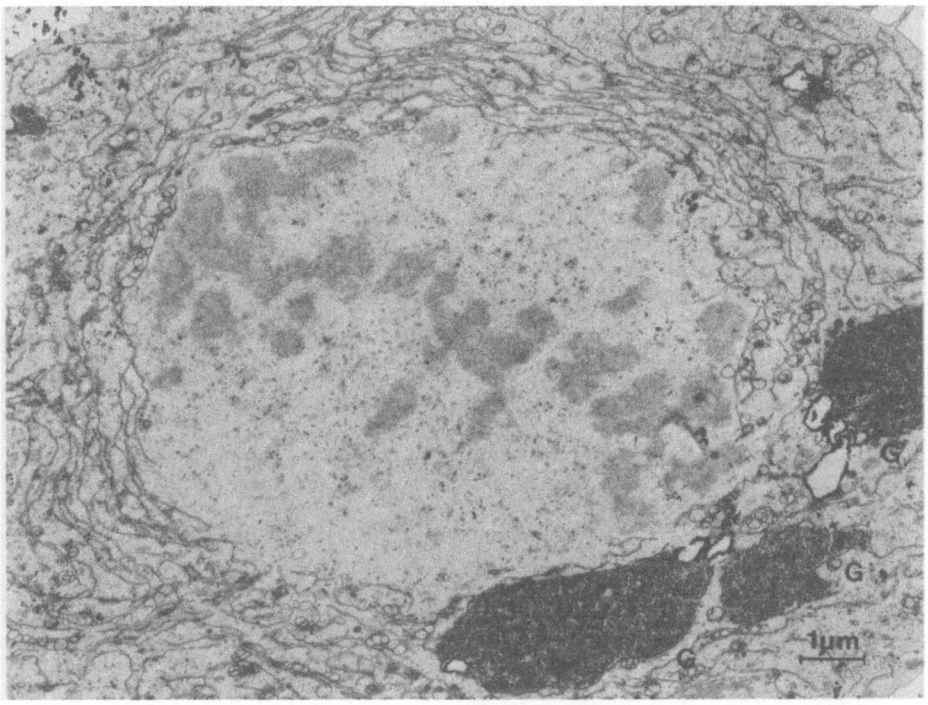

Fig. 7.   Cell in metaphase. The sheath of membranes around the mitotic apparatus is continuous by overlapping. Large accumulations of glycogen (G) cannot enter the spindle area and are excluded by membranous cisternae. The two poles are not in the plane of section (Courtesy, Dr. Moll).

Using an OsFeCN fixation (Hepler, 1980) the components of the Golgi apparatus appear electron opaque. Thus, vesicles of the Golgi apparatus can also be identified during mitosis since their opacity is retained and can be found again in the new Golgi apparatus after mitosis. It can be shown without any interruption that parts of the disintegrated Golgi apparatus remain intact and obviously also functionally active though their structural integrity is lost. The enzyme cytochemical studies on the membranous compartment of the mitotic apparatus (Schroeter et al., 1985) have confirmed these results also showing that at least some of the characteristic features of the Golgi apparatus are preserved during mitosis. It can be summarized that 1. membranous structures are integral components of the mitotic apparatus. 2. They assume a mitosis specific arrangement during the transition from interphase to mitosis. 3. They can serve as elements for the formation of compartments within the cytoplasm and the mitotic apparatus. These compartments enable the cell to maintain different but definite concentrations of certain compounds in different but neighboring regions of the same cell simultaneously or to change them rapidly. This function of the membranes of the mitotic apparatus becomes particularly clear when cells in interphase are fused with cells in mitosis. In this case, mitotic factors are unevenly distributed but later on these differences disappear (Ghosh and Paweletz, 1984). 4. Large particles like e.g. accumulations of glycogen (Fig. 7) are prevented from entering the mitotic apparatus. 5. While the cisternae can be used to store specific substances or to act as channels for a rapid transport the vesicles must be considered as easily moveable containers (Paweletz and Fehst, 1984b). 6. In another function, they probably serve as supports for MTs thus creating a strong counterfort for the forces producing movements of chromosomes by an intense association between these two structures. 7. Hepler and Wolniak (1984) have proposed that the membranous structures might also be involved directly in chromosome transport as MTs which are in contact with these membranes by means of cross-bridges could crawl along the cisternae and drag the chromosomes behind them to translocate them to the poles. 8. A series of investigations to find out the causes for aneuploidy during mitotic events revealed that membranes are also involved in an unequal distribution of the chromosomes (probably during prometaphase) and thus can be responsible for the development of an aneuploid genotype (for literature and discussion of this problem, see Paweletz and Schroeter, 1986, 1987).

DISCUSSION II or: THE MECHANISMS OF TRANSLOCATION OF MEMBRANES INTO AND WITHIN THE MITOTIC APPARATUS

The discussion about the functions of the membranes of the mitotic apparatus in the course of mitosis raises the question as to how the cisternae and vesicles can enter and penetrate the mitotic apparatus. Bajer (for literature see, Bajer and Molè-Bajer, 1972) could show in one of his impressive motion pictures that small particles within the spindle very often do not move in a saltatory manner (see Rebhun, 1972) but that the majority of them is rectilineally and orderly transported between poles and chromosomes with often alternating directions. It must be assumed that these particles which probably are at least partly identical with the vesicles described here are determined in their orientation and direction of movement by other cellular structures, e.g. MTs.

By an exact and careful ultrastructural analysis of the MTs and membranes of the mitotic apparatus, it can be clearly demonstrated that MTs and membranes are often in contact with each other (Fig. 8). This contact can be direct by MTs growing directly into the membranes (or originating from them or at their surface) or mediated by cross-bridges (Fig. 8). These contacts could be shown in several cases (Hepler et al., 1970; Paweletz and Finze, 1981; Paweletz and Fehst, 1984b).

These interrelationships do not occur at random but we must consider MTs to be responsible for the orderly arrangement and movement of vesicles (particles) and cisternae into and within the spindle (Paweletz and Fehst, 1984b). This mechanism could be proven for several different membranous structures of the cytoplasm (for literature, see Suchard and Goode, 1982).

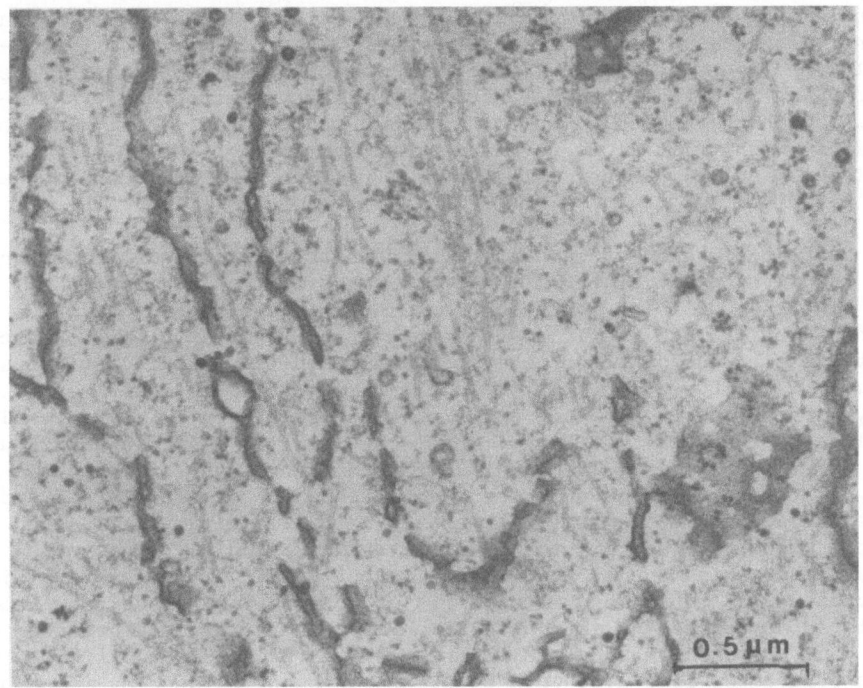

Fig. 8. Part of a cell in metaphase. Microtubules, cisternae of the ER-NE complex and vesicles are in close vicinity to each other.

For the transport of membranous structures several different ideas can be developed (for discussion of this problem, see Suchard and Goode, 1982). The MTs grow longer and therefore drag all structures connected with them behind. This is as well conceivable for a direct contact as also via cross-bridges. I have reported above (see also Paweletz, 1974) that outgrowing bundles of MTs produce folds and indentations of the nuclear envelope during prophase. It is, therefore, also easily intelligible that after disintegration of the nuclear envelope parts of it are translocated in the same way. This can be proven by the appearance of paired cisternae (remnants of the nuclear envelope) always in the neighborhood of MTs distributed all over the mitotic apparatus (for literature, see Paweletz, 1981).

Such a mechanism can translocate cisternae as well as vesicles. In the light of new results on the transport of vesicles in the cytoplasm (e.g. Koonce and Schliwa, 1985), another mechanism is conceivable especially for the vesicles in the mitotic apparatus. In the last year, two working groups reported independently of each other that a bidirectional transport of vesicles can be realized along one single MT (Koonce and Schliwa, 1985; Allen et al., 1985). While Koonce and Schliwa (1985) observed this phenomenon in finest cytoplasmic protrusions of a myxamoeba, the latter group found similar processes in the axon of the squid. Miller and Lasek (1985) demonstrated that cross-bridges between MTs and vesicles can mediate anterograde and retrograde transport along single MTs. In the meantime, an

MT translocator could be characterized (Vale et al., 1985a) which is called kinesin. Vale et al. (1985b) reported further that kinesin is only responsible for the transport in anterograde direction while another translocator which is not yet named functions during the retrograde translocation. Though these results up to now could be obtained in only two objects it is highly plausible to assume that the same or a very similar mechanism is active in other cell systems, too, as well in interphase as during mitosis.

This translocation of membranous structures is of special importance during the development of the mitotic apparatus of the fertilized sea urchin egg. In the living egg which contains yolk and pigment granules all over the cytoplasm and therefore is hardly transparent, a clear zone becomes visible in the center of the cytoplasm when the mitotic apparatus is formed. With proceeding time, this clear zone increases in size and roughly assumes the form of the spindle (see Paweletz et al., 1984). Electron microscopic investigations of this zone (Harris, 1975; Paweletz and Mazia, 1979; Paweletz et al., 1984) reveal that the light and electron opaque yolk and pigment granules have disappeared nearly completely from the clear zone while the mitotic apparatus is filled with vesicles and cisternae which give support to the microtubular spindle. It has often been asked how this "Entmischung" of different but nevertheless very similar structures, which would mean a transport of vesicles and cisternae into the mitotic apparatus and a translocation of yolk and pigment granules out of it, could be brought about. Harris (1975) has proven that different membranous structures like e.g. yolk granules can be orientated by means of MTs (or vice versa?). According to the newest ideas about the involvement of MTs in the bidirectional transport of vesicles in the cytoplasm, the phenomena can be explained as follows: The clear zone is only formed when at least one small microtubular aster is present in the neighborhood of the nucleus and the bipolarization process has finished. The MTs of the developing mitotic apparatus come in contact with neighboring membranous structures. While vesicles and short cisternae move into the mitotic apparatus, yolk and pigment granules are translocated outwards. This can occur along one single MT, provided that vesicles and cisternae on the one hand have receptors which are different from those of the granules on the other hand. While the receptors of the vesicles react with the translocator responsible for the anterograde transport, those of yolk and pigment granules bind to the translocator for the retrograde translocation. Such differences in the composition of the membranes of the different cellular structures are imaginable.

Taking again into consideration our question on the mechanisms for the transport of membranous structures into and within the spindle, we can now assume that at least vesicles can also be transported in the spindle in both directions along single microtubules. The most probable candidates for this mechanism are the free MTs of the spindle which are neither capped nor attached to a terminating structure (for definition, see Fuge, 1977). By temporal detachment of vesicles from one MT and reattachment of the same structure to a neighboring MT vesicle could also be laterally translocated. At the moment, this transport mechanism is being discussed for the transport of chromosomes, too.

It can now be stated definitely that various different cytoplasmic membranous structures can be transported within the cytoplasm by means of MTs (for literature and discussion, see Suchard and Goode, 1982). There is no reason to deny that membranous structures within the mitotic apparatus can also be translocated by MTs. Thus, the vesicles as easily movable containers gain a particular importance for the regulation of mitotic events.

In the last few paragraphs, I have been able to present a second transport mechanism which is active during mitosis and which has not been recognized in the past.

CONCLUDING REMARKS or: "MECHANICS OF CELL DIVISION" IS NO LONGER UP TO DATE

The two examples presented here in more detail about movements of cellular structures during mitosis and their respective mechanisms should show that besides the well known "classical" mechanisms of chromosome movement during prometa- and anaphase and the cleavage of the cytoplasm (cytokinesis), additional mechanisms are active of which investigations and explanations have been neglected in the past. From the preceding text, it has become clear that cell division represents the most complicated and complex process in the living cell which can be imagined. During all these events not only a cooperation of physical but also of chemical and biochemical parameters must be active, but the sum of all these parameters does not make a living dividing cell. In the meantime, it has been possible to analyse also molecular biological interrelationships but nevertheless it has not yet been possible and, let us hope, never will be possible to explain e.g. the mitotic events by means of simple physical, physico-chemical or chemical mechanisms.

As a life scientist, I consider the cell as a living unit with its own rules and laws. I am convinced that it will never be possible to elucidate all processes of the living cell by physics or chemistry. Consequently the term "mechanics of cell division" should no longer be used since it is no longer up to date. There are enough expressions such as mechanism or dynamics which are not so emotionally burdened and which can be used instead.

ACKNOWLEDGEMENT: The experimental work was skillfully performed by Ms. E.-M. Finze, Mrs. L. Doering prepared the micrographs and Ms. F. Schmitt and Mrs. U. Joa typed the manuscript. I am very grateful for all their help. Dr. E. Moll provided me with two of his micrographs. I gratefully acknowledge his support.

REFERENCES

Akkas, N., 1980, On the biomechanics of cytokinesis in animal cells, J. Biomechanics, 13:977.
Allen, R. D., Weiss, D. G., Hayden, J. H., Brown, D. T., Fujiwake, H., and Simpson, M., 1985, Gliding movement of and bidirectional transport along single native microtubules from squid axoplasm: Evidence for an active role of microtubules in cytoplasmic transport. J. Cell Biol., 100:1736.
Bajer, A., and Molè-Bajer, J., 1969, Formation of spindle fibers, kinetochore orientation and behavior of the nuclear envelope during mitosis in endosperm. Fine structural and in vitro studies, Chromosoma, 27:448.
Bajer, A. S., and Molè-Bajer, J., 1972, Spindle dynamics and chromosome movements, Int. Rev. Cytol., Suppl. 3, 34:1.
Boveri, T., 1900, Zellen-Studien. IV. Über die Natur der Centrosomen, Jena. Z. Naturwiss., 356:1.
Brenner, S., Pepper, D. A., Berns, M. W., Tan, E., and Brinkley, B. R., 1981, Kinetochore structure, duplication, and distribution in mammalian cells: Analysis by human auto-antibodies from Scleroderma patients, J. Cell Biol., 91:95.

Church, K., 1981, The architecture of and chromosome movements within the premeiotic interphase nucleus, in: "Mitosis/Cytokinesis", A. M. Zimmerman, and A. Forer, eds., Acadmic Press, New York, London, Toronto, Sydney, San Francisco.

Comings, D. E., 1968, The rationale for an ordered arrangement of chromatin in the interphase nucleus, Am. J. Human Genet., 20:440.

Comings, D. E., and Okada, T. A., 1970, Condensation of chromosomes onto the nuclear membrane during prophase, Exp. Cell Res., 63:471.

de Harven, E., and Bernhard, W., 1956, Étude au microscope électronique de l'ultrastructure du centriole chez les vertébrés, Z. Zellforsch., 45:378.

Euteneuer, U., and McIntosh, J. R., 1981, Polarity of some mobility-related microtubules, Proc. Natl. Acad. Sci. USA, 78:372.

Fuge, H., 1977, Ultrastructure of mitotic cells, in: "Mitosis - Facts and Questions", M. Little, N. Paweletz, C. Petzelt, H. Ponstingl, D. Schroeter, and H.-P. Zimmermann, eds., Springer-Verlag, Berlin, Heidelberg, New York.

Fuge, H., Bastmeyer, M., and Steffen, W., 1985, A model for chromosome movement based on lateral interaction of spindle microtubules, J. Theor. Biol., 115:391.

Ghosh, S., and Paweletz, N., 1984, Events associated with the initiation of mitosis in fused multinucleate HeLa cells, Chromosoma, 90:57.

Harris, P., 1975, The role of membranes in the organization of the mitotic apparatus, Exp. Cell Res., 94:409.

Hepler, P. K., 1980, Membranes in the mitotic apparatus of barley cells, J. Cell Biol., 86:490.

Hepler, P. K., McIntosh, J. R., and Cleland, S., 1970, Intermicrotubular bridges in mitotic spindle apparatus, J. Cell Biol., 45:438.

Hepler, P. K., and Wolniak, S. M., 1984, Membranes in the mitotic apparatus: Their structure and function, Intern. Rev. Cytol., 90:169.

Hughes, A., 1952, "The Mitotic Cycle", Academic Press, New York.

Koonce, M. P., and Schliwa, M., 1985, Bidirectional transport can occur in cell processes that contain single microtubules, J. Cell Biol., 100:322.

Kubai, D. F., and Ris, H., 1969, Division in the dinoflagellate Gyrodinium cohnii (Schiller). A new type of nuclear reproduction, J. Cell Biol., 40:508.

Lettré, H., 1961, Mitose und Dissoziabilität einzelner Mitoseschritte, Forsch. Fortschr., 35:39.

Lettré, H., and Lettré, R., 1959, A cytological problem: permanence of the chromosomal spindle fiber during interphase, Nucleus, 2:23.

Levine, L., 1963, "The Cell in Mitosis", Academic Press, New York, London.

Mazia, D., 1961, Mitosis and the physiology of cell division, in: "The Cell III", J. Brachet and A. E. Mirsky, eds., Academic Press, New York.

Mazia, D., 1978, Origin of twoness in cell reproduction, in: "Cell Reproduction: In Honor of Daniel Mazia", E. R. Dirksen, D. M. Prescott, and C. F. Fox, eds., Academic Press, New York, San Francisco, London.

Mazia, D., 1984, Centrosomes and mitotic poles, Exp. Cell Res., 153:1.

Miller, R. H., and Lasek, R.J., 1985, Cross-bridges mediate anterograde and retrograde vesicle transport along microtubules in squid axoplasm, J. Cell Biol., 101:2181.

Mitchison, T. J., and Kirschner, M. W., 1985, Properties of the kinetochore in vitro. II. Microtubule capture and ATP-dependent translocation, J. Cell Biol., 101:766.

Moll, E., and Paweletz, N., 1980, Membranes of the mitotic apparatus of mammalian cells, Eur. J. Cell Biol., 21:280.

Paweletz, N., 1967, Zur Funktion des "Flemming-Körpers" bei der Teilung tierischer Zellen, Naturwiss., 54:533.

Paweletz, N., 1974, Elektronenmikroskopische Untersuchungen an frühen Stadien der Mitose bei HeLa-Zellen, Cytobiol., 9:368.

Paweletz, N., 1981, Membranes in the mitotic apparatus. Mini-review, Cell Biol. Intern. Rep., 5:323.

Paweletz, N., and Fehst, M., 1984a, The vesicular compartment of the mitotic apparatus in mammalian cells, Cell Biol. Intern. Rep., 8:675.

Paweletz, N., and Fehst, M., 1984b, Are membranes of the mitotic apparatus translocated by microtubules? Cell Biol. Intern. Rep., 8:117.

Paweletz, N., and Finze, E.-M., 1981, Membranes and microtubules of the mitotic apparatus of mammalian cells, J. Ultrastruct. Res., 76:127.

Paweletz, N., and Mazia, D., 1979, Fine structure of the mitotic cycle of unfertilized sea urchin eggs activated by ammoniacal sea water, Eur. J. Cell Biol., 20:37.

Paweletz, N., and Mazia, D., 1987, The fine structure of bipolarization, in: "The Cell Biology of Fertilization"., H. Schatten, and G. Schatten, eds., Academic Press, New York.

Paweletz, N., Mazia, D., and Finze, E.-M., 1984, The centrosome cycle in the mitotic cycle of sea urchin eggs, Exp. Cell Res., 152:47.

Paweletz, N., and Risueño, M. C., 1982, Transmission electron microscopic studies on the mitotic cycle of nucleolar proteins impregnated with silver, Chromosoma, 85:261.

Paweletz, N., and Schroeter, D., 1974, Scanning electron microscopic observations on cells grown in vitro. II. HeLa cells in mitosis, Cytobiol., 8:229.

Paweletz, N., and Schroeter, D., 1986, On the fine structure of the mitotic apparatus of mammalian cells, in: "Genetic Toxicology of Environmental Chemicals, Part A", C. Ramel, B. Lambert, and J. Magnusson, eds., Alan R. Liss, New York.

Paweletz, N., and Schroeter, D., 1987, On the ultrastructure of the mitotic apparatus, in: "Progress and Topics in Cytogenetics. Aneuploidy - Incidence and Etiology", A. A. Sandberg, and B. K. Vig, eds., Alan R. Liss, New York.

Porter, K. R., and Machado, R., 1960, Studies on the endoplasmic reticulum. IV. Its form and distribution during mitosis in cells of onion root tip, J. Biophys. Biochem. Cytol., 7:167.

Rabl, C., 1885, Über Telltheilung, Gegenbaurs Morph. Jahrb., 10:214.

Rattner, J. B., and Berns, M. W., 1976, Centriole behavior in early mitosis of rat kangaroo cells (PtK$_2$), Chromosoma, 54:387.

Rebhun, L. J., 1972, Polarized intracellular particle transport: Saltatory movements and cytoplasmic streaming, Int. Rev. Cytol., 32:93.

Rhoades, M. M., 1961, Meiosis, in: "The Cell III", J. Brachet and A. E. Mirsky, eds., Academic Press, New York.

Rickards, G. K., 1975, Prophase chromosome movements in living house cricket spermatocytes and their relationship to prometaphase, anaphase and granule movements, Chromosoma, 49:407.

Rickards, G. K., 1981, Chromosome movements within prophase nuclei, in: "Mitosis/Cytokinesis", A. M. Zimmerman and A. Forer, eds., Academic Press, New York, London, Toronto, Sydney, San Francisco.

Roos, U.-P., 1973, Light and electron microscopy of rat kangaroo cells in mitosis. I. Formation and breakdown of the mitotic apparatus, Chromosoma, 40:43.

Sanger, J. M., Pochapin, M. B., and Sanger, J. W., 1985, Midbody sealing after cytokinesis. Cell Tiss. Res., 240:287.

Schatten, G., Maul, G. G., Schatten, H., Chaly, N., Simerly, C., Balczon, R., and Brown, D. L., 1985, Nuclear lamins and peripheral nuclear antigens during fertilization and embryogenesis in mice and sea urchins, Proc. Natl. Acad. Sci. USA, 82:4727.

Schrader, F., 1941, The spermatogenesis of the earwig Anisolabis maritima Bon. with reference to the mechanism of chromosome movement. J. Morphol., 68:123.

Schrader, F., 1953, "Mitosis: The Movement of Chromosomes in Cell Division", Columbia University Press, New York.

Schroeter, D., Ehemann, V., and Paweletz, N., 1985, Cellular compartments in mitotic cells: Ultrahistochemical identification of Golgi elements in PtK$_1$ cells, <u>Biol. Cell.</u>, 53:155.

Suchard, S. J., and Goode, D., 1982, Microtubule-dependent transport of secretory granules during stalk secretion in a peritrich ciliate, <u>Cell Mot.</u>, 2:47.

Taura, M., 1978, Origin and fate of paired cisternae in mitotic aortic cells of swine. <u>J. Electr. Microsc.</u>, 27:283.

Vale, R. D., Reese, T. S., and Sheetz, M. P., 1985a, Identification of a novel force generating protein (kinesin) involved in microtubule-based motility, <u>Cell</u>, 41:39.

Vale, R.D., Schnapp, B.J., Mitchison, T., Steuer, E., Reese, T. S., and Sheetz, M. P., 1985b, Different axoplasmic proteins generate movement in opposite directions along microtubules in vitro, <u>Cell</u>, 43:623.

Wassermann, F., 1929, "Die lebendige Masse. Wachstum und Vermehrung der lebendigen Masse", Verlag von Julius Springer, Berlin.

PROGRESS IN RESEARCH ON MITOSIS

J. Richard McIntosh

Dept. of Molecular, Cellular, and Developmental Biology
Univeristy of Colorado, Box 347
Boulder, Colorado, 80309

INTRODUCTION

The structure of the mitotic spindle (Bajer and Mole-Bajer, 1972; Heath, 1978; McIntosh, 1985), the action of the spindle on chromosomes (Nicklas, 1975; 1986), and the physiology of the spindle in vivo (Inoue, 1981) have all been studied and reviewed. From this work, the broad outline of mitotic events, both for the chromosomes and for the spindle fibers themselves, is well known. During prophase, as the chromosomes condense, the spindle begins to form in the cytoplasm by the initiation of microtubules (MTs) from the already duplicated centrosomes. The nuclear envelope then disperses, and the spindle MTs start to interact with the chromosomes at their kinetochores. Bundles of MTs, in cooperation with other, less well-defined components, become the spindle fibers that connect the kinetochores with the poles, and these fibers exert a pole-directed force on each kinetochore. Oppositely directed kinetochores on sister chromatids usually interact with fibers attached to opposite poles, and the two opposing forces exert a tension on the chromosome at its centromere. These forces are proportional in magnitude to the length of the spindle fibers (Hays, Wise, and Salmon, 1982), so the chromosomes are all gradually pulled to the spindle equator where they reside until the onset of anaphase.

Anaphase is initiated by the separation of the chromatids, a process that is not yet understood. The chromosomes then move nearer to the poles with a concomitant shortening of the pole-to-kinetochore fibers (anaphase A). Midway through this process, the poles usually start moving apart, increasing the distance by which the chromosomes will finally be separated (anaphase B). The latter process is accompanied by the elongation of an interdigitating system of MTs that connects the poles together. In late anaphase the chromosomes contract in toward the poles, and the nuclear envelop reassembles around the chromatin. The chromosomes then decondense (telophase) to re-establish the interphase condition. The cleavage furrow becomes established sometime during anaphase at the plane where the chromosomes used to be. It then acts to constrict the cell's equator, pinching the cell into two. The remains of the MT system that interconnected the poles is usually bunched by the furrow to form a compact MT bundle called the midbody. This bundle often persist through telophase while the centrosomes at the spindle poles are beginning to initiate an interphase MT complex in each of the two sister cells. The midbody slowly disappears during the subsequent interphase, allowing the sisters to separate and go their own ways.

During the last few years there has been considerable progress in developing a molecular understanding of the mechanisms for chromosome movement. The advances have come from a combination of structural and biochemical approaches. It has become clear from several lines of evidence that MTs are key for spindle structure and function. Spindle MTs have therefore served as a focus for much of the recent research on mitotic mechanism. Five questions have attracted the most attention: 1) What is the structure of the MT component of the fully-formed metaphase spindle? 2) How does that array of MTs form? 3) What are the changes in the structure of the MT array during anaphase? 4) What controls the stability and assembly properties of spindle MTs? 5) What are the motors that develop forces for the mitotic structural changes?

1. STRUCTURE OF THE MICROTUBULE COMPONENT OF THE METAPHASE SPINDLE

Spindle Microtubule Polarity

MTs are known to have a structural polarity (Amos and Klug, 1974), and two methods have been found to visualize this polarity in the spindles of lysed cells (Heidemann and McIntosh, 1980; Haimo and Telser, 1981). Studies with the two methods agree that all the MTs initiated from each spindle pole are of the same polarity; their "plus" or fast growing ends are distal to the pole from which they grew. Many MTs from each pole interdigitate near the metaphase spindle equator (Euteneuer and McIntosh, 1980), but the MTs that attach to the kinetochores do not extend that far from the spindle poles. Nonetheless, they too are oriented with their plus ends distal to the poles (Euteneuer and McIntosh, 1981), so the design of the MT arrangement in the metaphase spindle is simple (Figure 1a). This simplicity is preserved during anaphase, so the arrangement of MT polarities at telophase is little changed from that at metaphase (Figure 1b).

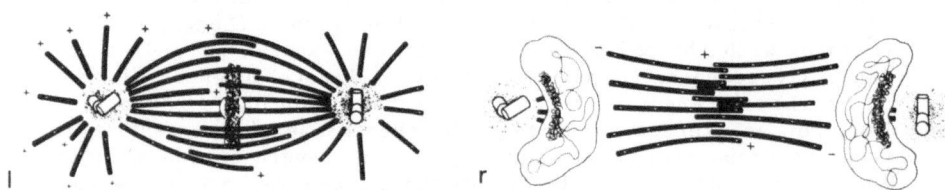

Figure 1. The polarity of spindle MTs at two different mitotic stages. Fig. 1a shows the arrangement at metaphase. The spindle poles are represented by paired cylinders that portray centrioles and stippling that depicts the pericentriolar material active in initiation of MTs. All the MTs have their plus ends distal to the centrosome from which they grew. Even the kinetochore MTs, whose pathways of formation are unknown, are oriented with their plus ends distal to the poles and proximal to the kinetochores. Fig. 1b shows that the midbody MTs that are left in the interzone by the separating chromosomes are oriented as if they were derived from some of the MTs from the metaphase spindle that did not end on chromosomes. This is probably their origin.

Spindle Microtubule Arrangement

The arrangement of the spindle fibers at metaphase has been studied by following the trajectory of individual MTs in three dimensions by serial thin sections and electron microscopy. (e.g. McIntosh et al., 1975; 1979; 1985; McDonald et al., 1979; Lafountain and Davidson, 1980; Tippit et al., 1980 a and b; 1983; 1984). The general design of the spindle seen in these reconstruction confirms that deduced from the studies of MT polarity; a spindle is constructed from two sets of MTs that interdigitate at the spindle equator (first described for midbodies by Paweletz, 1967). Some MTs end on chromosomes, but many do not, and the distribution of MT lengths is strikingly haphazard (Figure 2). The cross sectional arrangement of MTs in most spindles is similarly disordered.

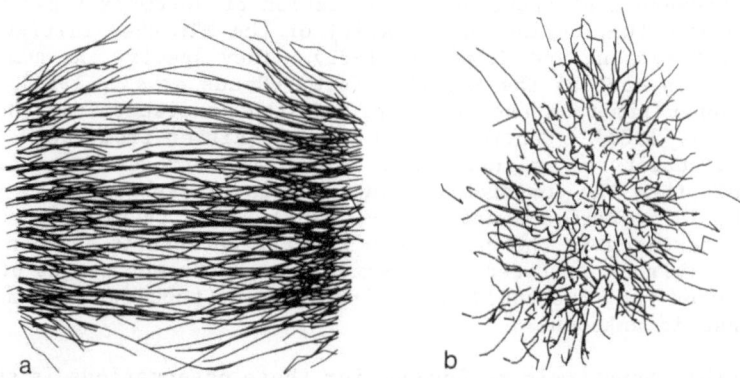

a   b

Figure 2. Longitudinal (Fig. 2a) and transverse (Fig. 2b) views of the MTs from an early anaphase spindle of <u>Dictyostelium</u>. Each line is the projection of a quartic polynomial fit to points lying on the cross-section of a single MT. All the MTs of the spindle are shown because these are projections, not sections. The projections have been distorted to make them easier to visualize. Spindle length is half what it should be for the cross-sectional dimensions shown.

2. ASSEMBLY OF SPINDLE MICROTUBULES DURING PROMETAPHASE

The <u>Morphology</u> <u>of</u> <u>Spindle</u> <u>Microtubule</u> <u>Assembly</u>

The formation of the spindle MTs has been described <u>in vivo</u> using both the light and the electron microscopes. With either polarization microscopy (reviewed in Inoue, 1981) or immunofluorescence (Figure 3), spindle MTs are seen to begin growing sometime during late prophase or early prometaphase, just as the nuclear envelope disperses. The growth is from two foci called centrosomes. Centrosomes vary in structure from one organism to another (reviewed in McIntosh, 1983), but they are always involved in the organization of the spindle, and they become the spindle "poles." Often the mitotic system of MTs forms at the very same time that the fiber array from the previous interphase is disassembling.

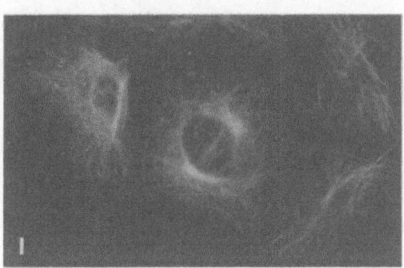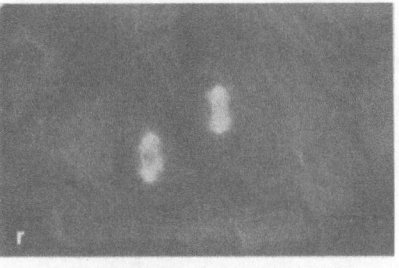

Figure 3. Immunofluorescence images of MTs in mammalian cells, strain PtK, as the cells enter mitosis and form a spindle. The importance of two foci in organizing the MTs of the spindle is obvious. These foci are the centrosomes.

The Role of the Centrosome in Spindle Assembly

At all stages of the cell cycle, the centrosomes serve as MT organizing sites, as defined by several criteria. They will initiate the polymerization of tubulin that is competent to assemble, even if the protein concentration is sufficiently far below the critical concentration that the solution is incapable of spontaneous initiation of assembly (Snyder and McIntosh, 1975). They define the polarity of the MTs they initiate (Bergen, et al., 1980; Heidemann and McIntosh, 1980). They specify the number of protofilaments in each of the MTs that forms (Evans et al., 1985), and they set an upper limit on the number of MTs that radiates from them (Brinkley et al, 1981). Centrosomes at mitosis differ from the corresponding structures in interphase. Mitotic centrosomes initiate 5 - 10 times more MTs than the comparable structures in interphase cells (Snyder and McIntosh, 1975; Kuriyama and Borisy, 1981). This change must be a result of differences in the centrosome itself, because it is seen using either lysed cells or isolated centrosomes from different stages of the cell cycle, even when they are challenged with the same tubulin. The mechanism for the change is unknown.

A plausible hypothesis to account for these observations is that the centrosome contains a defined number of objects that can bind tubulin and seed the assembly of a MT. While centrosomes do contain antigens that crossreact with some of the well known MT-associated proteins (reviewed in DeBrabander et al., 1985), there is no fine structural evidence for discrete structures in the region of the centrosome that initiates MTs (Figure 4). The chemistry of centrosomes has proven to be very difficult, and little can currently be said about the molecular mechanisms by which they work.

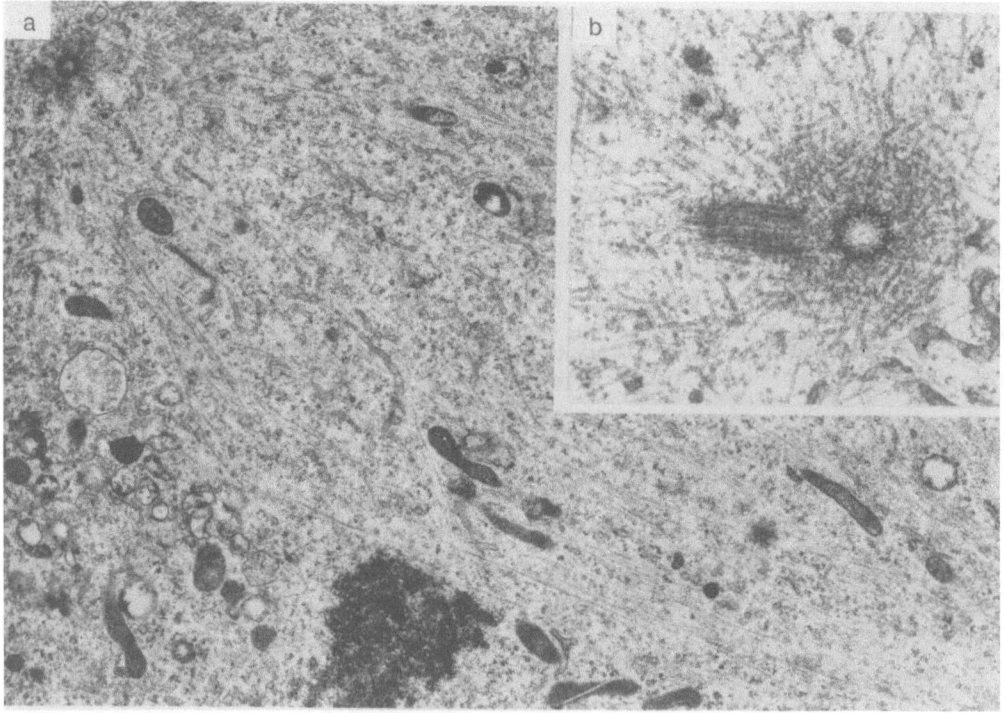

Figure 4. Electron micrographs of spindles from PtK cells. Fig. 4a is of a cell in prometaphase, and shows the still-disorganized arrangement of both MTs and chromosomes. Fig. 4b shows the centrosomal region at higher magnification. (This micrograph is kindness of Kent McDonald.)

The Role of Kinetochores in Spindle Assembly

During prometaphase MTs will also initiate at kinetochores under some experimental conditions. Cells blocked prior to mitosis with MT inhibitors such as colcemid or nocodazole will lack MTs. Upon washing out of the drug, MTs assemble and a functional spindle is formed (Debrabander et al., 1981; Witt et al., 1980). It is clear that during recovery from such inhibition, MTs assemble in the vicinity of the kinetochores. An analysis of the polarity of kinetochore-associated MTs in the spindles formed after reversal of such an inhibition shows that the MT polarities are set up in the normal, simple way (Euteneuer et al., 1983). The fine structure of the MTs near the kinetochores early during recovery, however, is more complex. The newly formed kinetochore-associated MTs are oriented almost at random, although they all lie in association with the "corona" of the kinetochore, i.e. the array of whispy microfibrils that protrudes from the outer dense plaque of the kinetochore proper (Ris and Witt, 1981). It looks as if the corona can potentiate the initiation of MTs, but does not define MT orientation (DeBrabander et al., 1981).

This impression has recently been confirmed in vitro where kinetochores will initiate MT growth from purified tubulin. The polarity of the resulting MTs is random with respect to the kinetochore (Mitchison and Kirschner, 1985 a and b). There are indications, however, that if ATP is present, either during or after MT initiation, the MTs will become bound to the kinetochores by their plus ends only. This behavior, which may result either from an active motion of the kinetochore along the MTs or from a release of incorrectly oriented MTs, seems to account for the pathway of spindle formation observed in cells recovering from the drug treatment. Whether kinetochore initiation of MTs is an important component of normal spindle formation, or whether kinetochore MTs are initiated only by the poles and subsequently captured by the kinetochores, remains to be determined. It is probably significant for spindle function that either pathway will result in the same arrangement of MTs, because MT polarity has been linked to MT function both in vivo (McNiven and Porter, 1984) and in vitro (Warner and Mitchell, 1980; Vale et al., 1985).

3. SPINDLE STRUCTURE CHANGES DURING ANAPHASE

Changes in Microtubule Length

The tracking of spindle MTs in three dimensions by electron microscopy has contributed several facts to our understanding of the structural changes that accompany spindle action. In an alga, a cellular slime mold, and a fungus the anaphase shortening of chromosome-to-pole MTs is accompanied by a shortening and drawing up to the pole of most of the other spindle MTs (McIntosh et al, 1985; Tippet et al, 1980 b; 1984). A similar situation probably pertains in mammalian cells, although the evidence here is less complete (McIntosh et al. 1975). Disassembly in anaphase is not, however, the pattern for all spindle microtubules; the MTs that interdigitate at the spindle equator lengthen during spindle elongation (Figures 5a and b). Probably these are the very MTs that interdigitated at metaphase, and something about their interactions with the MTs from the opposite pole distinguishes them from the non-interdigitating MTs that shorten during this time (McDonald et al., 1979).

The lengthening and shortening of spindle MTs during anaphase is of considerable functional significance. The chromosomes could not approach the poles if the intervening MTs did not shorten, and the spindle poles could not move apart if the MTs that connected them did not elongate. Some workers have proposed that it is the energy associated with MT polymerization and depolymerization itself that pulls and pushes on the chromosomes (reviewed in Inoue, 1981), while others have suggested that the polymerization events are permissive for chromosome movements, perhaps regulating their rate (e.g. Nicklas, 1985). At the moment there is not sufficient data to discriminate between these possibilities.

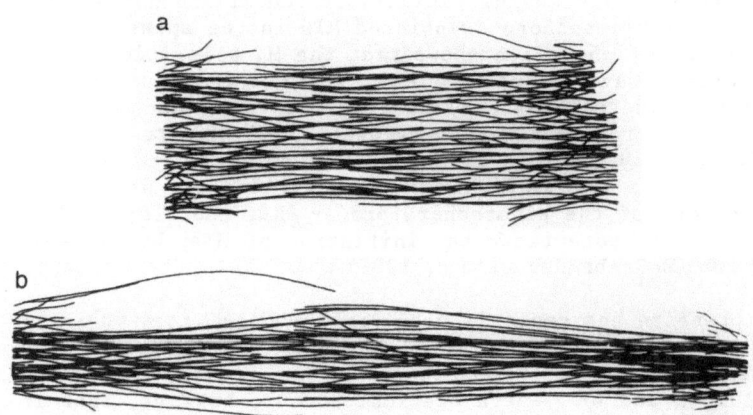

Figure 5. A mid anaphase (Fig. 5a) and an early telophase (Fig. 5b) spindle reconstructed from sections of <u>Dictyostelium</u>. Both spindles are made from interdigitating sets of MTs. The extents of MT interdigitation are not greatly different in the two reconstructions, but inspection shows that the interdigitating MTs of the later spindle are considerably longer than those of the earlier. The short MTs that cluster around the spindle ends are noticeably shorter than the corresponding MTs in Fig. 2. These spindles are distorted for display in the same way as that used in Fig. 2.

<u>Changes</u> <u>in</u> <u>Microtubule</u> <u>Position</u>

<u>Spindle</u> <u>Elongation</u>. Structural studies also show that during anaphase B in diatoms, the extent of MT interdigitation decreases, as if the MTs from the two poles slide apart like a muscle working backwards (Figure 6). This impression is supported by physiological evidence both <u>in</u> <u>vivo</u> (Leslie and Pickett-Heaps, 1983) and <u>in</u> <u>vitro</u> (Cande and McDonald, 1986).

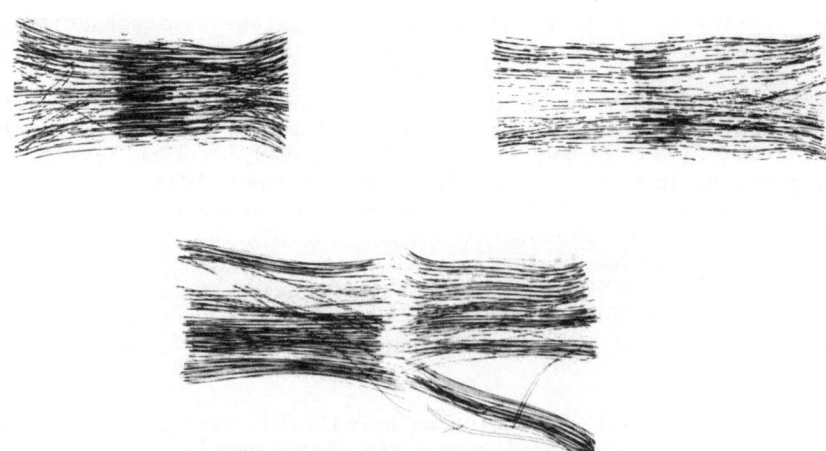

Figure 6. Reconstructions of the MTs from the central spindle of the diatom, <u>Diatoma</u> <u>vulgare</u>. The methods used were as for Figs. 2 and 4.

Examination of anaphase B in several cell types shows, however, that the situation is more complex than just a sliding apart of the two half spindles. Many spindles elongate during anaphase B by more than the extent to which their MTs interdigitate in metaphase. Further, measurement of MT length at different mitotic stages shows that the interdigitating MTs can elongate during anaphase (viz. Figs. 2, 5, and 6), so tubulin polymerization is also a part of anaphase B.

In some cells, such as the cellular slime mold shown in Figure 5, there is little or no anaphase change in the apparent extent of MT interdigitation, so one might argue that anaphase MT sliding is a rarity, confined perhaps to the diatoms, and the bulk of spindle elongation is achieved by MT polymerization. This possibility is unlikely, however, because in slime molds there is indirect evidence for the addition of tubulin subunits at the pole-distal ends of the interdigitating MTs during anaphase spindle elongation (McIntosh et al, 1985). These MT ends are frayed and uneven, as if they were the site of either MT assembly or disassembly. We have proposed that this is evidence of MT assembly, because these are the spindle MTs that elongate, while the MTs that shorten show only the blunt ends characteristic of normal MTs. If the tubulin is adding to these pole-distal ends, then the MTs of the <u>Dictyostelium</u> spindle must also be sliding apart to preserve the approximately constant extent of MT interdigitation at the spindle midplane (McIntosh et al., 1985).

Direct evidence for this mechanism of spindle elongation has recently come from an analysis of the elongation of the interpolar MT system in PtK cells. Fluorescent tubulin has been injected into living cells, then followed with a low light level video camera to trace its pattern of incorporation into polymer (Saxton et al., 1984). Tubulin adds to the elongating anaphase MTs at the region of their interdigitation near the spindle equator (Figure 7).

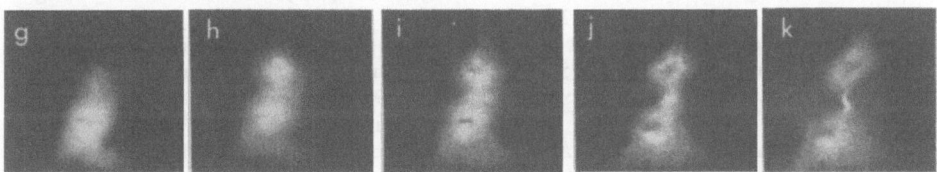

Figure 7. Incorporation of fluorescein-labeled tubulin into the anaphase spindle of a PtK cell in culture. The pictures show the spread of soluble fluorescent-labeled protein across the cell by diffusion, followed by the slower accumulation of fluorescence at the midregion of the midbody. These pictures are photographs taken from a video monitor that displayed the signal from an ISIT video camera. 64 consecutive frames (about 2 sec) were averaged by computer to improve the signal to noise ratio in each picture shown. We presume that this pattern of fluorescence development reflects assembly of tubulin into MTs. Pictures courtesy of W.M. Saxton.

Assembly of tubulin into MTs at their pole-distal ends where they interdigitate with MTs from the opposite pole (viz. Figs. 1b, 5, and 6) should, of course, only increase the extent of MT overlap, not elongate the spindle. In mammalian cells, however, the extent of overlap decreases or remains constant during anaphase (McIntosh and Landis, 1971; McIntosh et al, 1975), suggesting that the interpolar spindle MTs add tubulin at their pole-distal ends and simultaneously slide apart to effect spindle elongation. Direct evidence for this possibility is provided by experiments in which fluorescent interpolar MTs are marked by laser microirradiation to achieve local photobleaching. The changes in positions of the bleach marks with time shows that the two halves of the interzone spindle move away from the spindle midplane during late anaphase (Saxton and McIntosh, submitted).

Chromosome-to-Pole Motion. While MT behavior in anaphase B is now
fairly well understood, the corresponding information for anaphase A is
lacking. As the chromosomes approach the poles, the chromosomal MTs
shorten, as do most of the other MTs in the spindle, but the pathway for
this loss of subunits is not yet known. In principle, tubulin could
depolymerize from either the pole-distal or the pole-proximal MT ends; the
loss could even occur along the surface of the MT walls. We have tried to
distinguish these possibilities by using fluorescent tubulin and laser
photobleaching, but in the course of these experiments we found something
quite unexpected: during metaphase and anaphase the MTs in the regions
between chromosomes and poles turnover fast. The recovery is so fast that
we have not yet been able to follow a bleached mark long enough during
anaphase to determine whether a chromosome moves relative to its fiber as it
approaches the pole. It is therefore not yet possible to say much about the
mechanism by which the cell accomplishes anaphase A.

4. FACTORS REGULATING MICROTUBULE ASSEMBLY, DISASSEMBLY, AND TURNOVER

It is clear from the above account that there are changes in MT
polymerization and stability as the spindle forms and functions. Since the
disassembly of some MTs is essential for anaphase A and the elongation of
others MTs is a common part of anaphase B, the regulation of MT
polymerization is sure to be an important part of the mechanism for
chromosome motion.

Temporal Changes in the Rate of Microtubule Turnover in Vivo

When fluorescein-labeled (Saxton et al., 1984; Soltys and Borisy, 1985)
or biotin-labeled (Schultz and Kirschner, 1986) tubulin is injected into
interphase mammalian cells, it takes only a few minutes for the exogenous
protein to exchange with half of the endogenous polymer. Studies using a
laser microbeam to bleach the fluorescence of injected tubulin and a low
light level video system to follow the fluorescence redistribution after
photobleaching suggest a half time for interphase tubulin turnover of
something more than 200 sec (Saxton et al., 1984). Mitotic cells on the
other hand show a half time for redistribution of bleached tubulin
fluorescence of about 16 sec. A rapid turnover of spindle MTs is also
demonstrated by experiments that show how fast a spindle can dissolve when
the endogenous pool of subunits is inactivated (Salmon et al., 1984). There
is therefore an increase of more than 20-fold in the MT turnover rate in
going from interphase to mitosis.

Pathways of Microtubule Turnover

MT turnover at all stages of the cell cycle is rapid, and at mitosis it
is startlingly so. What is the pathway by which tubulin in solution can
exchange with protein in polymer to permit this surprisingly fast exchange?
The data available exclude several possibilities and leave a limited field
of acceptable hypotheses. The site of incorporation of labeled tubulin,
immediately following its injection into living cells, is always on the
centrosome-distal end of the cell's MTs; there is no observable addition at
either the pole proximal ends or on the walls of the MTs (Shultz and
Kirschner, 1986; Soltys and Borisy, 1985). Tubulin exchange into MTs in
vivo therefore seems to be an exchange process involving one MT end only.
The most likely model, a priori, for tubulin exchange between polymer and
solution is that the free MT end is simply exchanging subunits at random, as
in any chemical equilibrium. This attractive possiblity is excluded,
however, by the rapid rate of fluorescence exchange between monomer and
polymer. For half the subunits in spindle polymer to exchange by such a
process in just 16 sec, the mean MT length would have to be no more than

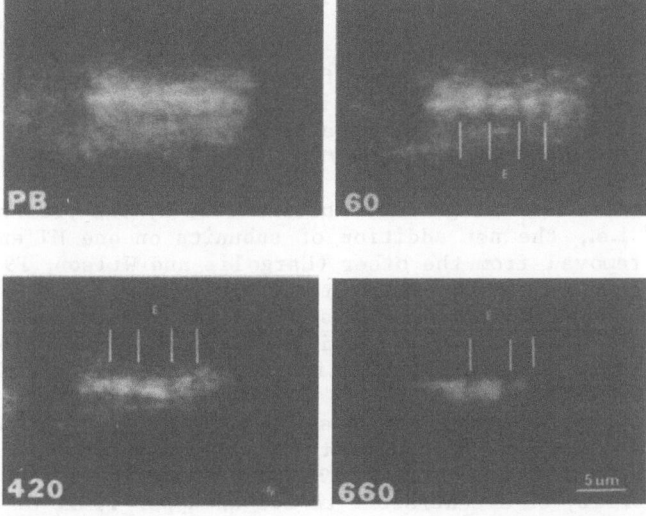

Figure 8.   Fluorescence in a late anaphase PtK cell that was injected with
fluorescent tubulin during prometaphase.   After anaphase was underway, the
spindle was photobleached with a 2 W argon laser whose beam was passed
through a Ronchi grating and then imaged on the specimen to effect a
periodic bleach pattern.   Such patterns made on spindles at this stage of
mitosis are rather stable, and one can follow the relative movements of the
bleach marks.   The bleached regions are probably good markers for positions
on the surfaces of MTs, and hence their motion apart reflects the relative
motion of the two halves of the forming midbody in PtK cells.   The results
have been quantified and show that the two halves of the midbody are sliding
apart at about 0.1 um/min at this time (Saxton and McIntosh, submitted).

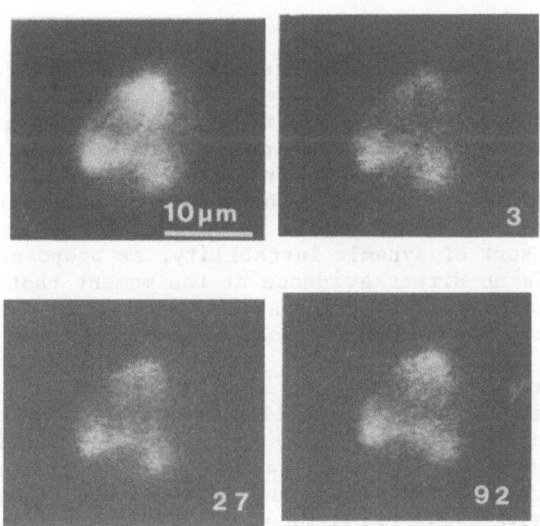

Figure 9. Representative images of tubulin fluorescence redistribution after
photobleaching.   The pictures were prepared as for Fig. 7.   The relative
times after photobleaching are marked.   The conclusion that tubulin turnover
is faster during mitosis than during interphase is obvious.

about 0.1 um, even assuming diffusion-limited rate constants. The structural studies cited above put the mean length more than an order of magnitude greater than this, depending on the spindle in question. A factor of 10 increase in length would mean a factor of 100 increase in mean half time, because the postulated exchange is essentially a diffusional process, and this value is far outside the range of acceptable rates of exchange established by the error in the observations.

The rapid rate of exchange could be achieved by the process of "treadmilling," i.e., the net addition of subunits on one MT end together with their net removal from the other (Margolis and Wilson, 1978). This model predicts that small bleached spots on clusters of parallel MTs should move in a direction defined by the MT's polarity. Direct examination of this hypothesis by fluorescence bleaching eliminates the possiblity as an explanation for fast turnover of mitotic MTs (Wadsworth and Salmon, 1986).

An important insight into the possible pathways for fast MT turnover _in vivo_ came recently from the _in vitro_ studies of centrosome-initiated MT assembly by Mitchison and Kirschner (1984 a and b). These workers not only confirmed the ability of a centrosome to set an upper limit on the number of MTs it will initiate, they also showed that the steady state number of MTs growing from a centrosome is a function of the tubulin concentration in solution. They proposed a model to account for this finding, based on the theory and experiments of Carlier and coworkers (Carlier and Pantaloni, 1981; Carlier et al., 1984). The model postulates a "cap" of tubulin subunits in stable association with the MT (Hill and Chen, 1984). The subunits in the cap are stable because their bound GTP has not yet been hydrolyzed. The length of this cap is presumed to depend on the rate of tubulin assembly, because GTP hydrolysis is thought to occur at a constant rate and the addition rate of GTP-tubulin would depend on the concentration of free subunits. Thus a population of MTs at steady state would consist of a large number of slowly growing polymers and a small group of MTs that have by chance just lost their cap of stable, GTP-subunits. The exposed GDP-subunits are thought to disassemble rapidly. In spite of the rather complex character of the resulting steady state, the model has some attractive features for explaining aspects of mitotic MT behavior (Kirschner and Mitchison, 1986).

The model has recently received qualitative support from the observation of MTs at steady state _in vitro_ with dark field light microscopy. Each MT displays dynamic fluctuations in length as a result of lengthy runs of subunit addition and loss at both of its free ends (Horio and Hotani, 1986). This behavior is inconsistent with the predictions of a model based on equilibrium subunit exchange at both MT ends, in which the expected fluctuations of length would be much shorter. It is consistent, however, with some sort of dynamic instability, as proposed by Mitchison and Kirschner. There is no direct evidence at the moment that the transition from a growing state to a shrinking state is related to the hydrolysis of GTP, but this is still a reasonable possibility (Williams et al., 1986).

It is important to note, however, that there is also compelling evidence for the reality of both treadmilling and annealing of MTs _in vitro_. Biochemical (Wilson and Farrell, 1986) and structural studies (Rothwell et al., 1985) have demonstrated that MTs can treadmill at steady state, and several lines of evidence show that annealing is surprisingly efficient (e.g. Rothwell et al., 1986; Williams et al., 1986). Recent work suggests that the factor which shifts MTs from dynamic instability to treadmilling _in vitro_ is the concentration of MT associated proteins. We expect, therefore, that MT behavior _in vivo_ will be modulated by the cell and may show differences that are important for cell physiology.

Factors Governing the Rate of Microtubule Turnover

The dynamic instability hypothesis accounts for the end-dependent exchange of MT subunits in vivo, for the unexpectedly rapid turnover of tubulin, and for the lack of bleach spot movement during tubulin fluorescence redistribution after photobleaching. It does not, however, explain why the mitotic MTs should turnover so much more rapidly than the MTs of interphase. Some insight into this difference has come from a study of the turnover of MT-associated proteins (MAPs) at different times in the cell cycle. MAPs that co-purify with tubulin through multiple cycles of polymerization may be labeled with fluorescein and injected into mammalian cells in culture, where they will bind to the existing array of MTs (Scherson et al., 1984). Fluorescence redistribution after laser photobleaching shows that these MAPs exchange with the interphase MTs to which they are bound with a half time of about 70 sec. In mitosis, on the other hand, the MAPs turnover with a half time of about 20 sec, within experimental error of the turnover rate for tubulin (Olmsted et al., 1987).

When mammalian cells are cooled to 26 deg. C, then the extent to which tubulin fluorescence will recover is reduced by 30 - 50%. MAP turnover, on the other hand, is unaffected by the temperature drop, in either mitosis or interphase. Thus MAP - MT interaction shows fast exchange in mitosis, whether or not the mitotic MTs are turning over quickly. Apparently the cell modulates the interaction between MAPs and MTs as a function of time in the cell cycle. This variation may be responsible for the significant increase in tubulin turnover rate when the cell goes into mitosis.

During metaphase and anaphase all the MTs connecting the chromosomes and poles turnover fast. The interzone of anaphase, on the other hand, contains MTs that turnover slowly. These are the interdigitating MTs that elongate during anaphase B. A bleach spot that covers about half the interzone reveals that the MTs immediately behind the separating chromosomes recover their fluorescence rapidly, while the interdigitating MTs that are the sole inhabitant of the region nearer to the spindle midplane turnover slowly, if at all (Figure 11).

This change in the rate of tubulin fluorescence redistribution is the reason that we were able to follow the behavior of MTs in anaphase B but not in anaphase A. It seems reasonable to suppose that the factors which make the interdigitating MTs slow to turnover also contribute to their ability to elongate while other spindle MTs are shortening. This distinction is not made by the MAPs so far investigated by fluorescence bleaching (MAP 2 and 4), because they show the rapid turnover characteristic of mitosis even in the mid region of the anaphase interzone where the tubulin turns over slowly (Olmsted et al., 1987). The factor most likely to govern the slow turnover of the interdigitating MTs is the material through which they interact to hold the two halves of the spindle together. Perhaps there are special MAPs the interconnect MTs of opposite polarity. Structural analysis of the relative positions of near neighbor MTs in this region show that MTs prefer to have near neighbors that emanate from the opposite pole (McDonald et al., 1979). These MTs lie at a sharply defined mean distance of 42 nm center-to-center, while the near neighbor MTs coming from the same pole lie essentially at random. There appears to be a mechanical connection between MTs of opposite polarity in the zone of MT interdigitation, as if there were a MAP that was a specific cross-linker for antiparallel MTs. This connection may stabilize the interdigitating MTs to subunit disassembly, reducing their rate of turnover and biasing their assembly equilibrium toward polymerization, even while the rest of the spindle is disassembling.

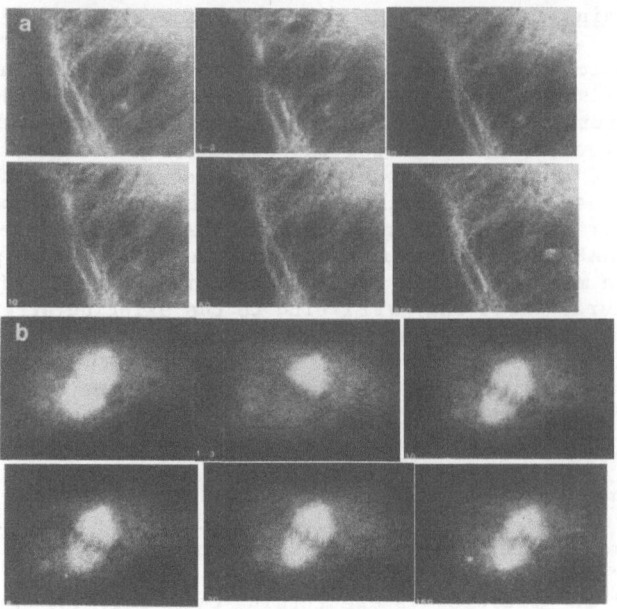

Figure 10. An interphase (Fig. 10a) and metaphase (Fig. 10b) PtK cell injected with fluorescent MTP 4 and allowed to equilibrate. The binding of the injected protein to the endogenous MTs is clear. Laser photobleaching is used to mark the proteins in one region of the cell, then the behavior of the protein is followed with a low light level video camera as in Fig. 8. Times are marked on the pictures. The fast turnover of mitotic protein relative to interphase protein is evident. (Courtesy of J.B. Olmsted).

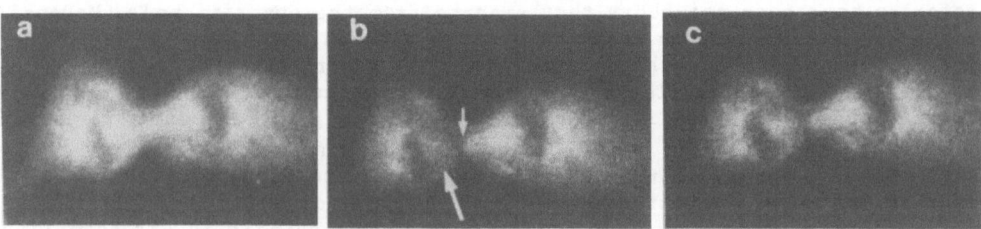

Figure 11. A photobleach and recovery of fluorescent tubulin during anaphase in PtK cells. (Courtesy of W.M. Saxton)

5. MOTORS THAT MOVE SPINDLE MICROTUBULES DURING ANAPHASE

Anaphase B seems to depend in part on a motor that makes microtubules slide. The dynein ATPase from flagella has long been recognized as capable of such an activity, but it is not yet clear whether a dynein-like enzyme is important for mitosis. Antibodies have been prepared against flagellar dynein and used for immunocytochemistry to see whether a related antigen is localized in the spindle. The results have been mixed, with some workers finding evidence for a spindle dynein (Mohri et al., 1976; Kobayashi et al., 1978) and others not (Zieve and McIntosh 1981). Much of this work has been done with eggs and embryos of marine organisms. This material has the disadvantage, however, that the embryos will form cilia after several rounds of cell division, so one can imagine that dynein is present in the eggs simply because it is a components of cilia. It might even be localized in the spindle simply because it binds adventitiously to the spindle microtubules. One really wants some sort of functional evidence for dynein's involvement in spindle action before taking it too seriously as a spindle protein, and such evidence for dynein is at present equivocal.

We have gone looking for motors that might be functionally significant in mitosis by following the logic that the relevant molecules are likely to bind to MTs in an ATP-sensitive manner. Using taxol to drive tubulin assembly in extracts of sea urchin eggs (reviewed in Vallee et al., 1984), we have identified a protein complex composed of at least two polypeptides with molecular wieghts of 130 KD and 60 KD. This complex is a MT dependent motor in the sense that it will induce MTs to glide over glass or latex beads to move over MTs in an ATP-dependent manner (Scholey et al., 1985; Porter et al., 1987).

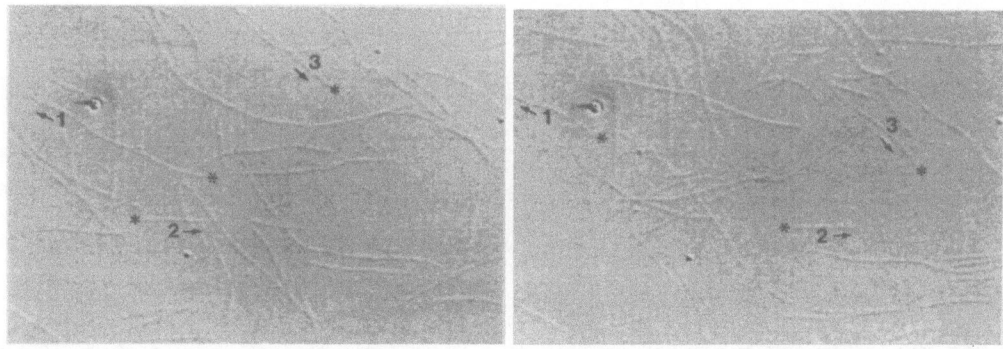

Figure 12. MTs moving over glass under the influence of the sea urchin egg protein and ATP. 28 sec elapsed between pictures. (Courtesy of M.E.Porter.)

The surface interacting with the MT is pushed toward the plus end of the fiber, so this sea urchin protein strongly resembles kinesin, the recently-described MT-dependent motor from squid axoplasm (Vale et al. 1985). Also like kinesin, the sea urchin egg protein does not have appreciable ATPase activity under the conditions we have used so far to measure it. Antibodies raised against the 130 KD polypeptide of sea urchin kinesin cross-react with the 110 KD polypeptide of squid, suggesting significant homology between the proteins, and both antibodies will stain the spindle of the sea urchin egg, as seen by immunocytochemisty (Scholey et al., 1985).

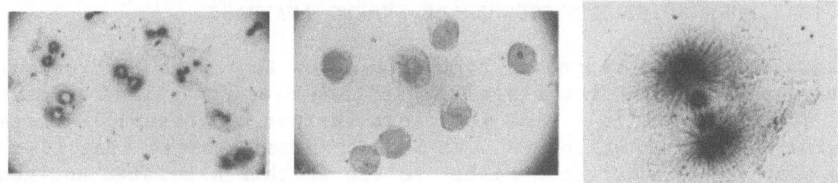

Figure 13.   Immunocytochemistry   of sea urchin embryos lysed in spindle-
stabilizing buffers and stained with the peroxidase conjugated indirectly to
affinity-purified antikinesin.   Fig. 11 a and b show reactions with the
immune serum, Fig. 11c depicts the staining seen with the preimmune.

We have begun to investigate the staining patterns seen with
antikinesin applied to mammalian cells. Antibodies raised against mammalian
neural kinesin in two different rabbits will react with crude kinesin
prepared from PtK cells. These antibodies show a weak staining of numerous
elongate objects in the interphase cytoplasm, perhaps mitochondria.   The
centrosome region of the cells is strongly stained at all stages of the cell
cycle.   The midbody of late anaphase and telophase cells shows some
staining, and there is a faint fibrous staining pattern in the prometaphase
and metaphase spindle in the region between the chromosomes and the poles.
The preimmune sera show a weak reaction with the elongate, mitochondria like
objects   (Figure 13).   Such patterns have been found in cells fixed in two
ways and with either IgG fractions or with affinity purified antibody.   The
patterns differ from the sea urchin spindle images in that the general
spindle staining is weaker and is more concentrated in particular places,
like the poles and the center of the midbody.

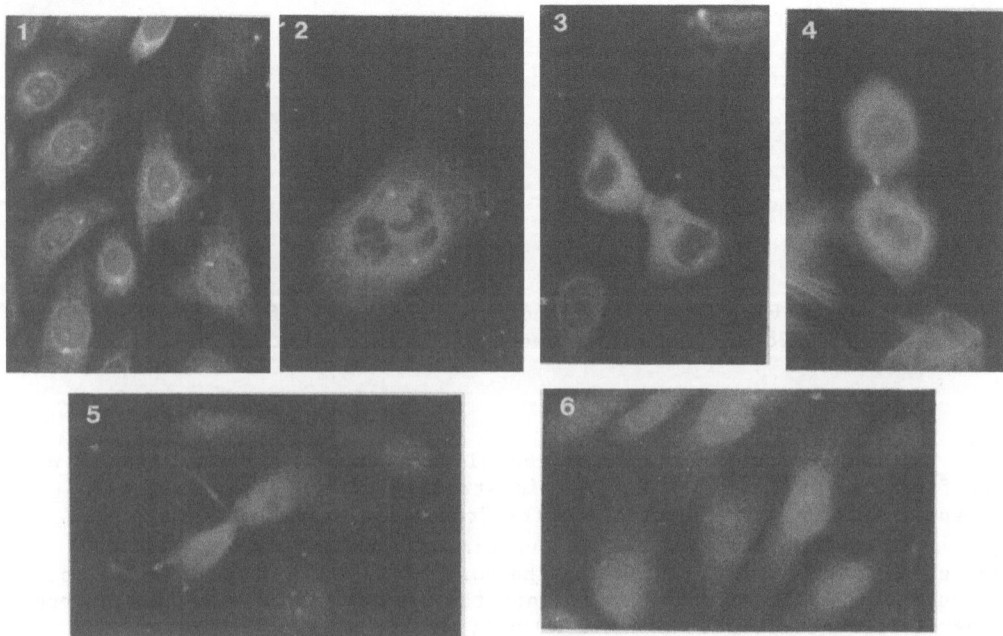

Figure 14.   Immunofluoresence of PtK cells fixed in cold methanol and
stained indirectly with antibodies raised against bovine brain kinesin.

Using kinesin's ability to move MTs over glass as an assay for its activity, we have determined that it has an aparent Km for ATP of 30 - 50 micromolar. The maximum speed of MT movement is about 0.5 um/min at 25° C, and axonemes that have been stripped of their dynein are moved at about the same speed as individual MTs, suggesting that the limitation on velocity is not the load of the fiber being moved but the kinetics of the motor doing the moving (Porter et al., 1987).

Kinesin's ability to promote motility of MTs is sensitive to the presence of the ATP analogue AMP-PNP; movement stops when the concentration of the inhibitor is higher than that of ATP. On the other hand, the movement is only half-maximally inhibited by 50 uM vanadate ions, while isolated flagella stop beating in 10 uM vanadate. Kinesin is also comparatively insensitive to the action of N-ethyl maleimide, a sulfhydryl reagent that inactivates both dynein and myosin at low concentrations. Kinesin is therefore a motor with some unique properties, and we can ask whether the properties of MT movements in mitosis correspond to them.

If kinesin were distributed thoughout the spindle, as it appears to be in sea urchins, and promoted the motion of the MTs relative to some matrix component, not yet defined, then the polarity of its action would induce the motion of all the spindle MTs away from the spindle equator. This direction of motion is indeed what is observed for both anaphase A and B. If kinesin were bound in the vicinity of the spindle poles, as it appears to be in mammalian cells, then it should serve to pull all the MTs in toward the poles, and it could in principle account for the pole-directed forces acting on the chromsomes during both prometaphase and anaphase A. There are, however, two problems with this picture. The forces acting on chromosomes during prometaphase are proportional to the distance between chromosome and pole (Hays et al., 1982). Kinesin could readily provide such a force if it were distributed throughout the spindle, as in the urchin spindle, but simple models of this kind with the kinesin all the pole, as it appears to be in the mammalian spindle, are harder to come by. Further, an investigation of MT disassembly during anaphase, using biotinylated tubulin microinjected during metaphase and electron microscopy of fixed cells to follow the loss of incorporated label, indicates that during anaphase A the kinetochore MTs lose subunits at the kinetochores, not at the poles (Mitchison and Kirschner, 1986). In such a situation, a kinesin motor localized at the poles would make no contribution to the forces pulling the chromosome to the pole. The localization of kinesin at the midregion of the anaphase and telophase spindle can be interpreted as an indication that a kinesin-matrix interaction forms the stuff that holds the two half spindles together where they interdigitate. If it worked as kinesin does _in_ _vitro_ it would promote MT sliding in the directions seen during anaphase B. One can argue, however, that kinesin is localized there simply because it walks to the plus ends of MTs and therefore accumulates in regions where plus ends are concentrated. One needs functional evidence to speak with more confidence about the role of the kinesin that is localized in the spindle.

Under just the right conditions, lysed mitotic cells will continue to move chromosomes after disruption of the cell's membrane. Cande and his co-workers have characterized two such model systems, one from mammalian cells and one from a diatom (Cande et al., 1981; Cande and McDonald, 1986). Both systems show that anaphase B is dependent on the presence of ATP and that 10 uM concentrations of vanadate ions block the movement. Vanadate sensitivity suggests that kinesin is not the motor for this process, but one must admit that a cell is so complex, that one could easily be fooled. The vanadate sensitivity could be imposed on a kinesin-dependent process through the action of some coupled process that is blocked by the ions. The evidence during anaphase A is more problematical, and there is some disagreement even on the ATP-requirement of this process. It is thus too soon to say what whether the physiological evidence dismisses the possiblity of a role for kinesin in anaphase.

We have recently isolated a novel protein that shares properties with both kinesin and dynein. Using the nematode worm, <u>Caenorhabditis elegans</u>, an organism that does not possess motile cilia at any stage of its life cycle, we have found a high molecular weight protein that shows ATP-sensitive binding to MTs (Lye et al., 1987). It is and ATPase makes MTs glide over glass in the presence of ATP with the same polarity as kinesin,

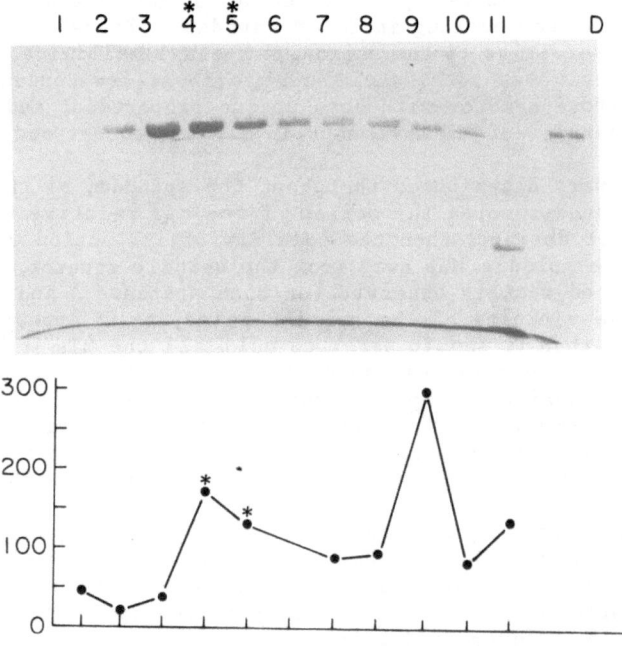

Figure 15. A sucrose gradient sedimentation of dynein-like protein isolated from a nematode worm. The ATPase sedimenting at 20 S corresponds with a polypeptide that coelectrophoreses with flagellar dynein. Stars indicate the fractions that could promote microtubule gliding over glass, as it Fig. 16 below.

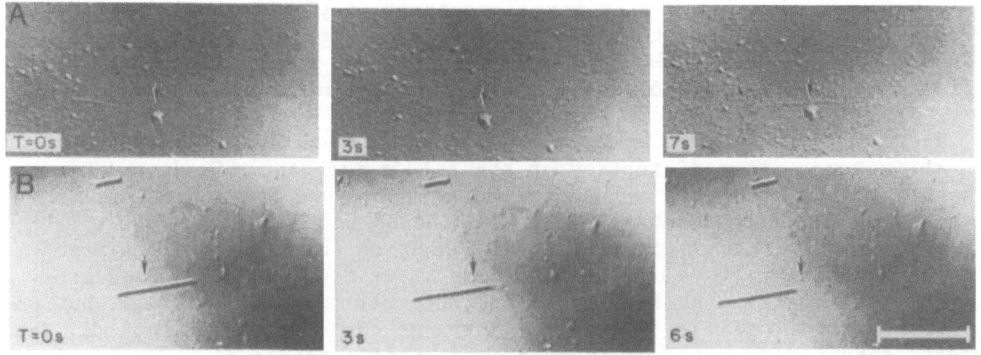

Figure 16. Axonemes gliding over glass in the presence of nematode dynein and ATP. The polarity of movement it from plus to minus, the same direction as results from kinesin.

so it too might be a mitotic motor. The sensitivity of this motor to vanadate and to other ATPase inhibitors closely resembles that of both dynein and the spindle motors seen in lysed cells.

This nematode enzyme is kinesin-like in its ability to effect ATP-dependent movement of MTs over glass. It is unlike kinesin in its physical properties, its possessing an ATPase activity, and its response to inhibitors. It is dynein-like in its sedimentation velocity, ATP-specificity, sensitivity to vanadate and NEM, and peptide molecular weight. It differs from dynein in its ability to move translocate MTs over glass. This property has never been described for axonemal dynein, and we have tried fairly hard to detect such an activity. The nematode protein may represent a class of dynein and kinesin like proteins that work as cytoplasmic microtubule translocators in a variety of cells, a possibility that we are currently exploring. We are also investigating the possibility that the nematode protein might be a spindle motor.

Collins and Vallee (1986) have recently described a microtubule-activated ATPase from sea urchin eggs that is distinct from either dynein or kinesin. Their enzyme is a good GTPase as well as ATPase and is insensitive to vanadate. It sediments at 10 S, but has been difficult to purify, so there is as yet no information about its localization in cells. This protein and the MT-dependent translocator identified in squid axoplasm (Vale et al., 1985 b) are additional possibilities as spindle motors, but more work is needed to clarify the issue.

6. THE PROGNOSIS FOR UNDERSTANDING SPINDLE MECHANICS

A complete and satisfactory understanding of mitosis will require a description of the forces and motions that occur within the spindle as essential parts of chromosome movement. Chromosome behavior can be more complex than is suggested by the above treatment. For example, when the chromosomes are gathered around a single spindle pole, they oscillate in and out from the pole (Bajer, 1983). This motion suggests that a pole can push on a chromosome as well as pull on it. Recent work implies that the pushing force acts over the entire surface of the chromosome, while the pole-directed pulling acts only on the kinetochore (Rieder et al., 1986). The spindle is thus quite versatile in its mechanical properties. There are even indications that forces from outside the spindle can have an effect on the events of mitosis, such as the elongation of the spindle (Aist and Berns, 1981). Clearly there is more to be done just in describing the mechanical activities of the mitotic machinery.

Understanding the production of mitotic forces will depend upon an identification of all the motors involved, of the fuels that the motors use, and of the factors that regulate the motor's actions. Finally, since the spindle forms for the occasion of mitosis and disassembles upon the completion of its task, it will be important to understand the pathways and controls for spindle fiber polymerization and depolymerization. This is a tall order. Clearly we are only a short distance along this long road. There is a good description of the principal forces produced by the cell to organize and segregate the chromsomes, but there are subtleties left to unravel. The movements of the spindle microtubules are beginning to be understood, but anaphase A is still a bit unsure. We can say that changes in the interactions between MAPs and MTs are a part of the regulation of spindle fiber stability, and there are some candidates for the motors that might push the spindle fibers around, but the chemistry of the spindle lags sadly behind our understanding of its structure. It is to be hoped that the leads in hand will take us in the right direction and that within a few years it will be possible to give difinitive answers to all the questions raised here.

Aist, J. R., and Berns, M. S., 1981, Mechanics of chromosome separation during mitosis in Fusarium, J. Cell Biol., 91:446-458

Amos, L. A. and Klug A., 1974, The arrangement of subunits in flagellar microtubules, J. Cell Sci., 14:523-549.

Bajer, A. S., 1983, Functional autonomy of monopolar spindles and evidence for oscillatory movement in mitosis, J. Cell Biol., 93:33-48.

Bajer, A., and J. Mole-Bajer, 1972, Spindle dynamics and chromosome movements. Int. Rev. Cytol., Suppl. 3.

Belar, K., 1929, Bertrage zur kausealanalyse der mitose. II. Untersuchangaenk an den spermatocyten von Chorthippus (Stenobothrus) lineatus panz, Arch. Entowmech, 118:359-484.

Bergen, L. G., Kuriyama, R., and Borisy, G. G., 1980, Polarity of microtubules nucleated by centrosomes and chromosomes of CHO cells in vitro, J. Cell Biol., 84:151-159.

Brinkley, B. R., Cox S. M., Pepper, D. A., Wible, L., Brenner, S. L., and Pardue, R. L., 1981, Tubulin assembly sites and the organization of cytoplasmic microtubules in cultured mammalian cells, J. Cell Biol, 90:554-562.

Cande, W. Z., and McDonald, K. L., 1985, In vitro reactivation of anaphase spindle elongation using isolated diatom spindles, Nature, 316:168-170.

Cande, W. Z., McDonald, K., and Meeusen, R. L., 1981, A permeabilized cell model for studying cell division, J. Cell Biol., 88:618-629.

Carlier, M. F., Hill, T, and Y. Chen, 1984, Interference of GTP hydrolysis in the mechanism of microtubule assembly: An experimental study, Proc. Natl. Acad. Sci., USA, 81:771-775.

Carlier, M. F. and Pantaloni, D., 1981, Kinetic analysis of guanosine 5'-triphosphate hydrolysis associated with tubulin polymerization, Biochemistry, 20:1918-1924.

Collins, C. A. and Vallee, R. B., 1986, A microtubule-activated ATPase from sea urchin eggs, distinct from cytoplasmic dynein and kinesin. Proc. Natl. Acad. Sci. USA, 83:4799-4803.

DeBrabander, M., Aerts, F., DeMey, J., Geuens, G., Moeremans, M., Nuydens, R., and Willebrords, R., 1986, Microtubule dynamics and the mitotic cycle: a model. in: "Aneuploidy: Etiology and Mechanisms", V. L. Dellarco, P. E. Voytec, and A. Holleander, eds., Plenum Press, New York.pp. 269-278

DeBrabander, M., Geuens, G., DeMey, J. and Joniau, M., 1981, Nucleated assembly of mitotic microtubules in living PtK cells after release from nocodazole treatment, Cell Motility, 1:469-484.

Euteneuer, U., and McIntosh, J. R., 1980, Polarity of midbody and phrogmoplast microtubules, J. Cell Biol., 87:509-515.

Euteneuer, U., and McIntosh, J. R., 1981, Structural polarity of kinetochore microtubule in PtK cells, J. Cell Biol., 89:338-345.

Euteneuer, U., Ris, H. and Borisy, G. G., 1983, Polarity of kinetochore microtubules in Chinese hamster ovary cells after recovery from a colcemid block, J. Cell Biol., 97:202-208.

Evans, L., Mitchison, T. J., and Kirschner, M. W., 1985, Influence of the centrosome on the structure of nucleated microtubules, J. Cell Biol., 100:1185-1191.

Haimo, L. T., and Telzer, B. R., 1981, Dynein-microtubule interactions. ATP-sensitive dynein binding and the structural polarity of mitotic microtubules, Cold Spring Harbor Quant. Biol. Symp. 46:207-218.

Hays, T. S., Wise, D., and Salmon, E. D., 1982, Traction force on a kinetochore at metaphase acts as a linear function of kinetochore fiber length, J. Cell Biol., 93:374-382.

Heath, I. B., 1978, "Nuclear Division in the Fungi", Academic Press, New York.

Heidemann, S. R., and McIntosh, J. R., 1980, Visualization of the structural polarity of microtubules, _Nature_ (London) 286:517-519.

Hill, T. L., and Chen, Y., 1984, Phase changes at the end of a microtubule with a GTP cap, _Proc. Natl. Acad. Sci., USA_ 81:5772-2776.

Horio, T and Hotani, H., 1986, Visualization of the dynamic instability of individual microtubules by dark-field microscopy, _Nature_ 321:605-607.

Inoue, S., 1981, Cell divison and the mitotic spindle, _J. Cell Biol._, 91:131s-147s.

Kirschner, M. W. and Mitchison, T. J., 1986, Beyond self assembly: from microtubules to morphogenesis, _Cell_ 45:329-342.

Kobayashi, Y., Ogawa, K., and Mohri, H., 1978, Evidence that the Mg-ATPase in the cortical layer of sea urchin eggs is dynein, _Exp. Cell Res._ 114:285-292.

Kuriyama, R., and Borisy, G. G., 1981, Microtubule-nucleating activity of centrosomes in Chinese hamster ovary cells is independent of the centriole cycle but coupled to the mitotic cycle, _J. Cell Biol._, 91:822-826.

La Fountain, J. R., and Davidson, L. A., 1980, An analysis of spindle ultrastructure during anaphase of micronuclear division in _Tetrahymena_, _Cell Motility_ 1:41-61.

Leslie, R. J., and Pickett-Heaps, J. D., 1983, Ultraviolet microbeam irradiations of mitotic diatoms: Investigations of spindle elongation, _J. Cell Biol._, 96:548-561.

Lye, R. J., Porter, M. F., Scholey, J. M., and McIntosh, J. R., 1987, Identification of a microtubule-based cytoplasmic motor in the nematode _Caenorhabditis elegans_, Submitted to _Cell_.

Margolis, R., and Wilson, L., 1978, Opposite end assembly and disassembly of microtubules at steady state _in vitro_, _Cell_, 13:1-8.

McDonald, K. L., Edwards, M. K., and McIntosh, J. R., 1979, Cross-sectional structure of the central mitotic spindle of _Diatoma vulgare_. Evidence for specific interactions between antiparallel microtubules, _J. Cell Biol._, 83:443-461.

McIntosh, J. R., 1983, The centrosome as an organizer of the cytoskeleton, in: "Spatial Organizational Eukaryotic Cells", J. R. McIntosh, ed., A. R. Liss, New York.

McIntosh, J. R., 1985, Spindle structure and mechanisms of chromosome movement in: "Aneuploidy: Etiology and Mechanisms", V. L. Dellarco, P. E. Voytec, and A. Holleander, eds., Plenum Press, New York. pp. 197-229.

McIntosh, J. R., Cande, W. Z., and Snyder, J. A., 1975, Structure and physiology of the mammalian mitotic spindle, in: "Molecules and Cell Movement", S. Inoue and R. E. Stephens, eds., Raven Press, New York.

McIntosh, J. R. and Landis, S. C., 1971, The distribution of spindle microtubule during mitosis in cultured human cells, _J. Cell Biol._, 49:468-497.

McIntosh, J. R., McDonald, K. L., Edwards, M. K., and Ross, B. M., 1979, Three dimensional structure of the central mitotic spindle of _Diatoma vulgare_, _J. Cell Biol._, 83:428-442.

McIntosh, J. R., Roos, U.-P., Neighbors, B., and McDonald, K. L., 1985, Architecture of the microtubule component of mitotic spindles from _Dictyostelium discoideum_, _J. Cell Biol._, in press.

McNiven, M. A., Wang, M., and Porter, K. R., 1984, Microtubule polatity and direction of pigment transport reverse simultaneously in surgically severed melanophore arms, _Cell_ 37:753-765.

Mitchison, T. J., and Kirschner, M. W., 1984, Microtubule assembly nucleated by isolated centrosomes, _Nature_, 312:232-236.

Mitchison, T., and Kirschner, M., 1984, Dynamic instability of microtubule growth, _Nature_, 312:237-242.

Mitchison, T. J. and Kirschner, M. W., 1985, Properties of the kinetochore in vitro I: microtubule nucleation and tubulin binding, J. Cell Biol., 101:755-765.

Mitchison, T. J. and Kirschner, M. W., 1985, Properties of the kinetochore in vitro II: microtubule capture and ATP-dependent translocation. J. Cell Biol., 101:766-777.

Mitchison, T. J. and Kirschner, M. W., 1986, Sites of microtubule assembly and disassembly in mitotic cells, Cell 45:515-527.

Mohri, H., Mohri, T., Mabuchi, I., Yazuki, I., Sakai, H., and Ogawa, K. 1976, Localization of dynein in the mitotic apparatus, Devel. Growth and Differen. 18:391-398.

Nicklas, R. B., 1986, Mitosis in eukaryotic cells: and overview of chromosome distribution in "Aneuploidy: Etiology and Mechanisms", V. L. Dellarco, P. E. Voytec, and A. Holleander, eds., Plenum Press, New York. pp. 183-195.

Nicklas, R. B., and Koch, C. A., 1969, Chromosome micromanipulation III. Spindle fiber tension and the reorientation of mal-oriented chromosomes, J. Cell Biol., 43:40-56.

Olmsted, J. B., Stemple, D. L., Saxton, W. M., Neighbors, B. W., and McIntosh, J. R., 1987, Dynamics of MAP4 and MAP2 in mitotic and interphase cells. Submitted to J. Cell Biol.

Paweletz, N., 1967, Zur funktion des "Fleming-Koerper" bie der Teilung Tierscher Zellen, Naturwissenschaften, 54:533-541.

Porter, M. E., Scholey, J. M., Stemple, D. L., Vigers, G. P. A., Vale, R. D. Sheetz, M. P., and McIntosh, J. R., 1987, Characterization of the microtubule movement produced by sea urchin egg kinesin, Submitted to J. Biol. Chem.

Rieder, C. L., Davidson, E. A., Jensen, L. C. W., Cassimeris, L., and E. D. Salmon, 1986, Oscillatory movements of monooriented chromosomes and their position relative to the spindle pole result from the ejection properties of the aster and half-spindle, J. Cell Biol. 103: 581-593.

Ris, H. and Witt, P., 1981, Structure of mammalian kinetochores, Chromosoma 82:153-170.

Rothwell, S. W., Grasser, W. A., and D. B. Murphy, 1986, End to end annealing of microtubules in vitro, J. Cell Biol., 102:619-627.

Salmon, E. D., McKeel, M., and T. Hays, 1984, Rapid rate of tubulin dissociation from microtubules in the mitotic spindle in vivo measured by blocking polymerization with colchicine, J. Cell Biol., 99:1067-1076.

Saxton, W. M., Stemple, D. L., Leslie, R. J., Salmon, E. D., Zavortink, M., and McIntosh, J. R., 1984, Tubulin dynamics in cultured mammalian cells, J. Cell Biol., 99:2175-2186.

Scherson, T., Kreis, T. E., Schlessinger, J., Littauer, U., Borisy, G. G., and Geiger, B., 1984, Dynamic interactions of fluorescently labeled MT-associated proteins in living cells, J. Cell Biol. 99:425-434.

Scholey, J. M., Porter, M. E., Grissom, P. M., and McIntosh, J. R., 1985. Identification of kinesin in sea urchin eggs and evidence for its localization in the mitotic spindle, Nature, 318:483-486.

Schultz, E. and Kirschner, M., 1986, Microtubule dynamics in interphase cells, J. Cell Biol., 102:1020-1031.

Snyder, J. A., and McIntosh, J. R., 1975, Initiation and growth of microtubules from mitotic centers in lysed mammalian cells, J. Cell Biol., 67:744-760.

Soltys, B. J., and Borisy, G. G., 1985, Polymerization of tubulin in vivo: Direct evidence for assembly onto microtubule ends and from centrosomes, J. Cell Biol., 100:1682-1689.

Tippit, D. H., Fields, C. T., O'Donnell, K. L., Pickett-Heaps, J. D., and McLaughlin, D. J., 1984, The organization of microtubules during anaphase and telophase spindle elongation in the rust fungus Puccinia, Eur. J. Cell Biol., 34:34-44.

Tippit, D. H., Pickett-Heaps, J. D., and Leslie, R., 1980 a, Cell division in two large pennate diatoms. III. A new proposal for kinetochore function during prometaphase, J. Cell Biol., 86:402-416.

Tippit, D. H., Pillus, L., and Pickett-Heaps, J. D., 1980 b, Organization of spindle microtubules in Ochromonas danica, J. Cell Biol., 87:531-545.

Tippit, D. H., Pillus, L., and Pickett-Heaps, J. D., 1983, Near-neighbor analysis of spindle microtubules in the alga Ochromonas, Eur. J. Biol., 30:9-17.

Vale, R. D., Reese, T. S, and Scheetz, M. P., 1985 a, Identification of a novel force generating protein (kinesin) involved in microtubule based motility, Cell, 41:39-50.

Vale, R. D., Schnapp, N. J., Mitchison, T., Steuer, E., Reese, T. S., and Sheetz, M. P., 1985, Different axoplasmic proteins generate movement in opposite directions along microtubule in vitro, Cell 43:623-632.

Vallee, R. B., Bloom, G. S., and Theurkauf, W. E., 1984, Microtubule associated proteins: subunits of the cytoplasmic matrix, J. Cell Biol., 99(1; part 2):38s-44s.

Wadsworth, P. and Salmon, E. D., 1986, Analysis of the treadmilling model during metaphase of mitosis using fluorescence redistribution after photobleaching, J. Cell Biol., 102:1032-1038.

Warner, F. D. and Mitchell, D. R., 1980, Dynein, Int. Rev. Cytol. 66:125-148.

Williams, R. L., Caplow, M., and McIntosh, J. R., 1986, Dynamic microtubule dynamics, Nature in press.

Wilson, L., and Farrell, K. W., 1985, Kinetics and steady-state dynamics of tubulin addition and loss at opposite microtubule ends; the mechanism of action of colchicine, Ann. N. Y. Acad. Sci., in press.

Witt, P. L., Ris, H., and Borisy, G. G., 1980, Origin of kinetochore microtubules in CHO cells, Chromosoma 81:483-505.

Zieve, G. W. and McIntosh, J. R., 1981, A probe for flagellar dynein in the mammalian mitotic apparatus, J. Cell Sci. 48:241-257.

MEASURING MECHANICAL PROPERTIES OF

CELL SURFACES AND THE RELATED THEORY

Richard Skalak

Bioengineering Institute
Department of Civil Engineering
and Engineering Mechanics
Columbia University
New York, New York 10027

INTRODUCTION

The measurement of cell surface property is made difficult by the fact that the cell size is extremely small and the membrane of the cell cannot usually be conveniently isolated. Therefore most techniques are indirect measures of cell surface properties and require a theory to interpret the measurements. The theories usually involve simplifying assumptions to allow solutions and this may introduce additional uncertainties. One of the principle difficulties is to separate the effect of the internal contents of the cell from that of the cell surface. In the case of red blood cells which are more simple in structure this can be done with some surity but in most other cells some uncertainty remains.

A variety of techniques have been developed for measuring mechanical properties of cell surfaces in which forces are applied to the surface of the cell and the deformation observed through a microscope. A principle technique of this type is the micropipette experiment in which a portion of the cell is sucked into a small pipette. This technique has been extensively used in red blood cells and leukocytes. Another technique developed more recently is cell poking. In this technique a compression is applied through a slender glass beam and the deformations are observed in the microscope. In this case the force is deduced from the deflection of the beam. Other techniques include whole cell compression or the use of a magnetic particle within the cell to which an external magnetic field can be applied to generate an internal force. These techniques and their related theory will be discussed in more detail below.

There are additional techniques probing properties of cells which do not involve local applied forces but rather rely on interactions with viscous flow fields or osmotic effects. Some of the earliest studies of red blood cells made estimates of the surface properties by use of osmotic swelling and estimates of the pressure developed. Another technique is to draw a cell entirely into a pipette, to then expel it and track its recovery as a function of time while freely suspended. Observations on both red blood cells and leukocytes have been made in this way. Cells may also be tested while they adhere to a rigid surface and are deformed by a fluid flow passing over them. The flow chamber technique of this type has been used for red blood cells. There are also instruments in which the deformations of freely suspended cells are observed. This is done directly in the rheoscope. In this instrument the counter rotating elements allow the observation of a single cell held stationary in space in a flow field of simple shear. A somewhat more indirect observation giving an average over a large number of cells is provided by the ektacytometer. In this instrument the diffraction of a laser light source by a suspension of cells is recorded under varying shear stress. This provides a measure of the mean properties of the cell. These techniques are also described in more detail in the paragraphs below.

MICROPIPETTE TECHNIQUES

The micropipette technique has been used to determine a variety of cell surface mechanical properties of several different types of cells. The micropipette aspiration technique was applied by Mitchison and Swann (1954) to the determination of the properties of sea urchin eggs. Rand and Burton (1964) adapted the technique for the study of red blood cell membrane. It has since been implied extensively to both red blood cell and leukocyte studies e.g. LaCelle (1969), Evans et al., (1976) Chien et al., (1981). Micropipette aspiration has been found useful to determine the shear elastic properties of red blood cell membranes as well as the area dilation modulus. It has also been used for determination of viscous properties by studying the time dependent response. In the study of leukocytes, the micropipette technique has been used to describe both short term and long term behavior of the cell as a viscoelastic body. In this case, the separation of surface properties and internal behavior allows a possibility of alternate interpretations as will be discussed below.

Consider first the use of the micropipette experiment to determine the shear properties of a membrane. This is possible when the membrane is not under an initial tension. This would be the case, for example, in a red blood cell in its normal state, but not in a spherically swollen cell under osmotically induced internal pressure. In the idealized analysis, the membrane is assumed to be initially flat and under no tension. In the experiment, a portion of the membrane is drawn into the pipette by a reduced pressure in the pipette as shown in Figure 1. As

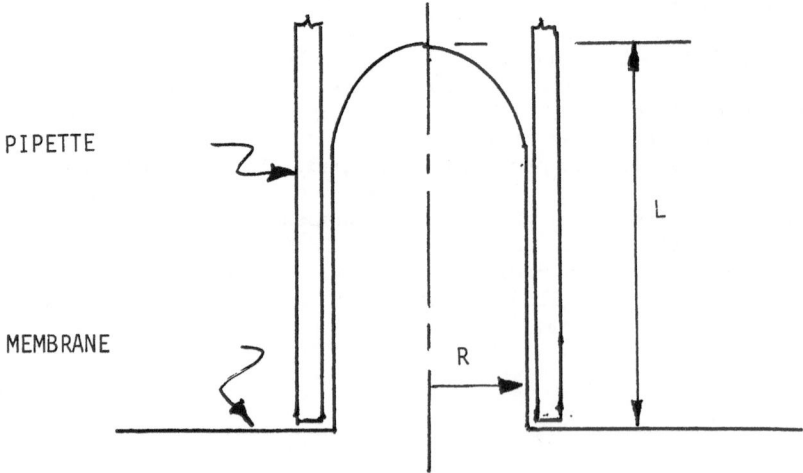

PIPETTE

MEMBRANE

L

R

Fig. 1.   Membrane drawn into micropipette to measure shear
           elasticity of the membrane. The membrane is assumed
           to be initially flat and flaccid.

material is drawn into the pipette, the membrane outside the pipette is drawn towards the center thus inducing a circumferential compression and a corresponding elongation in the radial direction. The extent of this radial movement will depend on the relative magnitudes of the shear modulus and areal stiffness of the membrane. It is possible that the portion of the membrane in the pipette is expanded due to the tension developed. However, in the case of the red blood cell membrane, for example, the areal modulus is much higher than the shear modulus so that the deformation shown in Figure 1 occurs at nearly constant surface area. In this case an analysis can be developed in which the shear modulus is determined from the pipette experiment. In this analysis it is assumed that the effect of the bending stiffness of the membrane is negligible. This is suggested by the fact that the membrane appears to bend readily at the tip of the pipette. It is also assumed that the membrane slides freely over the surfaces of the pipette so that no adhesion is involved. Details of the analysis are given in Evans and Skalak (1980). The final result is

$$\Delta P = \frac{\mu}{R} [(\frac{2L}{R} - 1) + \ell n \ (\frac{2L}{R})] \tag{1}$$

where L is the length of membrane in the pipette, R is the radius of the pipette and μ is the elastic shear modulus of the membrane; the pressure difference applied to the pipette is ΔP.   Equation 1 holds for $L > R$ and it is assumed that the curved portion of the membrane in the pipette is a hemisphere. The units of the shear modulus are dyne/cm rather than $dyne/cm^2$ as would be the case for a solid material. The membrane modulus may be regarded as the usual 3-dimensional modulus multiplied by the thickness of the membrane. However, since the membrane is anisotropic and its thickness is not readily determined, it is convenient to deal only with the total shear stiffness of the membrane, μ.

The areal elasticity of a membrane can be measured by the micropipette technique by first osmotically swelling the cells so that it is spherical in shape and under an initial isotropic tension. In the case it is assumed that the shear stresses are small and that the behavior is dominated by a constant isotropic tension T. When a small portion of the cell is drawn into the pipette the total surface area of a cell must increase, assuming the volume remains constant. Actually the volume of the cell may decrease slightly due to the exudation of water through the cell membrane. However, this is usually a negligible or preventable effect. The change of area due to a small movement L of the portion of the membrane in the pipette, Fig. 2, is given to first approximation by

$$\Delta A = 2 \pi R_p \Delta L \ (1 - R_p/R_c) \tag{2}$$

where R_p = the radius of the pipette and R_c is the radius of the spherical portion of the cell. The change of tension in the cell wall due to a decrement of pressure ΔP in the pipette is given by

$$\Delta T = \Delta P R_p /2 \ (1 - R_p/R_c) \tag{3}$$

The areal modulus K of the membrane is defined by

$$K = \Delta T \ (A/\Delta A) \tag{4}$$

where A is the total area of the cell.

The viscoelastic properties of a cell membrane can be determined using a micropipette by measuring the time history of the deformation and interpreting the results appropriately. Usually static tests are used to first determine the elastic components of the membrane behavior and the time dependent test may then be interpreted in terms of a membrane viscosity. This procedure has been developed in connection with measurements of the red blood cell membrane by Chien et al., (1978). The total stress in the membrane T_{ij} is usually assumed to be governed by a Kelvin model. Thus

$$T_{ij} = T_o \delta_{ij} + 2 \mu E_{ij} + 2 \eta V_{ij} \qquad (5)$$

where T_o is the isotropic stress associated with area change. E_{ij} is the Greens strain tensor and V_{ij} is the rate of deformation tensor. The coefficient η represents the membrane viscosity. In the case of the red blood cell membrane it was found that the viscosity coefficient varies in the aspiration and recovery phases of the motion. This behavior may be modeled by assuming that the preferred configuration of the membrane evolves in time when stress is maintained (Tozeren et al., 1984).

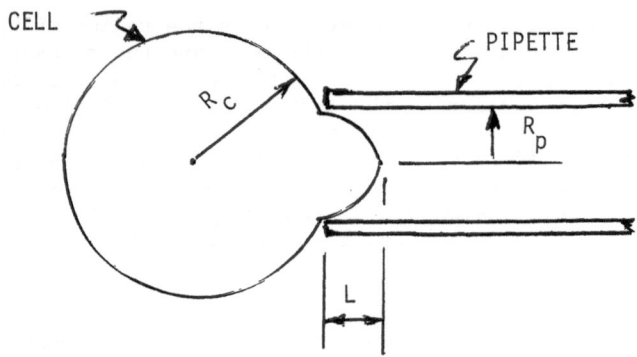

Fig. 2.   Micropipette experiment arranged to measure the areal elasticity of the red blood cell membrane. The cell is sphered osmotically prior to the application of the pipette suction pressure.

The viscoelastic properties of the red cell membrane may also be inferred from experiments in which the cell is deformed then released without any forces being applied. By tracking the time history of the recovery of the shape the viscosity of the cell membrane can be deduced. Hochmuth et al., (1979) have extended red blood cells by use of a pipette and by measuring a time constant of recovery upon release estimate the membrane viscosity as follows

$$\eta = t_c \mu \qquad (6)$$

where t_c is the exponential time constant of the observed recovery of normal discocyte shape.

Micropipette experiments have also been used to determine the properties of leukocytes. (Schmid-Schonbein et al., 1981). For small deformations the cell is considered as a viscoelastic sphere of uniform material. In this case the material behavior of the sphere is assumed to be that of a standard solid shown in Fig. 3. In this model the cytoplasm is assumed to be incompressible so the standard solid model applies only to the deviatoric stress and strain. It is found possible to fit the early stage of micropipette aspiration of leukocytes quite well with the standard solid model. However an alternative model which makes a different assumption as

to the cell structure can also fit the results. In this model the cell is
represented by a cortical shell with a Maxwell fluid as the interior
cytoplasm. The cortical shell is supposed to be a thin layer of the order
of 1μ m thick which is under an isotropic tension. This shell provides an
elastic behavior which returns the cell to a spherical shape after the
release of any deformation (Dong et al., 1985). This model may also be
useful in describing larger slow deformations in which the leukocyte is
drawn entirely into a pipette (LaCelle et al., 1982). For this purpose a
large deformation analysis is required.

The white blood cell is capable of some active motion. The typical
active phase involves the extension of pseudopods or protopods from the

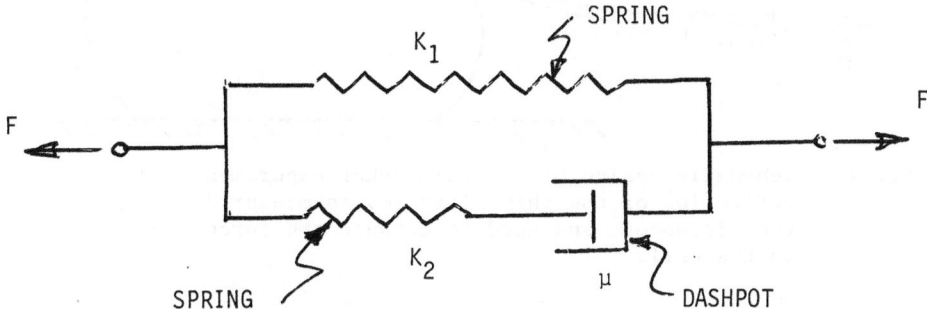

Fig. 3.   Standard solid model representing the shear behavior
of leukocytes. The cell is modeled as a homogeneous
incompressible viscoelastic sphere.

cell surface. Such pseudopods appear to be stiffer than the normal white
cell and to consist primarily of an actin gel (Stossel, 1982). It is not
clear whether the active motion of leukocyte involves a similar or different
mechanism from the   contraction that leads to cell division.

The micropipette technique has also been used to estimate the bending
resistance of the membrane of the red blood cell (Evans, 1979). In the case
of the red blood cell the bending elasticity is attributed to the cell
membrane. In other cells such as white blood cells or endothelial cells,
the membrane is ruffled and its bending stiffness is assumed to be
a negligible factor in the mechanical behavior of the entire cell. Even for
the red blood cell under large deformations the influence of the bending
resistance may be small. The micropipette experiment can be used to
estimate the bending modulus during initial deformations when the shear
strain energy is small. Another method (Evans, 1983) is to consider the
buckling of the membrane which occurs just outside of the pipette during
aspiration. If the shear modulus is assumed to be known, the pressure at
which buckling first occurs may be interpreted as a measure of the bending
stiffness.

CELL POKING EXPERIMENTS

The technique of cell poking is a relatively recent method in which the
deformability of cells is measured by applying a compressive force rather
than a tensile force as in the micropipette technique. The compressive
force is applied by a slender glass beam whose deflection is an indication
of the force applied. The deformation of the cell is viewed through the
microscope at the same time thus allowing a force-displacement history to
be developed. Depending on the cell and the geometry of the experiment,

theoretical analyses may be developed to interpret the measurements in terms of cell elasticity and viscoelastic behavior. The cell poker may provide information about the cell membrane properties in the case that it is applied to a cell with a relatively liquid interior such as a red blood cell. When applied to stiffer cells such as leukocytes or endothelial cells it may give information about the interior cytoplasm of the cells rather than surface properties which are less important in these cases.

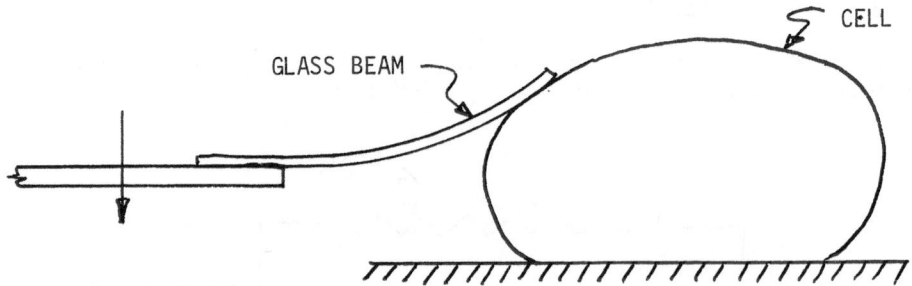

Fig. 4.   Schematic design of the cell poker experiment. The deflection of the thin glass rod is measured through the microscope and used to compute the force applied to the cell.

The basic design of the cell poker is shown schematically in Figure 4. A fine quartz fiber is glued to the end of a micropipette. When the fiber is moved downward the probe deforms and applies a force to the cell. The beam is mounted on a linear pizoelectric motor which can be driven with a programmable voltage wave form. Optical sensors are used to monitor the position of the glass beam and of the cell deformation. Details of the apparatus are given by McConnaughey and Petersen (1980).

The cell poking method has been used to measure the elastic area modulus of osmotically swollen human erythrocytes (Daily, et al., 1984). The results agree with the values determined by other methods. The theoretical prediction is based on a simple membrane model in which the surface area is minimized at constant cell volume. These results validate the cell poker as a method of measurement. The micropipette technique is probably a preferred method of measurement of the properties of cells in suspension and which are flacid such as normal red blood cells. Cell poking is probably better suited for making measurements on adherent and spread cells in culture. One advantage is that since a compression is applied the problem of adhesion between the cell and the poker tip is probably less problematical than in the micropipette technique.

The cell poking technique has been used to measure some properties of fibroblasts and modulated and unmodulated lymphocytes (Daily, 1985). It has also been used to measure the force of retraction and the stiffness of the active lamella on fibroblasts (Felder, 1984). It is observed that the active ruffles of fibroblasts tend to bend about a point near their base where the ruffle is attached to the substrate. These results indicate that the cell poking method may be useful to measure both passive and active properties of cells and should be a useful technique to discern differences in properties on different parts of a cell for example during mitosis.

MAGNETIC PARTICLES

An interesting method of testing the properties of cells is to apply an interior force as opposed to loading the exterior surface. Such an interior force can be produced by applying a magnetic field to a phagocytosed magnetic sphere. This technique

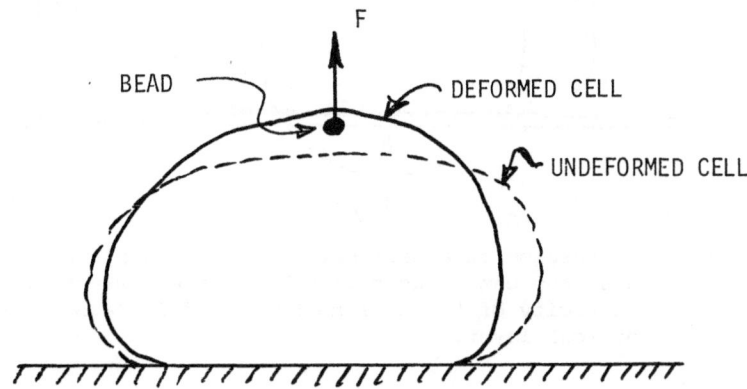

Fig. 5. A magnetic bead inserted into a cell is used to apply an internal force to the cytoplasm of the cell by application of an external magnetic field.

was introduced as a method of measuring the viscosity of cell protoplasm (Heilbronn, 1922). Hiramoto (1963, 1970) has used this procedure to determine properties of sea urchin eggs. In this case the membrane of the cell has a ruffled surface so that it may expand by unwrinkling. Thus when the cell surface is deformed by applying force to the magnetic sphere as shown in Fig. 5 the principal resistance probably comes from a cortical layer which may have an initial tension in it. In principle data obtained by the use of a magnetic particle under known applied forces should allow a discernment of whether the surface elasticity is due to a elastic behavior starting from zero stress or whether there is an initial isotropic tension (Evans and Skalak, 1981). Under large deformations it may be further possible to determine the contributions of shear elasticity and the bending stiffness. However, these possiblities by this technique do not appear to have been realized in any experiments to date.

The technique of using internal magnetic spheres to test the properties of cell cytoplasm has been carried on by Sato et al., (1983, 1985). In these studies the viscoelastic properties of the cell cytoplasm have been determined by studying the transient response to the forces applied through a magnetic field in various cells. This technique has the advantage that the part of the cell cytoplasm which is directly involved can be a part of the cell; variations of the structure within the cell and under different physiological conditions can be tested.

CELL COMPRESSION

If a cell is spherical and its volume is constant, then any deformation results in an increase of surface area. In particular if a spherical cell is compressed between two flat surfaces as shown in Fig. 6 there must be some increase in the surface area and it may be aniticipated that the dominant stress will be the membrane tension associated with the area dilation. Bending stresses may be significant in the early stages of the deformation and shear elasticity may play a role in the later stages. The interplay of these different kinds of reactions depends, of course, on the material properties and the effective thickness of the cell surface layer.

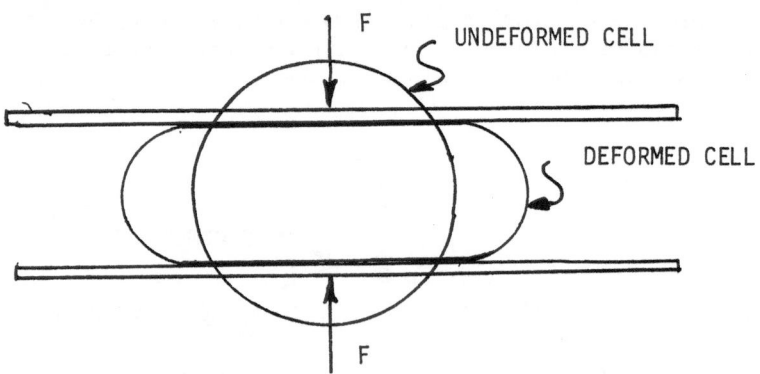

UNDEFORMED CELL

DEFORMED CELL

Fig. 6.   Compression of a cell between two flat plates.
This test can be used to measure the surface areal
elasticity of the cell membrane and its associated
cortical layer.

Hiramoto (1963) and Yoneda (1964) have developed experiments on sea
urchin eggs and a theoretical analysis assuming that the membrane of the
cell is under isotropic tension. In this case the total curvature must
be constant in the curved portion of the cell surface i.e.

$$(\frac{1}{R_1} + \frac{1}{R_2}) = C_o \tag{7}$$

where R_1 and R_2 are the principal curvatures and C_o is a constant. This
relation is sufficient to allow computation of the cell shape and to derive
the relation of the isotropic tension to the areal dilation from the
experimental data (Evans and Skalak, 1980). It is found that the sea urchin
eggs have a small initial tension and then an essentially linear relation
between the increase of tension and the increase in the surface area. The
initial component of isotropic tension is thought to be due to an active
tension in the cortical layer just below the plasma membrane.

The role of the bending stiffness of the membrane of a spherical shell
has been brought out by analysis of fluid filled spherical rubber shells by
Taber (1982, 1983). In the early stages of deformation the bending
resistance is important and gives rise to a force deflection curve which is
concave downward at first. As the membrane stresses become more significant
the typical curve turns to concave upward. These features may be
overshadowed by the nonlinear stress strain behavior of most biological
tissues which results in a steepening of the force deflection curve
(Lardner and Pujara, 1980).

Tests on cell compression have shown that the behavior of cells such as
lymphocytes will exhibit a hysteresis in the load deflection curve
indicating a viscoelastic or possibly plastic behavior (Daily, 1985). An
analysis of the problem of compression of a viscoelastic sphere of constant
volume between two plates is given for this purpose by (Daily, 1985). An
alternative model is that of an elastic prestressed spherical shell filled
with a viscoelastic incompressible liquid such as a Maxwell fluid. Such a
model has been developed by Dong et al., (1985) for purposes of describing
the behavior of leukocytes under pipette aspiration. This model is shown
to be able to represent the initial aspiration deformation and also the
recovery of spherical shape when the cell is released. These analyses
primarily assume samll strain theory and need to be extended to consider
finite deformations in order to represent biological experiments more
accurately.

The fact that cells generally will absorb water when placed in a hypotonic solution gives the opportunity to test the cell membrane by the internal pressures developed. By Starling's Law the flux of water continues until the sum of the hydrostatic pressure and the osmotic pressure is the same inside and outside the cell. If the cell membrane is sufficiently strong, an appreciable internal pressure can be built up within the cell. The osmosis is reversible so that if the cell is placed in an isotonic solution it recovers its initial volume, although some internal structures may be disrupted.

In the case of the red blood cell the membrane is initially smooth and placing the cell in a hypotonic solution produces a series of swollen shapes which result finally in a nearly perfect spherical shape. Any further reduction of the osmolarity of the external solution results in hemolysis. During osmotic swelling the surface area of the red blood cell has been shown to remain constant (Evans, 1977). The intermediate shapes are controlled primarily by the small bending stiffness as well as some contribution from the shear elasticity. However in the spherical form the primary stress is an isotropic tension which finally leads to lysis.

In the case of leukocytes or other cells such as endothelial cells, the cell membrane is initially ruffled and when the cell is osmotically swollen to a small extent these ruffles are partially unfolded which requires very little stress. The primary resistance in this case must be due to the elasticity of the cortical layer below the plasma membrane. A leukocyte can be osmotically swelled to the extent that its plasma membrane becomes smooth, taut and spherical. Again, a further reduction of the external osmolarity will lead to rupture of the cell membrane.

Osmotic swelling of red blood cells was utilized by Katchalsky, et al. in 1960 to estimate the viscoelastic properties of the red blood cell membrane. The pressure developed was estimated from osmotic considerations and the strain was estimated from cell volumetric measurements. The estimates gave a value of the apparent modulus of elasticity which corresponds to the areal stiffness of the membrane. The original techniques of Katchalsky, et al. do not seem to have been repeated, but with current equipment the measurements could probably be refined. Osmotic swelling has also been used to estimate the modulus of elasticity of the membrane of sea urchin eggs (Mela, 1967).

Osmotic swelling may have other uses besides direct measurement of the areal modulus of elasticity. In the case of the red blood cells the intermediate stages may be useful to evaluate the role of bending stiffness (Zarda et al., 1977). It may also be useful to pre-swell a cell prior to performing other experiments. This procedure has been used prior to applying micropipette suction to a cell to measure the areal modulus more precisely (Evans and Hochmuth, 1978).

It has also been shown (Sung et al., 1982) that the viscoelastic property of leukocytes measured by the micropipette technique vary with the osmolarity of the external solution. For sufficiently low external osmolarity the leukocytes become considerably less stiff and less viscous. On the other hand when the external osmolarity is increased, the water in the cell is partially extracted and the cells become much more stiff and viscous. The range of properties can be several decades. This large variation is not particularly associated with the surface or membrane properties of the cell, but rather is an indication of a change in the cytoplasm properties of the cell.

If a cell is attached to a rigid substrate, a viscous flow passing over the cell can exert sufficient stresses on it to deform the cell. Computation of the fluid stresses applied together with microscopic observation of the cell deformation allows estimating the mechanical properties of the attached cell. These principles are incorporated in the flow chamber experiments.

The flow chamber usually consists of two closely spaced glass plates, the spacing being of the order of 200 μm. The width of the channel is sufficient so that a two dimensional Poiseuille flow may be assumed in the gap between the plates. For a successful experiment the cells must become attached to one of the walls. Red blood cells which are allowed to settle on to a glass plate will become attached at one or more points in a short period of time. Then starting up the flow, the cells may be stretched out as indicated in Figure 7. Red blood cells are particularly deformable and

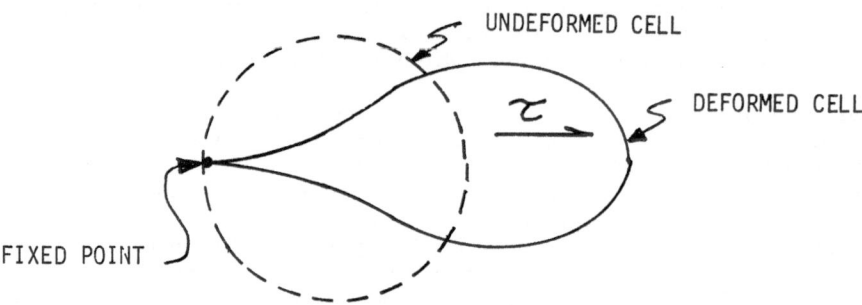

Fig. 7.   Flow chamber experiment in which a red blood cell adhering to a glass surface is extended by shear stresses exerted on the cell by the passing fluid flow.

may be elongated with stresses of an order of a few dynes/cm^2. The flow chamber experiment has not been used extensively for cells other than red blood cells.

For the analysis of the deformation of red blood cells in a flow chamber, the assumption is made that the wall shear stress of the Poiseuille flow is directly applicable to the plan surface area of the cell. The detailed stress distribution on a tethered red cell has not been computed. Assuming the stress applied to the cell is uniform and further that the response of the cell is a simple uniform axial extension on each cross-section perpendicular to the flow it is possible to derive an analytic solution for the shape of the cell in plan (Hochmuth et al., 1973, Evans, 1973). In the analysis it is assumed that the surface area of the red blood cell remains constant and that the bending resistance may be neglected. Although the volume of the cell is expected to remain constant this is not taken into account in the analysis because the theory assumes in effect that the cell consists of two flat circular sheets which are then deformed as shown in Figure 7. It is noteworthy that in spite of the several simplifying assumptions made for this analysis the results for the shear modulus are close to those derived from pipette aspiration experiments.

The flow chamber technique has been used to study the adhesion and disaggregation of red blood cells also. For this purpose situations are sought in which one cell settles directly over a second cell which is attached already to the glass surface. In passing a flow over this two cell stack the upper cell may be gradually peeled off by the fluid stresses applied to it. This allows estimating the adhesive stress between the two cells (Vayo, 1968). Other types of cells such as white blood cells or endotheleal cells have not been tested for their mechanical properties by the flow chamber technique although their adhesive properties may be tested in this matter. Most cells are too stiff to deform by the shear stresses available in a flow chamber. In the case of endothelial cells it has been shown that the surface shear stress effects the orientation and shape which the cells take (Dewey et al., 1985). In this case the cells are not deformed elastically but gradually change their shape by the reorganization of the cytoskeleton over a period of hours.

THE RHEOSCOPE

An interesting instrument which allows the microscopic examination of single cells suspended in a shear flow is the rheoscope in which counter rotating platens develop a shear field, one point of which is stationary in space. Cells at this level can be viewed through the microscope and appear to be stationary in space but are actually undergoing a rotation and/or so-called tank-treading motion (Schmid-Schonbein and Wells, 1969). Experimental observations show that depending on the viscosity of the suspending fluid, red blood cells may either rotate as rigid bodies or may undergo the tank-treading type of motion (Keller and Skalak, 1982) illustrated in Figure 8. In the tank-treading motion the orientation and shape of the cell are stationary in time, but the membrane rotates continuously like the tread on a tank. The interior fluid of the red blood cell which is tank-treading also undergoes a rotational motion. As the shear rate is increased, the frequency of the tank treading motion also increases and the red blood cell is elongated to a greater extent.

The rheoscope has been used to demonstrate various responses of red blood cells (Schmid-Schonbein, et al., 1983). For example diamide treated red blood cells show a reduced elongation and unusual buckling of the cell surfaces in the rheoscope while tank treading. This is attributed to cross-linking of the membrane proteins which is thought to produce an increase in the shear elasticity of the membrane. For normal red blood cells it appears that the viscosity of the membrane has a definite influence on the motion. An analysis of the dissipation in the membrane has been developed (Tran-Son-Tay, 1984) and allows an estimate of the viscosity of the membrane in this situation.

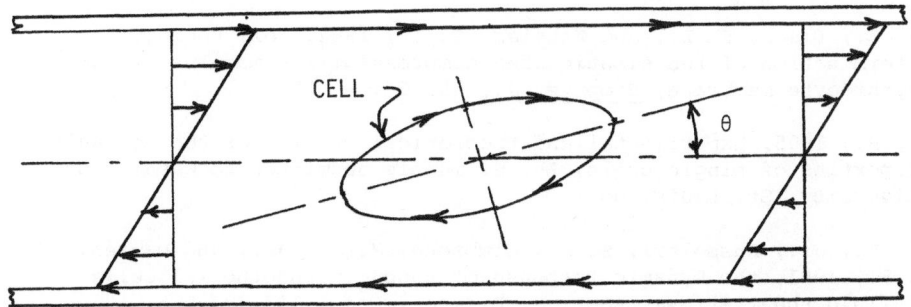

Fig. 8.   Schematic side view of a red blood cell undergoing tank treading motion in a rheoscope. The velocity distributions shown pertain to the suspending fluid at some distance from the cell. The shape of the cell is constant while the membrane tank-treads.

The rheoscope has not been used extensively to test cells other than red blood cells. Since white blood cells are relatively stiff and more or less spherical they tend to simply rotate in a shear field with very little deformation under the usual suspending fluids. The internal viscosity of leukocytes is very high compared to that of plasma or other fluids usually used to suspend cells in the rheoscope. In principle if an extremely high viscosity fluid were used to suspend white cells in the rheoscope, perhaps elongation during tank treading could be induced and used to determine gross cell properties. No such results have been reported. In low viscosity suspending fluids at high shear rates, leukocytes can be fragmented and lysed (Martin, et al., 1979).

The rheoscope might be useful to study the rheological properties of other cells also if very high viscosity suspending fluids were used. For example it is known that the deformability of circulating cancer cells affects their passage through the microcirculation and the formulation of metastasis (Weiss and Gloves, 1982). A suitable rheoscope test might allow mitotic behavior and rheology of different stages in mitosis might be explored in this manner.

CONCLUSION

The several methods described above collectively offer a wide variety of possible tests and information concerning the mechanical properties of individual cells. However each of them requires extensive, delicate equipment as well as skill and experience to produce reliable experimental results. Further elaborate theoretical treatments are required in order to extract basic parameters describing the properties of the individual cellular components. It does not seem practical for any one laboratory to develop expertise in all methods available. It is preferable to formulate critical biological questions carefully and then to select the particular method best suited to answer questions posed. There is still a need for simple inexpensive tests which can discern the microscopic mechanical properties of individual cells.

REFERENCES

Bull, B. S., and Brailsford, J. D., 1975, The relative importance of bending and shear in stabilizing the shape of the red blood cell, Blood Cells, 1:323.

Chien, S., Sung, K. L. P., Skalak, R., Usami, S., and Tozeren, viscoelastic properties of erythrocyte membrane, Biophys. J., 24:463-487.

Daily, B. B., Elson, E. L., and Zahalak, G. I., 1984, Cell Poking. Determination of the elastic area compressibility modulus of the erythrocyte membrane, Biophys. J., 45: 671 - 682.

Daily, B. B., 1985, Experimental and theoretical studies of the mechanical properties of single cells, Ph. S. Thesis submitted to Washington University, St. Louis, Mo.

Dewey, C. F., Jr., Bussolari, S. R., Gimbrone, M. A., Jr., and Davies, P. F., 1981, the Dynamic response of vascular endothelial cells to fluid shear stress, ASME. J. Biomech. Eng., 103:177-185.

Dewey, C. F., Jr., 1984, Effects of fluid flow on living vascular cells ASME J. Biomech. Eng., 106:31-35.

Dong, C., Schmid-Schobein, G. W., and Skalak, R., 1985, Rheological behavior of leukocytes, 1985 Biomechanics Symposium, Joint ASCE/ASME Mechanics Conference, Albuquerque, New Mexico, 68:163-166.

Evans, E. A., 1973, New membrane concept applied to the analysis of fluid shear and micropipette deformed red blood cells, Biophys. J., 13:941.

Evans, E. A., and Hochmuth, R. M., 1976, Membrane viscoelasticity, Biophys. J., 16:1.

Evans, E. A., Waugh, R. and Melnik, L., 1976, Elastic area compressibility modulus of red cell membrane, Biophys. J., 16:585-595.

Evans, E. A. and Waugh, R., 1977, Osmotic correction to elastic area compressibility measurements on the red cell membrane, Biophys, J., 20:307-313.

Evans, E. A., and Skalak, R., 1980, "Mechanics and Thermodynamics of Biomechanics," CRC Press, Boca Raton, Florida.

Evans, E. A., 1982, Bending versus shear rigidity of red blood cell membrane, Biophys. J., 43:27-30.

Felder, S., 1984, Mechanics of the motion of the leading active edge of animal cells attached to glass, Ph. D. Thesis submitted to Washington University St. Louis, Mo.

Heilbronn, A., 1922, Eine Neue Methode zur Bestimmung der viskositat lebender, protoplasten, Jahrbe. Wiss. Bot., 61:284-338.

Hiramoto, Y., 1963, Mechanical properties of sea urchin eggs. I. Surface force and elastic modulus of the cell membrane, Exp. Cell Res., 32:59.

Hiramoto, Y., 1963, Mechanical properties of sea urchin eggs. II. Changes in mechanical properties from fertilization to cleavage, Exp. Cell Res., 32:76.

Hiramoto, Y., 1970, Rheological properties of sea urchin eggs, Biorheology, 6:201.

Hochmuth, R. M., and Mohandas, N., 1972, Uniaxal loading of the red cell membrane, J. Biomech., 5:501.

Hochmuth, R. M., Mohandas, M. N., and Blackshear, P. L., 1973, Measurement of the elastic modulus for red cell membrane using a fluid mechanical technique, Biophys J., 13:747-762.

Hochmuth, R. M., Worthy, P. R., and Evans, E. A., 1979 Red Cell relaxation and the determination of membrane viscosity Biophys. J., 26:101-104.

Katchalsky, A., Kedem, O., Klibansky, C., and DeVries, A., 1960, Rheological considerations of the haemolysing red blood cell, in "Flow Properties of Blood and other Biological Systems", A. L. Copley, and G. Stainsby, eds., Pergamon Press, Oxford.

Keller, S. R. and Skalak, R., 1982, Motion of a tank-treading ellipsoidal particle in a shear flow, Journal of Fluid Mechanics, 120:27-47.

LaCelle, P. L., 1969, Alterations of deformability of the erthrocyte membrane in stored blood, Transfusion (Philadelphia), 9:238.

LaCelle, P. L., Bush, R. W., and Smith, B. D., 1982, Viscoelastic properties of normal and pathological human granulocytes and lymphocytes, in: "White Blood Cell. Morphology and Rheology as Related to Function," U. Bagge, G. V. R. Born, and P. Gaehtgens, eds., Martinus Nijhoff Publishers, The Hague.

Lardner, T. J., and Pujara, P., 1980, Compression of spherical cells, "Mechanics Today," S. Nemat-Nasser, ed., Pergamon Press, New York.

McConnaughey, W. B., and Petersen, N. O., 1980, The cell poker: An apparatus for stress-strain measurements on living, Rev. Sci. Instr., 51:575-580.

Mela, M. J., 1967, Elastic mathematical theory of cells and mitochondria in swelling process. I. The membranous stresses and modulus of elasticity of sea urchin eggs, Biophys. J., 7:95.

Mitchinson, J. M., and Swann, M. M., 1954, The mechanical properties of the cell surface. I. The cell elastmeter, J. exp. Bio., 31:443.

Petersen, N. O., McConnaughey, W. B., and Elson, E. L., 1982, Dependence of locally measured cellular deformability on position of the cell, temperature, and cytochalasin B, Proc. Natl. Acad. Sci. USA, 79:5327-5331.

Rand, R. P., and Burton, A. C., 1964., Mechanical properties of the red cell membrane. I. Membrane stiffness and intracellular pressure, Biophys. J., 4:115.

Sato, M., Wong, T. Z., and Allen, R. D., 1983, Rheological properties of living cytoplasma: endoplasm of physarium plasmodium, J. Cell Biol., 97; 1089-1097.

Sato, M., Schwarz, W. H., and Pollard, T. D., 1985, Mechanical properties of acathamoeba gelation protein and actin, J. Cell Biol. 101.

Schmid-Schonbein, G. W., Skalak, R., Sung, K. L. P., and Chien, S., 1982, Human leukocytes in the active state, in: "White Blood Cell Morphology and Rheology as related to Function," U. Bagge, G. V. R. Born, P. Gaehtens, eds., Martinus Nijhoff Publishers, The Hague.

Skalak, R., Tozeren, A., Zarda, P. R., and Chien, S., 1973, Strain energy function of réd blood cell membranes, Biophysical Journal, 13:245-264.

Skalak, R., 1973, Modelling the mechanical behavior of red blood cells, Biorheology, 10:228-238.

Skalak, R., Chien, S. and Schmid-Schonbein, G. W., 1984, Viscoelastic deformation of white cells: Theory and analysis, in: White Cell Mechanics, Basic Science and Clinical Aspects, I. Cellular and Membrane Mechanics of White Cells, H. J. Meiselman, M. A. Lichtman, P. L. LaCelle, eds., Alan R. Liss, Inc., New York.

Skalak, R., Schmid-Schonbein, G. W., and Chien, S., 1982, Analysis of white blood cell deformation, in: "White Blood Cells Morphology and Rheology as Related to Function," U. Bagge, G. V. R. Born and P. Gaehtgens, eds., Martinus Nijhoff, the Hague·

Stossel, T. P., Hartwig, J. H., Yin, H. L., Zaner, K. S., and Stendahl, O. I., 1981, Actin gelation and the structure of corical cytoplasm in: "Cold Spring Harbor Symposium on Quantitative Biology," 46:569-578.

Sung, K. L. P., Schmid-Schonbein, G. W., Skalak, R., Schuessler, G. B., Usami S., and Chien, S., 1982, Influence of physiochemical factors on rheology of human neutrophils Biophys. J., 39:101-106.

Taber, L. A., 1982, Large deflection of a fluid-filled spherical shell under a point load, ASME Journal of Applied Mechanics, 49:121-128.

Taber, L. A., 1983, Compression of fluid-filled spherical shells by rigid indenters, J. Appl. Mech., 50:717-722.

Tozeren, A., Skalak, R., Sung, K - L. P, and Chien, S., 1982, Viscoelastic behavior of erythrocyte membrane, Biophysical Journal, 39:23-32.

Vayo, M., 1986, Theoretical and experimental analysis of shear disaggregation of red cell rouleaux, Doctoral Dissertation, Columbia University, New York, New York.

Yoneda, M., 1964, Tension at the surface of sea urchin egg a critical examination of Cole's experiment, J. Exp. Biol., 41:893.

Yoneda, M., and Dan, K., 1972, Tension at the surface of the dividing sea urchin egg, J. Exp. Biol., 57:575.

Zarda, P. R., Chien, S., and Skalak, R., 1977, Elastic deformations of red blood cells, Journal of Biomechanics, 10:211-221

MECHANICS OF CELL DEFORMATION: RELATION TO CELL DIVISION

E. Evans and A. Yeung

Departments of Pathology and Physics
University of British Columbia
Vancouver, B.C. Canada V6T 1W5

INTRODUCTION

After a long growth phase where cells essentially double in size, nuclear division and cytokinesis (cell division) take place in a relatively short time for animal cells. The cleavage process involves assembly of actin filaments to form a contractile ring. The force of contraction in the ring must be sufficient to overcome the mechanical stiffness of the cell in order to bring about division. Several questions are immediately apparent: What are the levels of stress created in the cell during division? How large is the force of contraction in the cleavage ring? How do various structural elements of the cell resist cleavage? In the sections to follow, relevant features of the mechanics of cell deformation will be outlined and discussed in the context of these questions.

Animal cells have many common structural features: an outer plasma membrane envelope, subsurface cortical gel, internal granular or other organellar bodies, reticular elements, nuclei, etc. Even with this structural complexity, cells often remain spherical in suspension. Further, when stressed by external forces, many of these cells deform continuously, albeit lethargically, with no apparent limit to deformation. However, when the external forces are removed, the cells subsequently recover their original spherical shape. This behavior indicates that the cells can be modelled as complex liquid suspensions, encapsulated by cortical shells which are composites of plasmalemma and subsurface gel.[1] As such, the cell interior only limits the rate of deformation in response to applied forces whereas the cortical shell establishes the static or limiting balance of stresses. Also for motile cells, the active locomotory apparatus (a polymerizable actin filament solution) appears to be concentrated in the cortical layer adjacent to the cell surface [2,3]. Thus, mechanical abstraction of cells as complex liquid interiors surrounded by cortical shells can be a useful starting point for examination of cell deformation properties. It is emphasized that not all cells behave in this manner. When cells have structural connections that span the cytoplasm, the simple model will not be appropriate. In suspension, these cells are usually not spherical in shape. But the intrinsic mechanical properties of a cell must be determined by direct mechanical experiment and cannot be deduced from cell shape alone. Since a wide variety of cells do exhibit liquid-like response, we will develop the mechanical principles for the simplified view of cell structure, examine the major implications of the model based on known properties of cellular materials, and use the simple model to illustrate methods for analysis of the mechanics of cytokinesis.

MECHANICS OF CELL DEFORMATION

Identification of mechanical response of a cell as a liquid-like body is a major reduction in complexity; however, it remains to be determined which parts of the cell dominate the viscous resistance to flow and what are the properties of the cell cortex that oppose the excess pressure of the cytoplasm. The interior of the cell is a liquid "slurry" of suspended organelli and nucleus, which as a whole is non-uniform and may be non-Newtonian in character. Still, the contents only resist the <u>rate</u> of deformation and the body will exhibit a uniform isotropic stress when stationary (i.e. a uniform hydrostatic pressure). The static shape and stress distribution will be determined by the plasma membrane-cortical gel subject to uniform internal pressure. The dynamics of deformation of a cell with this type of structure will be determined by the properties of the cortical shell composite combined with the effective viscosity of the interior suspension.[1]

Analysis of cell deformation involves three independent and distinct developments which are the bases for the mechanics of continuous media: 1) quantitation of deformation and rate of deformation for changes in geometric shape (independent of substance and forces); 2) balance of forces (independent of deformation and substance); 3) material properties of the substance. Knowledge of any two of the three can be used to predict the third aspect, e.g. observation of deformation and rate of deformation in response to controlled forces can be analyzed to give the material properties (e.g. elastic and viscous coefficients). We will outline each of these components in the following paragraphs.

<u>Deformation and Stress Distribution in the Liquid Core</u>

The simplest approach to the mechanics of the liquid-like core is to model the response of the cell contents as a uniform Newtonian liquid with a constant viscosity. The equations of motion reduce to the "creeping" flow form - low Reynolds number limit- of the Navier-Stokes equations for a fluid.[5,6] Inertial forces (due to acceleration of the fluid) are completely negligible because of the small size of cells and the slow flow rates. Thus, the equations of mechanical equilibrium balance viscous forces against pressure forces. For example, consider axisymmetric flow with spherical boundaries; the equations of mechanical equilibrium are expressed in spherical polar coordinates (R, Θ, ϕ) by,

$$0 = \frac{1}{R^2} \cdot \frac{\partial(R^2 \cdot \sigma_{RR})}{\partial R} + \frac{1}{R \cdot \sin\theta} \cdot \frac{\partial(\sin\theta \cdot \sigma_{\theta R})}{\partial \theta} - \frac{\sigma_{\theta\theta}}{R} - \frac{\sigma_{\phi\phi}}{R}$$

$$0 = \frac{1}{R^2} \cdot \frac{\partial(R^2 \cdot \sigma_{R\theta})}{\partial R} + \frac{1}{R \cdot \sin\theta} \cdot \frac{\partial(\sin\theta \cdot \sigma_{\theta\theta})}{\partial \theta} + \frac{\sigma_{\theta R}}{R} - \frac{\cot\theta \cdot \sigma_{\phi\phi}}{R}$$

(1)

where $\sigma_{RR}, \sigma_{\theta\theta}, \sigma_{R\theta}(\sigma_{\theta R}), \sigma_{\phi\phi}$ are components of the stress field in the core (the first subscript defines the direction along which the force acts - either radial R, meridional Θ, or azimuthal ϕ - and the second subscript denotes the surface on which the stress is defined). For a Newtonian liquid, the constitutive behavior is represented by proportionality between stresses and rates of deformation,

$$\sigma_{RR} = -P+2\tilde{\eta}\cdot\frac{\partial v_R}{\partial R} \quad ; \quad \sigma_{R\theta} = \tilde{\eta}(\frac{1}{R}\cdot\frac{\partial v_R}{\partial \theta} + \frac{\partial v_\theta}{\partial R} - \frac{v_\theta}{R})$$

$$\sigma_{\theta\theta} = -P+2\tilde{\eta}(\frac{1}{R}\cdot\frac{\partial v_\theta}{\partial \theta} + \frac{v_R}{R}) \quad ; \quad \sigma_{\phi\phi} = -P+\frac{2\tilde{\eta}\cdot v_R}{R} \qquad (2)$$

where the hydrostatic pressure P is introduced because of fluid incompressibility,

$$0=\frac{1}{R^2}\cdot\frac{\partial(R^2\cdot v_R)}{\partial R}+\frac{1}{R\cdot\sin\theta}\cdot\frac{\partial(\sin\theta\cdot v_\theta)}{\partial \theta} \qquad (3)$$

The coefficient $\tilde{\eta}$ is the effective viscosity of the cell contents; v_R and v_θ are the velocity components along radial and meridional directions respectively. The solution for axisymmetric flow inside a spherical boundary is expressed in terms of infinite series expansions in angular harmonic functions (Y_n, Y'_n),[6]

$$v_R = -\sum_{n=2}^{\infty}[a_n\cdot R^{n-2}+c_n\cdot R^n]\cdot Y_{n-1}(\cos\theta)$$

$$v_\theta = \sum_{n=2}^{\infty}[n\cdot a_n\cdot R^{n-2}+(n+2)\cdot c_n\cdot R^n]\cdot Y'_n(\cos\theta)/\sin\theta \qquad (4)$$

$$P = -\tilde{\eta}\cdot\sum_{n=2}^{\infty}[\frac{2(2n+1)}{n-1}\cdot c_n\cdot R^{n-1}]\cdot Y_{n-1}(\cos\theta)+c_0$$

With the appropriate boundary conditions at the shell interface (for stresses or velocities), the coefficients a_n and c_n in these equations are fully specified.

The objective is to determine the dynamic stresses applied by the liquid core to the cortical shell. In this spherical example, the stresses are given by a dynamic pressure P' normal to the cortex and a shear stress σ_m tangent to the shell,

$$-P' = -P+2\tilde{\eta}\cdot\frac{\partial v_R}{\partial R} , \quad (R=R_0)$$

$$\sigma_m = \tilde{\eta}(\frac{1}{R}\cdot\frac{\partial v_R}{\partial \theta} + \frac{\partial v_\theta}{\partial R} - \frac{v_\theta}{R}) , \quad (R=R_0)$$

These fluid dynamic stresses are opposed by material forces in the shell. This approach has been used to model the flow response of the interior contents of blood phagocytes in micropipet aspiration tests.[10,14] Analysis of the creeping flow of a spherical liquid drop into a tube has shown that the stress applied normal to the sphere surface is essentially uniform over the spherical segment exterior to the tube. Hence for pipet aspiration, the pressure on the cortex can be approximated by a constant and an abrupt pressure drop inside the core is introduced at the pipet entrance in proportion to rate of entry.[1]

[The pressure drop at the pipet entrance is proportional to the entry flow rate scaled by the pipet radius $(\dot{L}/R_p)$,

$$\Delta P = \tilde{\eta} \cdot (\frac{\dot{L}}{R_p}) \cdot f(R_p/R_o)$$

multiplied by an algebraic function of the ratio of pipet size to cell diameter (R_p/R_o).]

In general, simultaneous solution of the equations of mechanical equilibrium for the cortical shell with those for the liquid core is required. The coupling between these sets of equations is provided by the dynamic stresses (P', σ_m) that act on the shell. The mechanics of shell deformation and force equilibrium will be discussed next.

Intrinsic Deformation and Rate of Deformation of the Cortex

Changes in overall shape or conformation of a shell can be viewed conceptually as the superposition of local deformations of "imaginary" differential elements of the shell. If the shell is thin in comparison to the radii of curvature that describe the contour of the shell, then the local deformation of each small element can be represented by three independent modes of deformation: 1) area dilation or condensation without changing shape or curvature of the element; 2) extension of the element without change in surface area or curvature; 3) changes in element curvature without change in element area or aspect ratio (the modes of deformation are illustrated in Fig. 1).[4] These geometric changes are quantitated by the fractional change in element area α; the in-plane extension ratio $\tilde{\lambda}$ (at constant surface density); and changes in element principal curvatures $(1/R_1)$ and $(1/R_2)$. Similarly, the intrinsic rates of deformation are defined by the rate of area dilation or condensation; the rate of extension at constant element area; and the rate of change of curvatures. Since deformation is defined locally, the small element of the shell surface can be considered "flat" with dimensions described by orthogonal surface coordinates that form a plane tangent to the shell surface. Thus, rectangular deformations of the element are characterized by the ratios (λ_1, λ_2) of instantaneous dimensions of the element to the initial dimensions in some reference state. Based on these (extension) ratios, the fractional change in area is given by,

$$a = (\lambda_1 \cdot \lambda_2) - 1 \tag{5}$$

and the measure of in-plane extension at constant area is given by a single extension ratio,

$$\tilde{\lambda} = \lambda_1/\lambda_2 \tag{6}$$

Simple rectangular deformation of an element embodies surface shear deformation which is maximal along lines in the surface at $\pm 45°$ to the extension axis; the maximum shear is represented by the Eulerian shear strain e_s,

$$e_s = |\tilde{\lambda}^2 - \tilde{\lambda}^{-2}|/4(1+a) \tag{7}$$

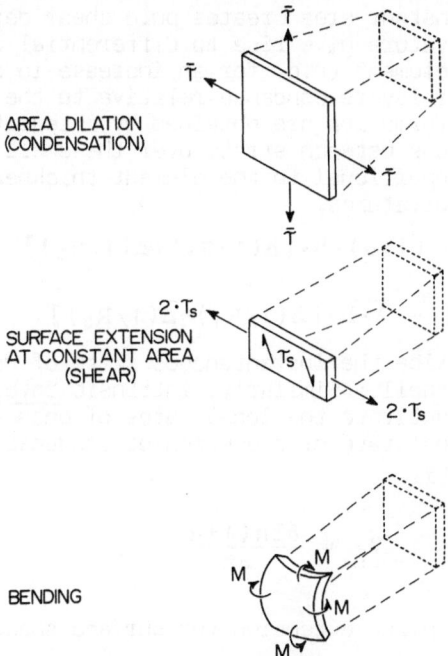

AREA DILATION
(CONDENSATION)

$2 \cdot T_s$

SURFACE EXTENSION
AT CONSTANT AREA
(SHEAR)

$2 \cdot T_s$

T_s

BENDING

M M M M

Fig. 1. Independent modes of shell deformation: 1) area dilation;
2) shear or extension at constant area; and 3) bending.
The force and moment resultants associated with each mode of
deformation are also shown.

Hence, extension at constant area creates pure shear deformations. Changes in element curvature give rise to differential deformations between strata of the element (e.g. for an increase in convexity, outer layers expand and inner layers condense relative to the mid-plane). Measures of bending deformation are obtained by integration of the differential deformations between strata over the shell thickness; these bending strains are proportional to the element thickness times the changes in principal curvatures,

$$|\delta a| \sim (1+a)\cdot h\cdot [\Delta(1/R_1)+\Delta(1/R_2)]$$

$$|\delta\tilde{\lambda}| \sim \tilde{\lambda}\cdot h\cdot [\Delta(1/R_1)-\Delta(1/R_2)]$$

(8)

Equations (5)-(8) describe the instantaneous state of deformation at a local position on the shell. Similarly, intrinsic <u>rates</u> of deformation can be derived that quantitate the local rates of change of element geometry. The fractional rate of expansion or condensation of the element area is given by,

$$V_a = \frac{\partial \ln(1+a)}{\partial t}$$

(9)

The maximum rate of in-plane extension (or surface shear rate) is given by,

$$V_s = \frac{\partial(\ln \tilde{\lambda})}{\partial t}$$

(10)

Finally, the rates of bending deformations are given by,

$$|\delta V_a| \sim h.\frac{\partial[(1/R_1)+(1/R_2)]}{\partial t}$$

$$|\delta V_s| \sim h.\frac{\partial[(1/R_1)-(1/R_2)]}{\partial t}$$

(11)

In the previous discussion, deformation was evaluated in a domain small enough so that the material element could be considered flat and locally uniform. This approach leads to intensive quantitation of shape changes, i.e. the variables are point definitions which do not depend on the size or shape of the shell as a whole. However, deformation and rate of deformation variables may vary over the shell contour. The continous variation is derived from differential geometry. For example with axisymmetric deformations, the shell geometry can be described by curvilinear coordinates (s,Θ,ϕ) and (s_0, Θ_0, ϕ) for the instantaneous (deformed) and initial (reference) contours respectively as shown in Fig. 2. The distance along a meridian (contour line) is defined as "s"; the coordinate "Θ" is defined as the angle between the outward normal to the surface and the axis of symmetry; the coordinate "ϕ" is the azimuthal angle. By transformation, these intrinsic coordinates are related to the spatial coordinates (r, Z, ϕ) for the instantaneous geometry and (r_0, Z_0, ϕ) for the reference contour. Extension ratios, which define local changes in element geometry, are given by the ratio of differential lengths (deformed:undeformed) along a meridian and around a latitude circle for the surface elements as follows,*

$$\lambda_m = \frac{ds}{ds_0} \Rightarrow \lambda_1 \quad ; \quad \lambda_\phi = r/r_0 \Rightarrow \lambda_2$$

(12)

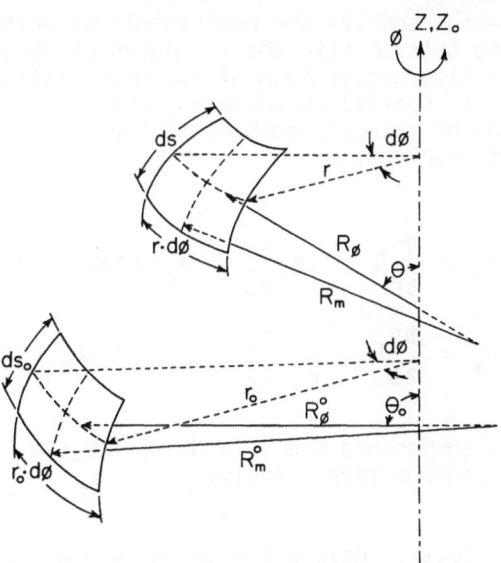

Fig. 2. Initial and deformed elements of an axisymmetric shell.
Extension ratios for deformation (deformed: undeformed);
$\lambda_m = ds/ds_0$, $\lambda_\phi = r/r_0$.

These extension ratios specify the positional variation of surface deformation for the axisymmetric shell. Rates of deformation are derived from local time derivatives of extension ratios as given before in Eqs. (9)-(11). In spatial coordinates, the rates of deformation are also given in terms of velocity components that characterize motion of points on the shell contour,

$$V_a = \frac{\partial v_s}{\partial s} + \frac{v_s}{r} \cdot \frac{\partial r}{\partial s} + v_n \cdot (1/R_m + 1/R_\phi)$$

$$2 \cdot V_s = \frac{\partial v_s}{\partial s} - \frac{v_s}{r} \cdot \frac{\partial r}{\partial s} + v_n \cdot (1/R_m - 1/R_\phi)$$

(13)

where the velocity components are v_s and v_n, tangent to the contour and normal to the contour respectively.

---

   * The subscripts (m, ϕ) replace the index values (1,2) since the former represent the principal directions in the case of axial symmetry.

Shell Mechanical Equilibrium

The general approach to the mechanics of thin shells is to cumulate the actions of forces that are distributed in the shell by integration over the shell thickness. This yields equivalent force and moment resultants which act at a mathematical surface that replaces the shell.[4,7] Figure 3 schematically illustrates the effective reactions of the shell that oppose the dynamic stresses (p', σ_m) applied normal and tangent to the shell surface.

Fig. 3. Cumulated stresses in a shell: in-plane force resultants (τ_n tension, τ_s shear); transverse shear resultant Q_m; bending moment resultant M.

Integration of the "lateral" stresses over the shell thickness yields forces per unit length along edges of the element that act in the plane of the shell, normal and tangent to the edge (i.e. tension and shear force respectively). Deviation of lateral stresses from the mean values of stress gives rise to force couples that cumulate into moment resultants which act around the edge of the element. Finally, the shear stress component parallel to the outward normal cumulates into the transverse shear resultant which acts to "cut" the shell as shown in Fig. 3.

Examination of the in-plane force resultants shows that these forces can be partitioned into two independent parts: a mean or isotropic force resultant $\overline{\tau}$; and a deviatoric or maximum shear force resultant τ_s (that acts maximally along lines at $\pm 45°$ to the axes for purely tensile or compressive stresses as shown in Fig. 1). Similary, moment resultants can be decomposed into isotropic and deviatoric parts as well.

$$\overline{\tau} = (\tau_m + \tau_\phi)/2 \quad ; \quad \tau_s \equiv (\tau_m - \tau_\phi)/2$$

$$\overline{M} = (M_m + M_\phi)/2 \quad ; \quad \widetilde{M} \equiv (M_m - M_\phi)/2$$

In general, the sum of all forces on the shell is equal to the inertial force due to acceleration. Because of the small scale of biological structures and low accelerations, inertial forces are totally negligible. Hence, mechanical equilibrium is established when the sum of forces on the body is equal to zero. Equations of mechanical equilibrium are not equivalent to thermodynamic statements of equilibrium. For example in mechanical equilibrium, forces may be dissipative (non-conservative) as well as elastic (conservative). Thermodynamic equilibrium is based on reversibility of the physical and chemical processes that occur at the atomic and molecular level during deformation and, consequently, only represents elastic forces.

For an axisymmetric shell (illustrated in Fig. 2) subject to a dynamic pressure differential P' and a tangential traction or shear stress σ_m along the meridian of the shell contour, two differential equations must be satisfied for the local balance of forces,

$$P' = \overline{\tau} \cdot (1/R_m + 1/R_\phi) + \tau_s \cdot (1/R_m - 1/R_\phi) - \frac{1}{r} \cdot \frac{\partial(r \cdot Q_m)}{\partial s}$$

$$0 = r \cdot \sigma_m + r \cdot \frac{\partial(\overline{\tau} + \tau_s)}{\partial s} + 2 \cdot \tau_s \cdot \frac{\partial r}{\partial s} + \frac{(r \cdot Q_m)}{R_m} \tag{14}$$

Likewise, the balance of moments for the shell element is given by,

$$r \cdot Q_m = r \cdot \frac{\partial \overline{M}}{\partial s} + \frac{1}{r} \cdot \frac{\partial(r^2 \widetilde{M})}{\partial s} \tag{15}$$

Here, the equations of equilibrium have been written in terms of surface-isotropic and maximum shear force resultants plus moments to demonstrate the explicit dependence of mechanical equilibrium on these independent actions. The stresses applied to the cortex are obtained from simultaneous solution of the equations for "creeping" flow of the liquid core (as discussed previously).

General Constitutive Equations for Thin Shell Materials

The principles of thermodynamics can be used to derive phenomenological recipies for the storage and dissipation of energy when work is done on a material by displacement of external forces.[4] The differential work associated with deformation of an element of the shell can be expressed per unit surface area of the shell as a local density function,

$$d\widetilde{W} = \bar{\tau} \cdot da + 2 \cdot \tau_s \cdot (1+a) \cdot d(\ln \widetilde{\lambda})$$
$$+ \bar{M} \cdot (1+a) \cdot d[(1/R_m) + (1/R_\phi)] + \widetilde{M} \cdot (1+a) \cdot d[(1/R_m) - (1/R_\phi)] \tag{16}$$

Here, it is seen that the work is made up of contributions from dilation and extension in the plane of the shell plus a contribution from bending or curvature changes, all given by products of intensive forces multiplied by conjugate deformations. For reversible thermodynamic processes, work can be attributed to conservative forces which are derivatives of a potential function that depends only on the independent modes of deformation. For a shell whose properties do not depend on directions tangent to the shell (i.e. isotropic in the plane of the surface), an elastic potential can be postulated by,

$$(d\widetilde{F})_T \simeq K \cdot d(a^2/2) + \mu \cdot d[(\widetilde{\lambda}^2 + \widetilde{\lambda}^{-2})/2] + \bar{B} \cdot d[(1/R_m + 1/R_\phi)^2/2]$$
$$+ \widetilde{B} \cdot d[(1/R_m - 1/R_\phi)^2/2] \tag{17}$$

where the coefficients in this equation are elastic properties: area compressibility modulus K; the surface extension (shear) modulus ; and bending or curvature elastic modulus B. From the first law of thermodynamics for a reversible process, work is equal to the change in the elastic potential which leads to proportionality between material forces and deformation. Thus, mean or isotropic force resultants are proportional to area dilation or condensation,

$$\bar{\tau} = K \cdot a \tag{18}$$

Shear forces in the plane of the shell are proportional to extensional deformation,

$$\tau_s = \mu \cdot (\widetilde{\lambda}^2 - \widetilde{\lambda}^{-2})/2 = 2 \cdot \mu \cdot e_s \tag{19}$$

Bending moments are proportional to changes in curvature,

$$\bar{M} = \bar{B} \cdot \Delta(1/R_m + 1/R_\phi)$$
$$\widetilde{M} = \widetilde{B} \cdot \Delta(1/R_m - 1/R_\phi) \tag{20}$$

In contrast with elastic behaviour, irreversible processes create non-conservative forces that depend on rate of deformation as well as the instantaneous deformation state.[4] For example, a solid shell structure will be characterized by elastic levels of stress at static equilibrium, but the time dependent response of the shell to applied forces will involve greater stresses because of viscous dissipation within the material. Further, shell materials may exhibit liquid-like or plastic behaviour which is characterized by continuous flow in response to prolonged exposure to stress when the magnitude of the stresses are sufficiently large. For such non-conservative stresses, mechanical power is dissipated in the material as heat and the density of mechanical power is given by the local time derivative of the work density,

$$\frac{\partial \widetilde{W}}{\partial t} = \overline{\tau} \cdot \frac{\partial a}{\partial t} + 2 \cdot \tau_s \cdot (1+a) \cdot \frac{\partial (\ln \widetilde{\lambda})}{\partial t} + \overline{M} \cdot (1+a) \cdot \frac{\partial [(1/R_m)+(1/R_\phi)]}{\partial t}$$
$$+ \widetilde{M} \cdot (1+a) \cdot \frac{\partial [(1/R_m)-(1/R_\phi)]}{\partial t} \qquad (21)$$

For ideal liquids, dissipation of mechanical power is a quadratic function of the rates of deformation. Hence, force and moment resultants are simply inelastic and proportional to rates of deformation,

$$\overline{\tau} = \kappa \cdot \frac{\partial \ln(1+a)}{\partial t} \quad ; \quad \tau_s = 2\eta \cdot \frac{\partial (\ln \widetilde{\lambda})}{\partial t}$$
$$(22)$$
$$\overline{M} = \overline{\nu} \cdot \frac{\partial [(1/R_m)+(1/R_\phi)]}{\partial t} \quad ; \quad \widetilde{M} = \widetilde{\nu} \cdot \frac{\partial [(1/R_m)-(1/R_\phi)]}{\partial t}$$

where proportionality is established by coefficients of viscosity for the appropriate type of deformation. In general, material behaviour can only be approximated by solid or liquid-like constitutive relations and usually is characterized by more complicated rheological equations that must be defined empirically by experiment.

A reasonably general representation for the mechanical response of materials is given by constitutive relations for three ideal regimes: solid, semi-solid, and plastic or liquid.[4] Solid-like materials are represented by viscoelastic equations that model parallel superposition of elastic and viscous processes as originally envisioned by Kelvin,

$$\tau_s = \mu \cdot [(\widetilde{\lambda}^2 - \widetilde{\lambda}^{-2})/2 + 2 \cdot t_e \cdot \frac{\partial (\ln \widetilde{\lambda})}{\partial t}] \qquad (23)$$

given here for the in-plane shear resultant. The key feature for this type of behaviour is that the dynamic response is characterized by a time constant t_e, which is the ratio of the coefficient of viscosity to the elastic modulus (η_e/μ). The transition from solid to liquid-like behaviour depends on the magnitude and duration of forces applied to the material. This intermediate regime is characterized by creep and relaxation phenomena, i.e. the material exhibits slow continuous deformation when the force is held constant whereas the material responds elastically to rapid changes in applied force. This behaviour can be modelled by the serial addition of elastic (conservative) and inelastic (plastic) processes as originally developed by Maxwell. Here, the in-plane extensional constitutive relation for creep and relaxation is represented by,

$$\frac{\partial(\ln\tilde{\lambda})}{\partial t} = \frac{1}{\mu}\left[\frac{1}{2\sqrt{(\tau_s^2/\mu^2+1)}}\cdot\frac{\partial\tau_s}{\partial t} + \frac{\tau_s}{2\cdot t_c}\right] \qquad (24)$$

where the characteristic time constant, t_c, is given by the ratio of a different coefficient of viscosity to the elastic modulus (η_c/μ). As such, the characteristic material time constants for viscoelastic solid and semi-solid material regimes represent different molecular relaxation processes. Finally, the simplest form of material transition from solid to liquid behaviour is that of an ideal plastic modelled by elastic solid behaviour below a yield threshold and liquid behaviour above the yield,

$$\frac{\partial(\ln\tilde{\lambda})}{\partial t} = (\tau_s - \hat{\tau}_s)/(2\cdot t_p) \ , \ \tau_s > \hat{\tau}_s \qquad (25)$$

as introduced by Bingham in 1922 and generalized ten years later by Hohenemser and Prager. Two material properties are necessary to characterize the behaviour of a plastic body: a yield force resultant and a viscosity ($t_p = \eta_p/\hat{\tau}_s$). For a perfect liquid, the yield threshold is identically zero. For both semi-solid and plastic types of material behaviour, deformations of the material are irrecoverable in that the shell is left permanently deformed after the forces are removed. The constitutive relations given above outline general regimes of material behaviour but are not unique. However, these relations characterize the essential features of mechanical response of materials and transitions in behavior; the time constants (t_e, t_c, t_p) are especially useful as guidelines for prediction of the appropriate response.

Equations for material behaviour, i.e. the relation between intensive force (and moment) resultants and intensive deformation and rate of deformation variables, are used with the equations of mechanical equilibrium to predict the evolution of shell geometry in response to applied forces. The necessary ingredients are the material properties (elastic and viscous coefficients) for the actual shell materials and the effective viscosity of the liquid core. In the following paragraphs, a brief overview of the known properties of synthetic (phospholipid) bilayer and cell membranes will be given as well as a discussion of the behaviour anticipated for plasma membrane-cortical gel composite materials. In the context of these properties and constitutive relations, a simple model for the mechanics of cell division will be examined.

Material Properties of Lipid Bilayer and Red Blood Cell Membranes

From the viewpoint of mechanics, membranes are simply <u>very</u> thin shells so the effects of moment resultants (due to bending ridigity) on the mechanical equilibrium can be neglected in comparison to the contributions from force resultants in the plane of the shell. In biology, the term membrane usually represents a molecularly thin structure that separates one region of a cell or subcellular body from another and is often synonomous with the concept of a plasmalemma or lipid bilayer structure. Because of molecular thinness, bilayer membrane structures must in general be supported by stiff adjacent materials. The unifying feature of all bilayer membranes is the preferential assembly of the lipid amphiphiles into two dimensional, condensed liquid bilayers which form tight-cohesive chemical insulators. Subsurface or superficial layers provide "scaffolding" and support for the liquid bilayer; these differ from cell to cell but are essential for maintenance of cell shape and strength. For simple capsules like red blood cells, the lipid bilayer is supported by an adjacent network of filamentous and globular proteins which is strongly associated with the bilayer. As such, the red cell envelope is a thin tri-lamellar membrane composite which is smooth in contour and exhibits recoverable elastic deformations.[4] On the other hand, cells like blood phagocytes appear to possess a gelatinous layer adjacent to the ruffled outer plasmalemma which is forced onto the spherical form. The difference between smooth and ruffled configurations is more readily understandable after examination of the specific properties characteristic of phospholipid bilayer membranes.

Lipid bilayer membranes can exist either as thin solid- or liquid-materials where the polar head groups are anchored to the water interfaces and the hydrocarbon polymer chains form a thin double layer between the polar interfaces.[8,9] In biological cells, these hydrocarbon chains are in the fluid state so the cell surfaces behave as two dimensional liquids (i.e. $\mu \equiv 0$). Because of the relatively large energies required to expose the hydrocarbon interior to water, the bilayer surfaces have very small area compressibility (i.e. $K \sim 10^2$ dyn/cm); thus, the molecules remain closely packed even when subjected to tensions that lead to rupture. A spherical bilayer capsule with fixed internal volume is essentially rigid because any deformation of the sphere requires an increase in area whereas deflation of the sphere leaves the capsule completely flaccid (easily deformable) stabilized in shape by the extremely small bending rigidity (i.e. $B \sim 10^{-12}$ dyn-cm) of the bilayer. Bilayer bending energies only become comparable to surface dilation energies when the membrane radii of curvature are less than 10^{-6}cm.[4] For Example, osmotic deflation of a vesicle (shown in Fig. 4) allows the vesicle to be aspirated by a micropipet with very low suction pressure ($>10^{-6}$ atmosphere) to an extent determined by vesicle area and reduced volume. When this limit is reached, large suction pressures are required to produce small displacements of the vesicle projection into the pipet which leads to vesicle rupture.[8,9]

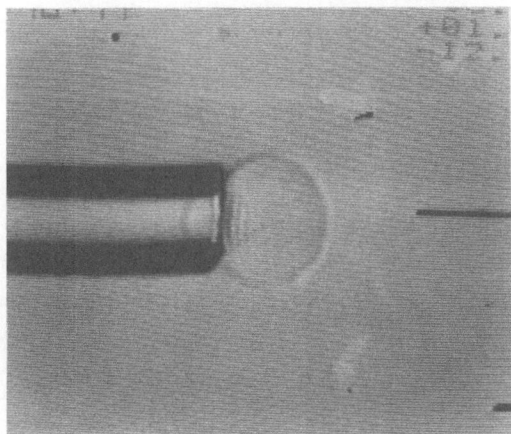

Fig. 4. Single-walled (phospholipid bilayer) vesicle aspirated with low suction pressure to the extent permitted by the fixed area envelope and internal volume. The vesicle diameter is about 2×10^{-3} cm.[8]

On the other hand when compressed, the bilayer readily folds and wrinkles unless subsurface structure exists to provide a stiff support for the membrane. Hence, the plasmalemma of cells offers no resistance to deformation until the surface area is required to dilate upon which the bilayer membrane becomes rigid to expansion. The bilayer is simply a fixed area envelope that limits the expansion and distribution of the interior contents. Since the interior contents of cells have a high osmotic activity, pressures created by cell deformation are insufficient to change cell volume which adds a second important constraint to cell deformation. Departure from the constant plasmalemma surface area and constant cell volume requirements can only occur if there is transport of materials out of the cell interior or incorporation of amphiphilic molecules into the plasmalemma. It is important to note that large amounts of surface may be present in the wrinkled and ruffled conformation of the plasmalemma. For example, micropipet aspiration of blood phagocytes has shown that more than 110% excess surface area (over that of a sphere of equivalent volume) is in the form of surface ruffles.[10] When the ruffles are pulled smooth, the cell envelope becomes tight and rigid; further expansion leads to cell lysis.

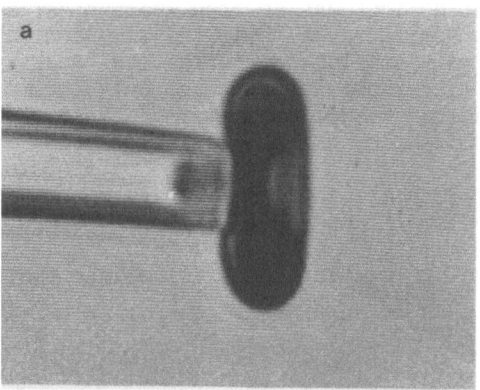

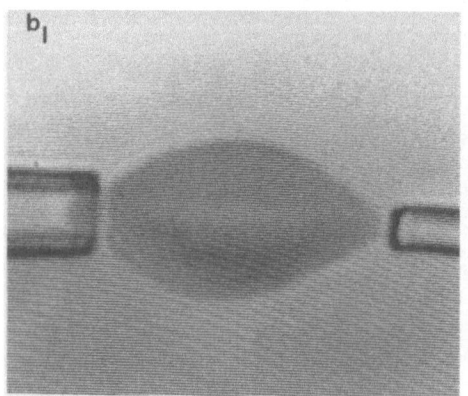 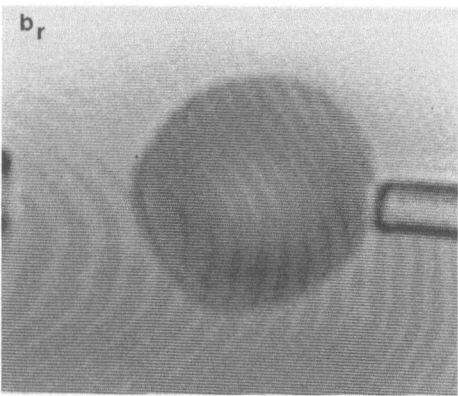

Fig. 5. (a) Human red blood cell held fixed by the suction pressure. The major dimension of the red cell is about 8 x 10^{-4} cm.[12] (b) Elastic recovery of a red blood cell from end-to-end extension. The recovery time is on the order of 0.1 sec.[4]

The red blood cell is unique in that the surface contour of the plasmalemma is kept smooth by the subsurface protein meshwork. The red cell membrane composite normally behaves as an elastic solid material which again has a very small area compressibility (K $\sim 10^2$-10^3 dyn/cm) but with a measureable non-zero extensional modulus ($\mu \sim 10^{-2}$ dyn/cm) and a small bending rigidity similar to lipid bilayers.[4,11-13] The extensional rigidity accounts for the static resistance to deformation (shown in Fig. 5a) as well as the elastic recovery of the red blood cell from large deformations (shown in Fig. 5b). It is important to note that the red cell membrane extensional (shear) modulus is four orders of magnitude lower than the area compressibility modulus. Bending rigidity of the red cell is much smaller yet in its effect but is sufficient to maintain the red cell membrane contour in a smooth form until compressional stresses cause the surface to "buckle" or fold.[13] Red cell membranes also exhibit a wide range of dynamic and plastic transition properties which are important measures of material structure and viability.[4] However, it is usually sufficient to represent the red cell as a viscoelastic body with a solid membrane structure, characterized by a response time t_e equal to 0.1 sec and the elastic coefficients just described; the internal hemoglobin solution offers negligible dynamic resistance to deformation at normal cell concentrations ($\sim$32 g/dl).

Properties of a Plasmalemma - Cortical Gel Composite Shell with Liquid Core

For a plasmalemma-cortical gel composite, the plasmalemma is simply a ruffled envelope that bounds the gel on one side but offers no resistance to deformation until the ruffles are pulled smooth where further expansion is not possible without rupture. Hence within the limits of this constraint, the deformation of the cortical composite is determined by the properties of the subsurface gel layer. In general, there are a minimum of six parameters (three elastic properties K, μ, B; and three viscous properties κ, η, ν) for the gel which may be of comparable magnitude unlike the properties of simple bilayer or red cell membranes where there are orders of magnitude differences between parameters. The simplest case is to assume that the gel is isotropic (in the three local dimensions) like an amorphous, bulk material. Then, the following relations can be used to reduce the number of parameters to single elastic and viscous properties,

$$K \approx 3 \cdot \mu \; ; \; B \approx K \cdot h^2/12$$

$$\kappa \approx 3 \cdot \eta \; ; \; \nu \approx \kappa \cdot h^2/12$$

(26)

where h is the thickness of the gel. For most cells, there is also a net passive contractility in the cortex which is represented by a constant mean stress $\bar{\tau}_0$. An example of this behaviour is shown in Fig. 6 which presents a sequence of video micrographs of a blood phagocyte subjected to micropipet aspiration and then released.[10,14] Here, the continuous flow of the cell into the pipet was proportional to suction pressure in excess of a small threshold; when the pressure was lowered to the threshold value, the flow ceased and the aspirated portion of the cell inside the pipet remained stationary. The shape of the cell exterior to the pipet remained close to spherical throughout aspiration; and when the cell was released from the pipet, it slowly but eventually recovered its original spherical form. Even with the simplified approximation for gel properties, solution of the equations of mechanical equilibrium coupled with the constitutive relations requires complicated numerical computations.[1]

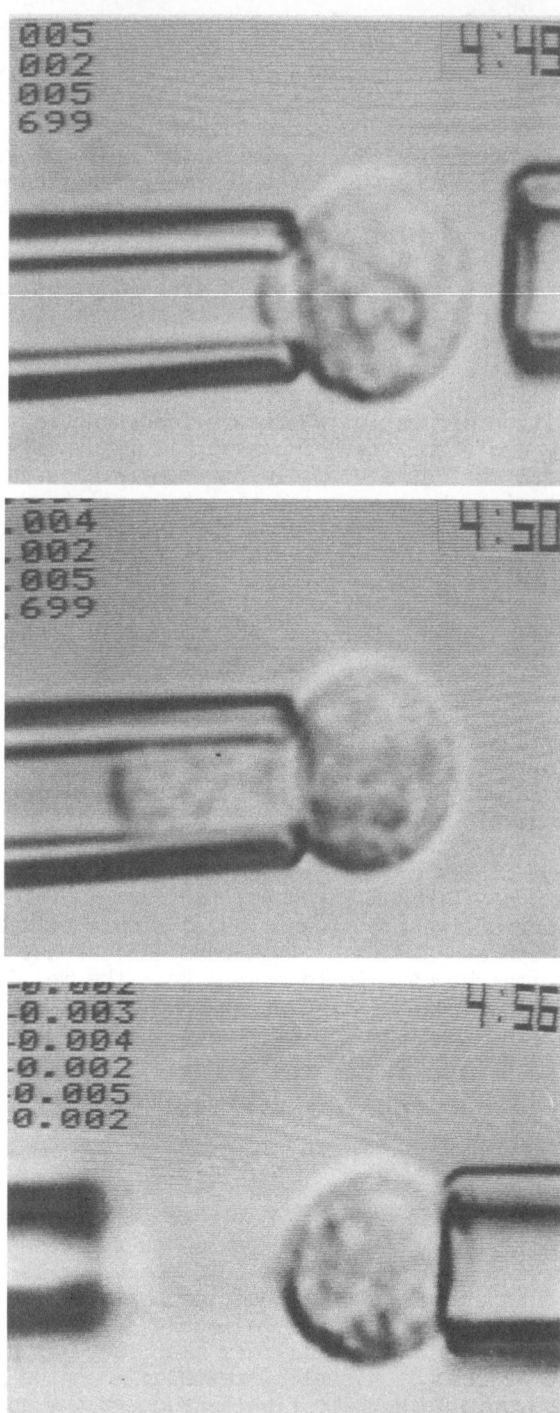

Fig. 6. Continuous aspiration of a blood granulocyte when the suction pressure exceeds the small threshold established by the contractile stress in the cell cortex and subsequent recovery to its original spherical shape.[10,14]

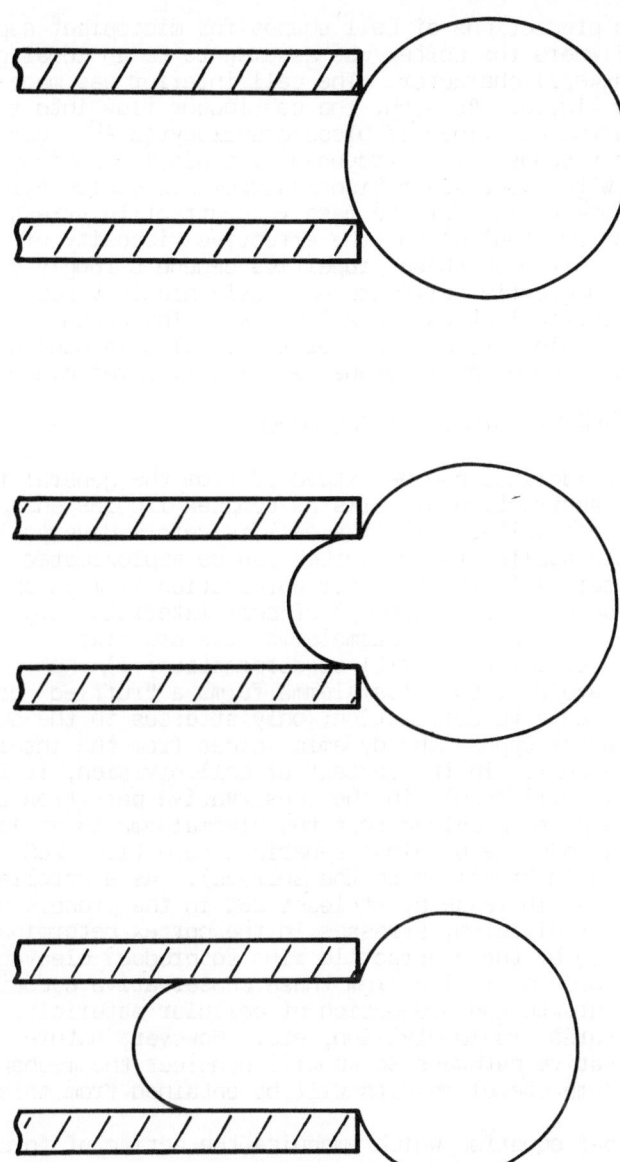

Fig. 7. Shapes predicted by numerical computation for the cell modeled
as a viscous liquid encapsulated by cortical gel.

Figure 7 shows predictions of cell shapes for micropipet aspiration of a spherical cell where the cortex was assumed to be an ideal gel but with semi-solid (Maxwell) character. The cell interior was modelled as a highly viscous liquid. As such, the continuous flow into the pipet closely models the behaviour of blood granulocytes.[10] Correlation of the model of a viscous core surrounded by a plasma membrane - cortical gel composite with pipet aspiration measurements yields values on the order of 3×10^{-2} dyn/cm for the passive contractile stress and 10^3-10^4 dyn-sec/cm^2 (poise) for the effective viscosity of the cell contents. Although these properties depend strongly on temperature and activity, phagocytic cells are obviously highly viscous and possess low levels of cortical stress when inactive. The dynamic rigidity (resistance to rapid deformation) for these cells is many orders of magnitude greater than for membrane capsules like red blood cells.

MECHANICS OF CELL DIVISION: SIMPLE MODEL

Important guidelines can be extracted from the general features of the mechanical abstraction of cells as complex liquids encapsulated by cortical shell composites: 1) if a cell is deformed very slowly, the dynamic pressure applied to the cortex can be approximated as uniform over the surface; 2) if the time for deformation is much shorter than the rate of production and exchange of cell materials (e.g. growth), both the cell volume and the plasmalemma area are fixed and thereby establish a constraint (or limit) to deformation; 3) when this restriction is avoided, the plasmalemma forms a "ruffled" boundary and offers no resistance to deformation; only stresses in the subsurface (or superficial) cortex oppose the dynamic forces from the interior and exterior of the cell. In the context of cell division, it is expected that cytokinesis will result in the conservative partition of the cell volume into two parts provided that the plasmalemma is at least 26% greater in area than the original spherical form (i.e. 26% excess of plasmalemma area in "ruffles" on the surface). As a corollary, the area of the cortex must increase by at least 26% in the process of division. For slow rates of division, stresses in the cortex determine the level of force required by the contractile ring to produce cleavage. Many factors could cause deviation from these conservative deductions: non-negligible growth and production of cellular materials, stiff internal structures, rapid division, etc. However, nature usually prefers conservative pathways so we will consider the mechanics of the simple model; some useful results will be obtained from this exercise.

The principal equation which embodies the action of forces that govern division is derived from the balance of cortical stresses against the force in the contractile ring at the cleavage section (illustrated in Fig. 8). For a differential segment of the ring, mechanical equilibrium is given by,

$$\frac{f_c}{r} = 2 \cdot \tau_m \cdot \cos\theta \tag{27}$$

where f_c is the "line" force in the contractile ring; τ_m is the local cortical stress that acts normal to the ring; Θ is the angle between the shell contour and the cleavage section. The factor of two reflects the action of both segments of the dividing cell. We must now define the constitutive behaviour of the shell and analyze deformation.

We will assume that the dividing cell can be represented by two intersecting spherical segments. In this case, the cortical stress is uniform over each segment surface.[4] For many cells, the stress in the cortex will include a net contractility $\bar{\tau}_0$ plus a dynamic or time dependent stress that increases with the rate of deformation. Taking this as the minimal definition of the response of the cortex, we express the cortical stress as,

$$\tau_m = \bar{\tau}_0 + \frac{\kappa}{(1+a)} \cdot \frac{da}{dt} \tag{28}$$

where κ is a coefficient for viscous resistance to area dilation. The coefficient κ not only represents dissipation in the cortex itself but also can be used to represent dissipation or viscous effects in the core as well (for slow flow where the dynamic pressure on the cortex is essentially uniform over the segment surface).

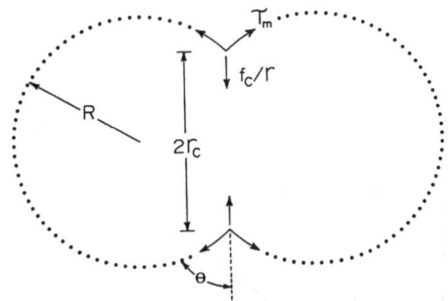

Fig. 8. Symmetric spherical segments formed by contraction of a ring around the midsection of the cell.

Deformation of the initial sphere to form symmetric spherical segments at constant volume requires area dilation. The area dilation depends only on the extent of cleavage and is given explicitly by an algebraic function of the radius of the cleavage section r. This function is plotted in Fig. 9 where the size of the cleavage section has been normalized by the initial radius of the cell before division, $\tilde{r} = r/R_0$. Therefore, the stress in the cell cortex can be expressed as a function of the rate of closure of the cleavage section,

$$\tau_m = \bar{\tau}_0 \cdot [1 + \frac{t_D}{(1+a)} \cdot \frac{da}{d\tilde{r}} \cdot \frac{d\tilde{r}}{dt}] \quad ; \quad t_D \equiv \kappa/\bar{\tau}_0 \tag{29}$$

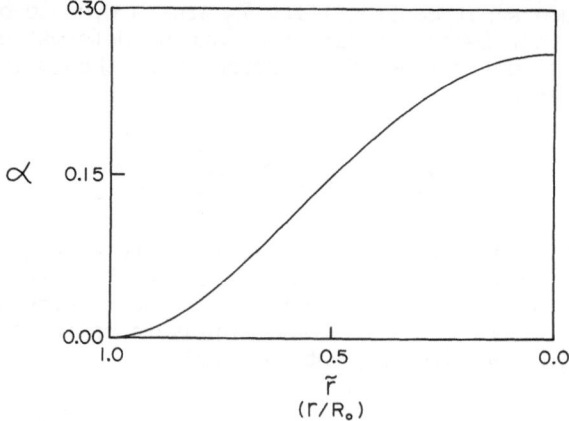

Fig. 9. Uniform area dilation of the shell segments versus extent of cleavage.

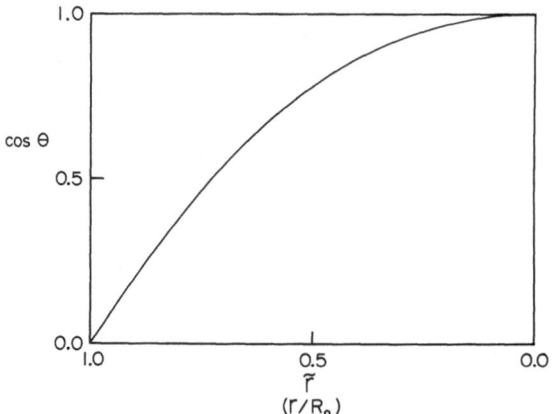

Fig. 10. Cosine of the angle formed between the shell contour and the cleavage section versus the extent of cleavage.

180

Deformation of the sphere to form two segments at constant volume also defines the angle of intersection between the spherical surfaces for a symmetric cleavage profile. Figure 10 shows the dependence of $\cos\Theta$ on the size of the cleavage opening. With this result and the previous prescription for the cortical stress, the contractile force in the ring is reduced to a function of the size of the ring <u>and</u> the rate of contraction,

$$f_c = (\bar{\tau}_0 \cdot R_0) \cdot [1 + \frac{t_D}{(1+a)} \cdot \frac{da}{dr} \cdot \frac{d\tilde{r}}{dt}] \cdot \cos\theta \qquad (30)$$

Two parameters completely characterize the dynamics of division for this simple model: the net stress or contractility of the cell cortex $\bar{\tau}_0$ multiplied by the cell dimension R_0 before division; the viscous coefficient for the surface dilation κ divided by the shell contractility $\bar{\tau}_0$, i.e. the dynamic response time. How do these parameters affect the dynamics of cell division?

Consider two different processes for division: 1) driven by a constant force in the contractile ring; 2) driven by a force sufficient to produce a constant <u>rate</u> of cleavage. Figure 11 shows the predictions for size of the cleavage section versus time when the force in the contractile ring has various constant values. Here, two key observations are made: there is a minimum force of contraction ($0.8\bar{\tau}_0 \cdot R_0$) below which cleavage will not be completed; for contractile forces above this threshold, the time for cleavage is a definite value determined by the viscous dissipation through the ratio $t_D/(f_c/\bar{\tau}_0 \cdot R_0 - 0.8)$. On the other hand for a constant rate of cleavage, the force generated by the contractile ring is variable and depends on the size of the cleavage section as shown in Fig. 12. Again, there is a threshold value for the contractile force since cleavage at the slowest possible rate of closure requires a non-zero force that must exceed $0.8\bar{\tau}_0 \cdot R_0$.

The simple model for mechanics of dividing cells was based on perfectly spherical contours. This assumption dictates that the shell meridional stress is uniform over each shell segment and that the stress depends only on area deformation (and rate). For a general gel-like structure, stresses will be produced by both area dilation and shear. In this case, the shell will deviate from a spherical contour; the shape will be flattened in the vicinity of the cleavage section because of shear stresses in the shell (which are large in this region). The same equation for the force in the contractile ring holds, i.e. Eq. (27); but the meridional stress will vary over the shell contour <u>and</u> will include both dilatory and shear components,

$$\bar{\tau} = \bar{\tau}_0 + \kappa_1 \cdot V_a + K_2 \cdot \int_0^t V_a \cdot e^{-K_2 \cdot (t-t')/\kappa_2} \cdot dt'$$

$$\tau_s = 2 \cdot \eta_1 \cdot V_s + 2\mu_2 \cdot \int_0^t V_s \cdot e^{-\mu_2 \cdot (t-t')/\eta_2} \cdot dt'$$

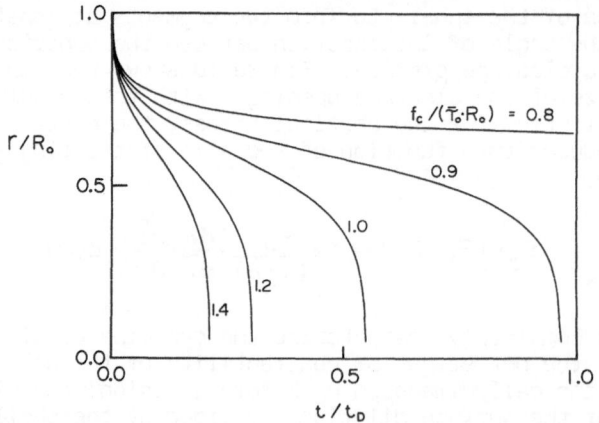

Fig. 11. Extent of cleavage versus time for fixed levels of force in
the contractile ring. A minimum level of contractile force
in the ring is required to produce complete division.

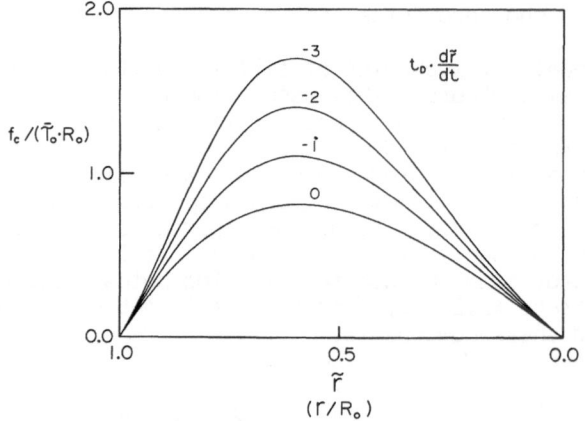

Fig. 12. Force in the contractile ring versus extent of cleavage for
various rates of contraction.

The dynamic stress contributions in a gel-like material approach the limiting stress representations,

$$\bar{\tau} \Rightarrow \bar{\tau}_0 + (\kappa_1 + \kappa_2) \cdot V_a \qquad ; \qquad \tau_s \Rightarrow 2(\eta_1 + \eta_2) \cdot V_s \qquad (31)$$

Fig. 13. Shapes predicted by numerical computation for the cell modeled as a viscous liquid encapsulated by a cortical gel.

Mechanical equilibrium is established by the balance of these stresses against the excess pressure of the cytoplasm, Eqs. (14). Because the equations are non-linear, solution necessitates numerical computation.[1] Figure 13 shows the shapes predicted for division of a cell with a gel-like cortex. Even for this more realistic (and more complicated) model, the key features of the division process remain the same as for the simple model: i.e. a minimum level of force of contraction is required for complete division and the contractile force depends strongly on rate of cleavage. However, the shapes are better approximations to real cell geometries; the form of the dimensionless functions given in Figures 11 and 12 are also altered somewhat.

COMMENTS

Our purpose has not been to rigorously analyze the mechanics of cell division but rather to indicate methods and approaches that can be used in the biophysical study of cell division. Further, we have endeavored to outline important and relevant features of cell deformability and cell material properties. Again, we emphasize that direct test of material properties are required for each cell type and the best test of any cell division theory is in situ measurement of cell stresses. Even for the naive model introduced here, it is clear that the intrinsic mechanisms of cell division are directly related to the dynamic stresses produced in cells by cleavage.

ACKNOWLEDGEMENT

This research was supported in parts by the U.S. National Institutes of Health, grant HL26965, and the Medical Research Council of Canada, grant MT7477.

REFERENCES

1. Evans, E. and Yeung, A., 1986, Limiting models for passive flow of liquid-like spherical cells into micropipets: (1) Uniform viscous droplet with constant interfacial tension; (2) Plasma membrane-cortical gel composite surrounding an inviscid fluid interior, Biophys. J. (submitted).

2. Southwick, F.S. and Stossel, T.P., 1983, Contractile protein in leukocyte function, Sem. in Hematology 20:305.

3. Valerius, N.H., Stendahl, O., Hardwig, J.H. and Stossel, T.P., 1981, Distribution of actin-binding protein and myosin in polymorphonuclear leukocytes during locomotion and phagocytosis, Cell 24:195.

4. Evans, E.A. and Skalak, R., 1980, "Mechanics and Thermodynamics of Biomembranes," CRC Press Inc., Boca Raton, Florida.

5. Landau, L.D. and Lifshitz, E.M., 1959, "Fluid Mechanics," Pergamon Press, London.

6. Happel, J. and Brenner, H., 1973, "Low Reynolds Number Hydrodynamics," Noordhoff International, Leyden.

7. Flügge, W., 1966, "Stresses in Shells," Springer-Verlag, New York.

8. Evans, E. and Needham, D., 1986, Giant vesicle bilayers composed of mixtures of lipids, cholesterol and polypeptides: Thermomechanical and (mutual) adherence properties, Faraday Disc. Chem. Soc. 81:(in press).

9. Evans, E. and Needham, D., 1986, Physical properties of lipid bilayer membranes: cohesion, elasticity, and (colloidal) interactions, J. Phys. Chem. (submitted).

10. Evans, E., Kukan, B. and Yeung, A., 1986, Mechanics of a plasma membrane-cortical gel composite: Model for passive deformability of blood phagocytes, Biophys. J. (submitted).

11. Evans, E. and Waugh, R., 1977, Osmotic correction to elastic area compressibility modulus of red cell membrane, Biophys. J. 20:307.

12. Waugh, R. and Evans, E.A., 1979, Thermoelasticity of red blood cell membrane, Biophys. J. 26:115.

13. Evans, E.A., 1983, Bending elastic modulus of red blood cell membrane derived from buckling instability in micropipet aspiration tests, Biophys. J. 43:27.

14. Evans, E. and Kukan, B., 1983, Large deformation recovery after deformation, and activation of granulocytes. Blood 64:1028.

SURFACE CHANGES IN RELATION TO CYTOKINESIS AND OTHER STAGES DURING THE

CELL CYCLE

C.A. Pasternak

Department of Biochemistry
St George's Hospital Medical School
Cranmer Terrace, London SW17 ORE, UK

SUMMARY

Cytokinesis in exponentially-growing mastocytoma cells is accompanied by a reduction in microvilli, suggestive of a redistribution of existing plasma membrane to form the cytokinetic furrow; the actin filaments inside those microvilli that unfold may become the actin filaments of the contractile ring. Such a mechanism appears also to partially account for the formation of the cytokinetic furrow and the contractile ring in Drosophila blastoderm embryos undergoing cellularization.

During the cell cycle of Lettre cells, membrane potential remains virtually constant, despite an increase in the amount of Na^+/K^+-ATPase on which it is dependent. This is consistent with the view that changes in transmembrane signalling do not underlie the progression of continuously-dividing cells through the cell cycle.

INTRODUCTION

The plasma membrane of dividing cells becomes altered both in shape and in composition during each cell cycle. The most obvious change in shape occurs during cytokinesis, when the cytokinetic furrow forms to produce two daughter cells. Is this membrane formed by synthesis of new material, or is it found by a redistribution of existing membranes? Some years ago we proposed that for a somatic cell growing in suspension culture, the latter mechanism is the operative one [Knutton et al., 1975]; the evidence underlying that suggestion will briefly be reviewed. Because cytokinesis in the early stages of embryonic development is a more dramatic one than in somatic cells, in the sense that the generation of up to hundreds of times more new membrane is involved, it is clearly of interest to determine whether a redistribution of existing membrane occurs in that situation also. Recent evidence, as shown below, indicates that in the case of insect embryos undergoing cellularization at the blastoderm stage, both synthesis and redistribution take place.

The trigger for cells to undergo division has been much debated. In the case of quiescent somatic cells, the plasma membrane has been implicated in several ways. First, extracellular signals such as epidermal growth factor (EGF), platelet-derived growth factor (PDGF) and other mitogenic proteins exert their action through an initial binding to specific receptors at the cell surface. Second, that binding event has been shown to induce an alteration of cytoplasmic components through a change in plasma membrane function: alterations in phosphoinositide turnover, in ion transport, in membrane potential, in enzymes such as protein kinase C and nucleotide cyclase, have all been demonstrated.

In the case of malignant cells undergoing repeated cell division, similar mechanisms appear to operate: that is, external signals continually stimulate cells to further rounds of division (a) because the receptors and/or the transducing mechanisms have become amplified and (b) because the dividing cells secrete more growth factor, which taken acts in a local or 'autocrine' manner. Do the transducing mechanisms show changes during the cell cycle, as they do during the stimulation of quiescent cells, or are they permanently 'set' at a higher level, with the actual trigger for successive rounds of cell division being produced within the cells? In the case of early embryo- genesis the trigger is most likely within the cells (once fertilization has occurred), so the answer to the question is an important one in relation to the analogy that has been drawn between the cell cycle during embryogenesis, and the cell cycle in continuously dividing malignant cells. We have suggested that whatever membrane changes occur in somatic cells undergoing repetitive division, they do not directly control the progression from one state of the cell cycle to the next [Pasternak, 1979]. On the other hand one plasma membrane enzyme, namely Na^+/K^+-ATPase, was shown to increase rather dramatically during the cell cycle of an established line of rapidly growing mastocytoma cells [Graham et al., 1973]. Since Na^+/K^+-ATPase directly generates the membrane potential in another line of rapidly growing cells, namely Lettré cells [Bashford and Pasternak, 1984], and since membrane potential is an important parameter in the control of ion movements across the plasma membrane, we have examined Na^+/K^+-ATPase and membrane potential simultaneously during the cell cycle of Lettre cells. The results presented below show that while Na^+/K^+-ATPase increases during the Lettre cell cycle as it does in mastocytoma cells, membrane potential remains virtually constant. This observation is therefore compatible with our proposal that basic functions of the cell surface, such as membrane potential, are not involved in controlling the progression from one phase of the cell cycle to the next.

CYTOKINESIS

Malignant Cells

The amount of plasma membrane that is required when a spherical cell divides into 2 daughter cells is approx. 40% of its surface area just prior to cytokinesis [Pasternak, 1976]. The evidence that this membrane arises by redistribution of existing plasma membrane is based on three observations made with continuously dividing mastocytoma (P815Y) cells. First, the amount of plasma membrane, measured by its major components (proteins, phospholipid and cholesterol), exactly doubles between the start of a new cell cycle (early G1) and the onset of cytokinesis (late G2) [Pasternak, 1979]; it is important to note that this material is isolated from a plasma membrane fraction, not from an intracellular precursor pool fraction [Graham et al., 1973]. Second, morphological assessment by scanning electron microscopy shows that there are approximately 3 times as many microvilli on a late G2 cell as

on an early Gl cell. Since the size of microvilli is the same on a late
G2 cell as on an early Gl cell, the additional microvilli are equivalent
to a 3-fold increase in surface area (from approx. 110 to approx.
300μm^2). At the same time, the volume of the cell doubles, generating
approximately 60% more surface area (from approx. 225 to approx.
350μm^2). When these two components contributing to the amount of plasma
membrane are added together (showing an increase from 335 to 650μm^2
 between early Gl and late G2), they account precisely for the doubling
in amount assessed by chemical measurement [Knutton et al., 1975].
Third, scanning electron microscopy of the occasional cell caught
undergoing cytokinesis shows the poles of the two forming daughter cells
to be relatively devoid of microvilli, whereas the cytokinetic furrow is
still villated (Fig. 1), indicative of a redistribution process as
indicated in Fig. 2. Since microvilli contain bundles of actin
filaments, the possibility that their unfolding liberates such filaments
in order to form the actin-rich contractile ring, makes an attractive
hypothesis [Pasternak, 1976]. Quantitative measurement of the
distribution of F actin before and during cytokinesis would be
instructive in this regard.

A scheme such as that depicted in Fig. 2 ascribes a very dynamic
role to microvilli. Yet microvilli at the luminal surface of epithelial
cells lining the intestine or the renal tubule, - the so-called brush
border, - are clearly rather stable structures. What is the evidence
that microvilli in other cells undergo redistribution of their membrane
in the manner indicated in Fig. 2? First, cells are able to increase
their surface area at the expense of microvilli when swelling as a
result of osmotic changes (e.g. Knutton et al., 1976); the rapidity of
this process, and other considerations, clearly rule out the possibility
that new membrane synthesis, matched by loss of microvilli as vesicles
into the surrounding medium, accounts for the increase in cell size.
Second, a redistribution of membrane at the expense of microvilli occurs
during the capping of lymphocytes: the cross-linking of antibody to
antigenic determinants on the plasma membrane [Taylor et al., 1971], -
or merely exposure of cells to hypertonic medium [Yahara and Kakimoto-
Sameshima, 1977], - forms patches that remain villated; during the
capping process, these microvilli disappear and are replaced by lamellar
'folds' at the cell cap [Yahara, 1982]. The redistribution of membrane
during this process bears a certain similarity to that postulated to
occur during cytokinesis (Figs. 1 and 2).

Embryonic cells

Since early embryos are surrounded by a thick protective coat, the
surface structure of cleaving cells cannot be visualized unless the coat
is first removed. The finding that mild proteolysis [Epel, 1970] or
treatment with dithiothreitol [Epel et al., 1970] is able to do this
without damage to the underlying cells in the case of sea urchin
embryos, has led to the realization that such cells are as highly
villated as any somatic cell [Mazia et al., 1975; Eddy and Shapiro,
1976; Schroeder, 1978; Pasternak, 1979]. So far, correlations between
composition and morphology of plasma membrane, of the kind described
above in the case of cultured mastocytoma cells, does not appear to have
been carried out. What detailed studies have been made, while confirming
the presence of microvilli in the cytokinetic furrow of cleaving sea
urchin eggs, postulate a rather static role for these structures
[Schroeder, 1982].

In the case of cytokinesis in Drosphila embryos, some evidence
for the operation of the type of mechanism outlined in Fig. 2 has

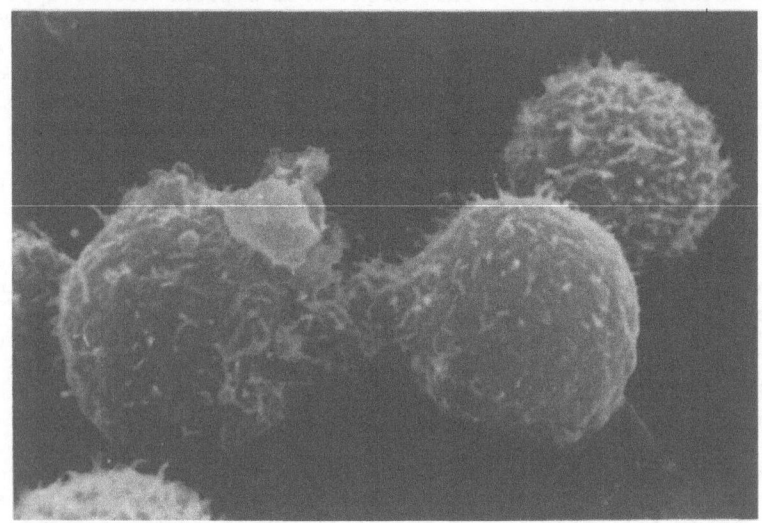

Fig. 1. P815Y mastocytoma cell undergoing cytokinesis; a highly-
villated G2 cell is seen in the upper right corner.
Reproduced from Knutton et al., 1975, by copyright permission
of The Rockefeller University Press.

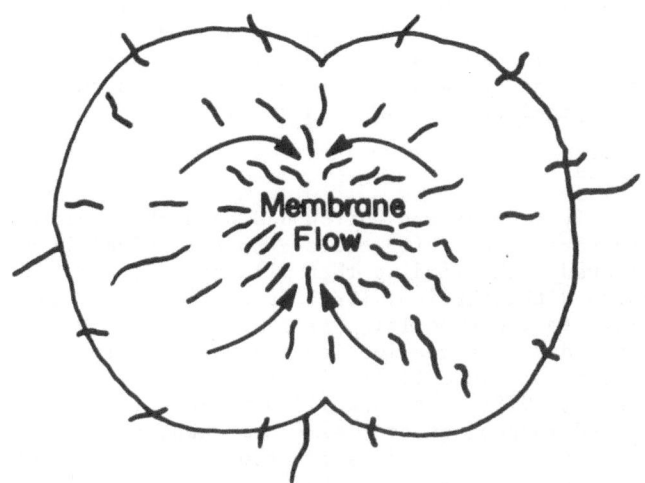

Fig. 2. Redistribution of membrane during cytokinesis of P815Y
cells

emerged. During the transition of syncitial to cellular blastoderm, some thousands of nuclei in the peripheral cytoplasm elongate; at the same time plasma membrane grows inwards around each nucleus to form thousands of discrete cells. It has been suggested [Fullilove and Jacobson, 1971] that this process occurs in two stages: a relatively slow process, during which new plasma membrane is assembled de novo, followed by a relatively fast process, during which microvilli that were originally at the surface of the blastoderm, disappear in order to provide the plasma membrane for the remainder of each elongating cell [Turner and Mahowald, 1976]. In other words cytokinesis in this situation occurs partially by synthesis of new membrane and partially by a mechanism akin to that of Fig. 2.

Direct support for this suggestion by following the transition from syncitial to cellular blastoderm by scanning electron microscopy is not available because it has so far not proved possible to remove the protective vitalline coat from unfixed preparations without damage to the underlying embryo. Instead, a novel technique [Warn and Magrath, 1983] has provided indirect evidence in favour of the model proposed by Fullilove and Jacobson [1971] and Turner and Mahowald [1976]. Briefly, fluorescently-labelled phalloidin, – which specifically stains F actin but not G actin [Wulf et al., 1979], – is introduced to the surface of Drosophila blastoderms at various developmental stages. Fluorescence microscopy of stained preparations at different levels of focus allows visualization of successively deeper layers of blastoderm. The results show that during the early stages of cellularization the blastoderm surface shows intense, diffuse staining, indicating the presence of F actin, – presumably in the form of microvilli, – randomly distributed over the surface of each developing cell. As the process of cellular-ization continues, and each contractile ring spreads deeper into the interior of the blastoderm, the surface of the blastoderm becomes progressively less stainable, while the contours around each developing cell stain intensely, indicating the presence of F actin in the contractile ring [Warn and Magrath, 1983]. In this instance, then, the correlation between the unfolding of microvilli to provide extra plasma membrane, and the concomitant movement of actin filaments from within the microvilli to the site of contractile ring formation, is rather compelling.

MEMBRANE POTENTIAL AND Na^+/K^+-ATPase

The membrane potential of Lettre cells, an established line of malignant cells grown ascitically, is around -50 to -60mV [Bashford and Pasternak, 1984]. Unlike excitable cells of nerve or muscle that contain voltage-sensitive ion channels, the membrane potential of Lettre cells is set directly by the electrogenic Na^+ pump (Na^+/K^+-ATPase). Inhibitors of the enzyme such as ouabain depolarize Lettre cells, while ionophores that increase cytoplasmic Na^+, a substrate of the enzyme, hyperpolarize Lettre cells [Bashford and Pasternak, 1984, 1986]. This behaviour is not a special feature of malignant cells since normal neutrophils, for example, show similar features [Bashford and Pasternak, 1985]. It might therefore be wondered to what extent membrane potential reflects the total amount of plasma membrane Na^+/K^+-ATPase in cells. Since it had been shown [Graham et al., 1973] that the activity of this enzyme more than doubles during the cell cycle of the P815Y mastocytoma cells referred to earlier, it seemed worthwhile to measure the enzyme at different stages of the cell cycle of Lettre cells, and to compare the findings with the membrane potential exhibited by the same cells.

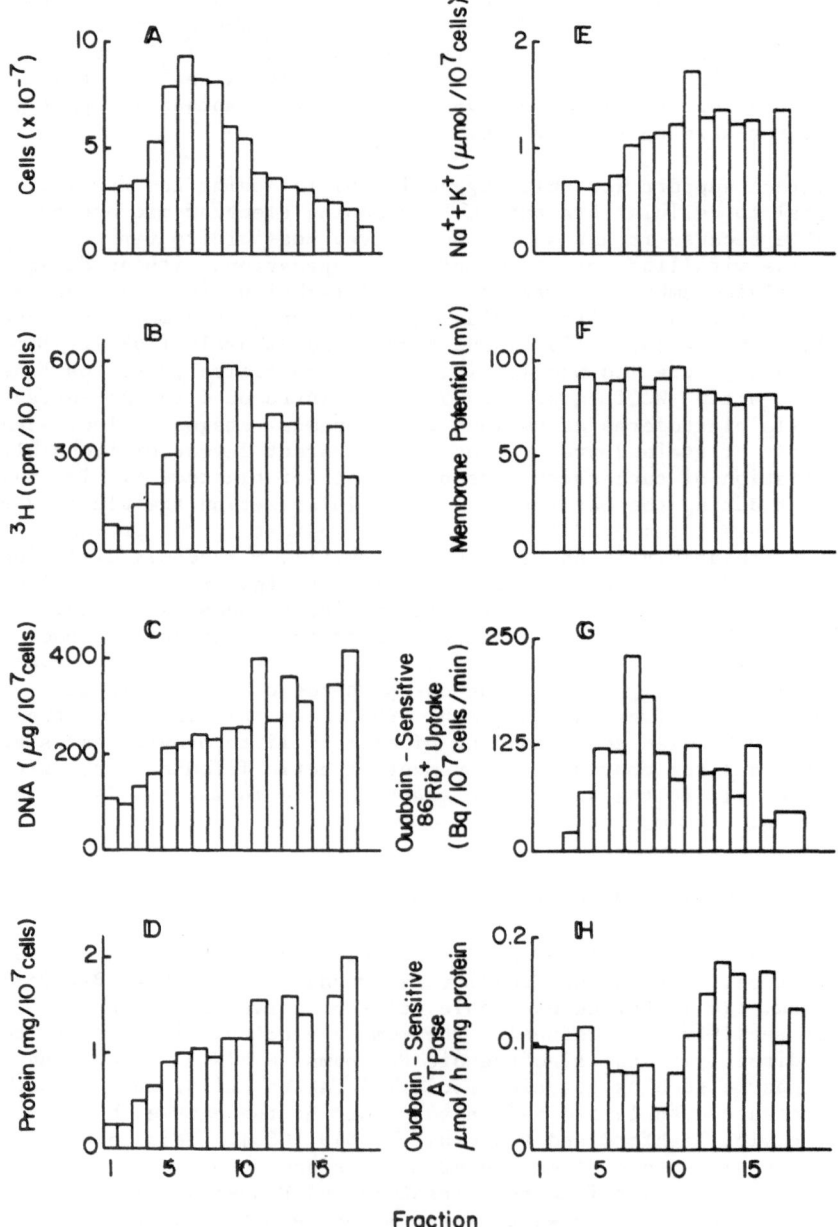

Fig. 3. See opposite page for legend

Fig. 3. Assay of Lettré cells at different stages of the cell cycle.

Approximately 2 x 10^9 Lettré cells, prelabelled in hepes-buffered saline [Impraim et al., 1980] with 100μC of [^{3}H]-thymidine for 15 minutes at 37°, were loaded on to a gradient of 5-10% Ficoll containing 10% metrizoate (900ml) and centrifuged on an MSE type A zonal rotor for approx. 20 minutes at 600 rpm [Warmsley et al., 1970]. Fractions were pumped out of the rotor, washed free of Ficoll and suspended in tris-buffered saline (150mM NaCl, 10mM tris-HCl, pH 7.4).

Cell number (panel A) was determined with a Coulter counter. The content of ^{3}H (panel B), total DNA (panel C) and total protein (panel D) was assessed by conventional techniques following precipitation by ice-cold trichloroacetic acid. Total Na^+ and K^+ (panel E) was determined according to Bashford et al. [1985a] and membrane potential (panel F) according to Bashford et al. [1985b]. Ouabain-sensitive $^{86}Rb^+$ uptake (panel G) was measured by exposing cells to $^{86}Rb^+$ +/- 2mM ouabain and measuring isotope in cell pellets as described by Bashford et al. [1985a]. Ouabain-sensitive ATPase (i.e. Na^+/K^+-ATPase) was measured as described by Graham et al. [1973].

Fractions 1-3 probably correspond to early G1 or G0 cells, or are contaminated with host-derived non-Lettre cells. Fractions 4-6 correspond to later G1 cells. Fractions 7-10 correspond to S cells. Fractions 11-16 to a mixture of S and G2 cells, and fractions 17-18 to predominantly G2 cells.

The results shown are taken from a single experiment, carried out in collaboration with C.L. Bashford, J.M. Graham, J.A. Vincent and R.B.J. Wilson, and are in general typical of several experiments performed over a month with the same line of Lettré cells.

Rapidly-growing Lettre cells were removed from the peritoneal cavity of mice that had been inoculated with 10^7-10^8 Lettre cells a week previously. Following a brief incubation with [^{3}H]thymidine in order to label cells in S phase, cells were separated according to size by zonal centrifugation [Pasternak,1973]. Fig.3 panels A and B show that the cell peak does not coincide with the peak of [^{3}H]thymidine incorporation; that is, nearly half the cells do not incorporate [^{3}H]thymidine and are smaller than those, - in S phase, - that do. Panel D shows that total cell protein increases along the gradient, and panel E shows the same for the total content of Na^+ plus K^+ (which reflects the cytoplasmic water content [Bashford et al., 1985]), with an approximate doubling in amount between fractions 3-6 and fractions 11-17; that is, of course, the anticipated result for cells separated according to size. The important point is that the content of DNA (panel C) also approximately doubles between fraction 4 and fractions 11-17. This shows that larger cells are indeed further in the cell cycle than smaller cells; in other words variation in size reflects physiological age of cells, not merely a random heterogeneity of size among the population. It will be noted that some cells (fractions 1-3) are particularly low in DNA and protein; whether these are a population of non-cycling G0 cells, or whether they represent some contamination with non-Lettre cells, remains to be established. Certainly some of their properties are different from those of the rest of the population.

Measurement of total Na^+/K^+-ATPase, that is the ouabain-sensitive ATPase (panel H), shows an approximate doubling between fractions 1-10 and fractions 11-17. If the low value in fraction 9 is real, the sharp rise between that fraction and fraction 13 is reminiscent of the sharp increase in plasma membrane Na^+/K^+-ATPase seen with P815Y cells [Graham et al., 1973], and indicates, together with other evidence, that total Na^+/K^+-ATPase in Lettre cells at different stages of the cell cycle indeed reflects the activity of the plasma membrane Na^+/K^+-ATPase. What is important to note is that the membrane potential (panel F) remains virtually constant between fractions 3-17; if anything, it decreases slightly in the larger (G2) cells. In other words, although the membrane potential of Lettre cells depends directly on the activity of the plasma membrane Na^+/K^+-ATPase, the precise level at which it is set depends on other factors. A major one of these is the extent to which a balancing leak of anions out of the cell accompanies the outward pumping of a Na^+ ion (an electrogenic Na pump in the absence of such a balancing leak would hyperpolarize a cell to the point at which the bilayer collapses) [Bashford and Pasternak, 1984, 1986]. Interestingly, the ouabain-sensitive uptake of $^{86}Rb^+$, which is a reflection of the activity of the Na^+ pump in intact cells, shows a pattern (panel G) more like that of membrane potential (panel F) and not at all like that of Na^+/K^+-ATPase measured in extracts (panel H). The low uptake in fraction 3 may be due to the presence of G0 or non-Lettre cells (it is similar in extent to $^{86}Rb^+$ uptake by human peripheral lymphocytes [C.L. Bashford and C.A. Pasternak, unpublished observations]). The high uptake in fractions 7 and 8, and the low uptake in fractions 11 onwards, probably reflect experimental variation; they are not a consistent feature from one experiment to another. It is pertinent to note that in Chinese hamster lung cells, membrane potential measured by intracellular electrophysiological recording also shows a maximum during S and G2, with lower values in G1 and just prior to mitosis [Sachs et al., 1974].

The conclusion to be drawn from these results is that membrane potential reflects the 'texture' of the plasma membrane and that this, - like the composition of the constituent phospholipids [Pasternak, 1979] or the distribution of intramembranous particles [Knutton, 1976],

- remains constant throughout the cell cycle, in contrast to the
amount of plasma membrane, which gradually doubles between G1 and G2.
This conclusion is consistent with the view [Pasternak, 1979] that
whatever triggers are responsible for the progression of continuously-
dividing cells through the cell cycle, including the initiation of
cytokinesis, they are of intracellular origin, and that a change in
extracellular signalling across the plasma membrane is not involved.

ACKNOWLEDGEMENTS

I am grateful to Dr R.M. Warn for much helpful discussion, to Drs
J.M. Graham and C.L. Bashford for collaborative experiments, to Mrs V.
Marvell and Mrs B. Bashford for help in preparation of this article and
to the Cancer Research Campaign and Cell Surface Research Fund for
financial assistance.

REFERENCES

Bashford, C.L., and Pasternak, C.A., 1984, Plasma membrane potential of
    Lettre cells does not depend on cation gradients but on pumps,
    J.Membr.Biol., 79:275.
Bashford, C.L., and Pasternak, C.A., 1985, Plasma membrane potential of
    neutrophils generated by the Na+ pump, Biochim.Biophys.Acta, 817:
    174.
Bashford, C.L., and Pasternak, C.A., 1986, Plasma membrane potential of
    some animal cells is generated by ion pumping, not by ion gradients,
    TIBS, 11:113.
Bashford, C.L., Alder, G.M., Gray, M.A., Micklem, K.J., Taylor, C.C.,
    Turek, P.J., and Pasternak, C.A., 1985a, Oxonol dyes as monitors of
    membrane potential: The effect of viruses and toxins on the plasma
    membrane potential of animal cells in monolayers, J.Cell.Physiol.,
    123:326.
Bashford, C.L., Micklem, K.J., and Pasternak, C.A., 1985b, Sequential
    onset of permeability changes in mouse ascites cells induced by
    Sendai virus, Biochim.Biophys.Acta, 814:247.
Eddy, E.M., and Shapiro, B.M., 1976, Changes in the topography of the
    sea urchin egg after fertilization, J.Cell.Biol., 71:35.
Epel, D., 1970, Methods for removal of the vitelline membrane of sea
    urchin eggs. II Controlled exposure to trypsin to eliminate
    post-fertilization clumping of embryos, Exp.Cell Res., 61:69.
Epel, D., Weaver, A.M., and Mazia, D., 1970, Methods for removal of the
    vitelline membrane of sea urchin eggs. I Use of dithiothreitol
    (Cleland Reagent), Exp.Cell Res., 61:64.
Fullilove, S.L., and Jacobson, A.G., 1971, Nuclear elongation and cyto-
    kinesis in Drosophila montana, Dev.Biol., 26:560.
Graham, J.M., Sumner, M.C.B., Curtis, D.M., and Pasternak, C.A., 1973,
    Plasma membrane assembly: sequence of events during the cell cycle,
    Nature, 246:291.
Impraim, C.C., Foster, K.A., Micklem, K.J., and Pasternak, C.A., 1980,
    Nature of virally-mediated changes in membrane permeability to small
    molecules, Biochem.J., 186:847.
Knutton, S., Sumner, M.C.B., and Pasternak, C.A., 1975, Role of micro-
    villi in surface changes of synchronized P815Y mastocytoma cells,
    J.Cell Biol., 66:568.
Knutton, S., 1976, Structural changes in the plasma membrane of
    synchronozed P815Y mastocytoma cells, Exp.Cell Res., 102:109.
Knutton, S., Jackson, D., Graham, J.M., Micklem, K.J., and Pasternak,
    C.A., 1976, Microvilli and cell swelling, Nature 262:52.

Mazia, D., Schatten, G., and Steinhardt, R., 1975, Turning on of activities in unfertilized sea urchin eggs: Correlation with changes of the surface, Proc.Nat.Acad.Sci.USA, 72:4469.

Pasternak, C.A., Synchronization of mammalian cells by size separation, in: "Methodological Developments in Biochemistry", Vol. 4, A.I. Laskin, and J.A. Last, eds., Marcel Dekker Inc., New York (1973).

Pasternak, C.A., 1976, The cell surface in relation to the growth cycle, J.Theor.Biol., 58:365.

Pasternak, C.A., The cell surface and the cell cycle, in, "The Biochemistry of Cellular Regulation", Vol. IV, P. Knox, ed., CRC Press Inc., Boca Raton, Florida (1979); note that the word 'Cell' should replace 'All' in both parts of Fig. 9.

Sachs, H.G., Stambrook, P.J., and Ebert, J.D., 1974, Changes in membrane potential during the cell cycle, Exp.Cell Res., 83:362.

Schroeder, T.E., 1978, Microvilli on sea urchin eggs: a second burst of elongation, Dev.Biol., 64:342.

Schroeder, T.E., 1982, Interrelations between the cell surface and the cytoskeleton in cleaving sea urchin eggs, in: "Cell Surface Reviews", Vol. 7, G. Poste, and G.L. Nicolson, eds., North Holland Publishing Co., Amsterdam.

Taylor, R.B., Duffus, W.P.H., Raff, M.C., and Petris, S., 1971, Redistribution and pinocytosis of lymphocyte surface immunoglobulin molecules induced by anti-immunoglobulin antibody, Nature New Biol, 233:225.

Turner, F.R., and Mahowald, A.P., 1976, Scanning electron microscopy of Drosophila embryogenesis. I The structure of the egg envelopes and the formation of the cellular blastoderm, Dev.Biol., 50:95.

Warmsley, A.M.H., Phillips, B., and Pasternak, C.A., 1970, The use of zonal centrifugation to study membrane formation during the life cycle of mammalian cells. Synthesis of 'marker' enzymes and other components of cellular organelles, Biochem.J., 120:683.

Warn, R.M., and Magrath, R., 1983, F-actin distribution during the cellularization of the Drosophila embryo visualized with FL-phalloidin, Exp.Cell Res., 143:103.

Wulf, E., Deboben, A., Bautz, F.A., Faulstich, H., and Wieland, T., 1979, Fluorescent phallotoxion, a tool for the visualization of cellular actin, Proc.Nat.Acad.Sci.USA, 76:4498.

Yahara, I., 1982, Transmembrane control of the mobility of surface receptors by cytoskeletal structures, in: "Structure, Dynamics and Biogenesis of Biomembranes", R. Sato, and S-I. Ohnishi, eds., Plenum Press, New York.

Yahara, I., and Kakimoto-Sameshima, F., 1977, Ligand-independent cap formation: Redistribution of surface receptors and thymocytes in hypertonic medium, Proc.Nat.Acad.Sci.USA, 74:4511.

POLYMERS AND RUBBER ELASTICITY: THERMODYNAMICS OF LARGE

DIMENSIONAL CHANGES IN AMORPHOUS SYSTEMS

Burak Erman

School of Engineering
Boğaziçi University
Bebek 80815, Istanbul, Turkey

INTRODUCTION

Polymeric molecules constitute a large fraction of the living material. They may be in the form of a single macromolecule dissolved in a suitable fluid, or may exist as a three dimensional topological network swollen – to equilibrium – with the surrounding fluid. In either case, the polymer-solvent system constitutes a semi-open thermodynamic system where the solvent molecules of much smaller size may enter or leave the space pervaded by the polymeric molecules. The size of the polymer-solvent system changes upon transport of the solvent. At equilibrium, the size of the polymer-solvent system is determined by the equality of the activity of the solvent in the polymer-solvent system to that in the surrounding region, which in most cases is the pure solvent. The solvent activity depends predominantly on   i) the constitution of the polymeric chains and the network, and   ii) the thermodynamic interaction between the solvent molecules and the polymer. Presence of solvent in a polymeric network dilates the configurations of the macromolecules constituting the network. On the other hand, the connectivity of the network chains opposes dilation. The balance between these two opposing effects determines the equilibrium degree of swelling of the network.

In a "good solvent", the polymeric network (or the single chain) exists in a highly swollen state. The quality of the solvent may be changed by modifying the factors governing the interaction between the solvent and polymer. The pH and the ionic strength of the solvent form the two most common factors. In a "poor solvent", the elastic potential of the network chains dominates, and the concentration of solvent in the polymeric medium at equilibrium may be vanishingly small. The ratio of volumes of the polymeric system in the highly swollen state to that in the unswollen state may be as high as 1000.

The transition between a highly swollen to the unswollen state by changing the quality of the solvent may be continuous or discrete. The latter may be of major interest in biological systems. Throughout the various stages of swelling or deswelling, various mechanical properties of the polymer-solvent system display large-scale variations. The equilibrium osmotic compressibility and the permeability of the system are two properties that change significantly with solvent content. Consequently, the mechanical properties of the polymer-solvent system in the highly swollen state and the "collapsed" unswollen state, show discrete differences and have to be accounted for in mechanical calculations.

Considerable improvements have been made over the last decades in the understanding of the behavior of polymer-solvent systems. Foundations of the molecular theory of rubber elasticity and solutions of polymeric systems have been outlined by Flory and collaborators[1-4]. Experimental and theoretical analyses have shown that various thermodynamic processes may be related to the constitution of the macromolecular system and to experimental conditions. In the following paragraphs, the progress in the field is summarized and the thermodynamics of finite dimensional changes in a network-solvent system is analyzed.

THE NETWORK-SOLVENT SYSTEM

The polymeric network is a three dimensional structure formed by linking individual chains. The most common method of linking the molecules is by covalently joining few reactive groups along the chains by a cross-linker molecule or by radiation. The sites of reactive groups may be randomly distributed along the chains, or they may be situated at specific sites. Among the latter group are chains with reactive groups at the ends. Cross-linking in this manner produces end-linked networks. The linearity of the individual molecules, although common, is not required. Star or branched polymers may also be cross-linked to form a three dimensional network.

A network chain is defined as the polymer that joins two junctions. Depending on the degree of cross-linking, network chains may range from 10 to over 1000 skeletal bonds, the length of a bond being in the order of 0.1 nm. The former value of bonds corresponds to a highly cross-linked, "stiff" network of very limited extensibility. The modulus of such a network shows rapid increase upon deformation. A network with about 1000 or more bonds between junctions, on the other hand, is very soft and shows "gell-like" behavior. Representative chain lengths in networks are in the range of 400-500 bonds[5]. The average molecular weight of a network chain corresponding to this range of bonds is in the order of 10^4. The junctions are ordinarily tetrafunctional where four chains are attached at each cross-link. The junctions may, however, have values of three or more, depending on the structure of cross-linker.

The structure of a network may be characterized by a few parameters. Thus, networks may be considered to consist of ν chains whose ends are joined to multifunctional junctions of any functionality $\phi > 2$. The number of junctions is μ. In an "imperfect" network, some of the chains are joined to junctions at only one end, the other end being free (mono functional). The network consists of ξ cycles. ξ is termed as the cycle rank, which equals to the minimum number of chains to be cut to reduce the network to a tree. The topological structure of a network may conveniently be described in terms of graph-theory language[7].

The number of chains and junctions are not independent of the cycle rank and the junction functionality. Two relationships exist among the four variables:

$$\nu = \mu\phi/2$$

$$\nu = \xi/(1 - 2/\phi)$$

(1)

If junctions of different functionalities are present, (with $\phi \geq 3$), then ϕ may be construed as the average functionality.

The chains between the junctions may be of varying flexibility. The basic requisite of the rubbery state of a network is the presence of sufficient degree of mobility of the segments of the chains. The only postulate of the molecular theory of rubbery elasticity is this mobility.

The end-to-end length of a chain in a network is assumed to be Gaussian according to theory. Although this assumption is valid for sufficiently long chains of any constitution, it may not be approximated by short chains. For networks with chains of about 10-20 bonds, corrections for the non-Gaussian nature of chains are required. The following treatment is confined to longer Gaussian chains.

The root mean square end-to-end length $<r^2>^{1/2}$ for the chain of 400 bonds in an undistorted network is in the order of 7 nm. A spherical domain of this radius contains about 40-50 junctions. The junctions are not stationary in the structure and fluctuate. Instantaneously, the root mean square fluctuations, $<(\Delta r)^2>^{1/2}$ of junctions in a tetrafunctional network from their mean positions is about 4-5 nm. Thus, fluctuations of chain dimensions and junction positions are substantial in a polymeric network[5].

The molecular theory of elasticity of polymeric networks treats the problem in two steps. First, the fluctuations of junctions are assumed to take place without any consideration of the entanglements to which the chains and junctions are subject as a result of molecular interpenetration. In this picture, the chains cross each other in a phantom-like manner, without any effect of the exclusion of volume exerted by the neighboring molecules. Such a network is aptly termed[5] as the "phantom network". The effect of physical entanglements is brought into the picture at a second stage, where the junctions are constrained to move in smaller domains than would be obtained in the phantom structure. The severity of entanglements in a real network may be represented by the ratio,

$$\kappa = <(\Delta R)^2_{ph}>/<(\Delta R)^2_r> \tag{2}$$

where the numerator represents the mean squared fluctuation of junctions from their mean locations in the hypothetical phantom network. The denominator represents the mean squared dimensions of the constraint domains surrounding the junctions in the real network. Defined in this manner, κ varies between 0 and ∞, the former being obtained in the absence of constraints and the latter in severely constrained networks where the fluctuations are totally supressed. The parameter κ, measuring the constraints on fluctuations of junctions due to the surrounding chains in which they are embedded, should depend on the degree of interpenetration of the network structure[8]. The degree of interpenetration is measured by the number of junctions occuring within the domain pervaded by the ϕ chains emanating from a given junction. Thus

$$\kappa = I<r^2>_o^{3/2}(\mu/V^o) \tag{3}$$

where I is a numerical constant, $<r^2>_o$ is the mean squared length of an unperturbed network chain, and μ/V^o is the number of junctions per unit volume, V^o denoting the volume of the network in the state of reference. According to the theory, the length of a chain is expressed by the number x_c of segments it contains, a segment being defined as the portion of a chain whose volume equals that of the solvent molecule in the network-solvent system. Thus

$$x_c = n\overline{v}M_\ell/V_1 \tag{4}$$

where n is the number of chemical bonds of the chain, V_1 is the molar volume of the solvent, $\overline{v}$ is the specific volume of the polymer and M_ℓ is the molecular weight per bond. The number of network chains and junctions are also related to x_c. For a tetrafunctional network, this relation reads as

$$\mu = \nu/2 = v_2^o v^o N_A / 2 x_c V_1 \qquad (5)$$

where v_2^o is volume fraction of polymer in the polymer-solvent system during cross-linking, and N_A is the Avogadro number. Calculations and use of experimental evidence lead to an expression for κ for tetrafunctional networks as[6]

$$\kappa = (1/4) p v_2^o x_c^{1/2} \qquad (6)$$

where p is a dimensionless parameter defined by

$$p = (C_\infty \ell^2 / \bar{v} M_\ell)^{3/2} V_1^{1/2} N_A \qquad (7)$$

Here, C_∞ is the characteristic ratio of the chain, ℓ is the bond length, and M_ℓ is the molecular weight per unit bond. p depends only on characteristics of the generic type of polymer and on the molar volume of the diluent.

The final parameter that contributes to the network property is the number, i, of ionizable groups per chain.

The complete network is thus characterized by the four independent parameters ξ, μ, κ, and i. κ depends on the generic nature of the polymer and the solvent, ξ and μ are network parameters which are fixed when the network is constructed, and i depends on the degree of ionization of the chains.

The degree of swelling at equilibrium in the network-solvent system may be expressed by the volume fraction, v_2, of polymer. This quantity may be related to the linear dilation ratio, λ, by

$$\lambda = (V/V^o)^{1/3} = (v_2^o/v_2)^{1/3} \qquad (8)$$

where V and V^o are the prevailing volume of the network and the volume in the state of reference at which the network is constructed, respectively, and v_2 and v_2^o are the corresponding volume fractions of polymer.

THE TOTAL FREE ENERGY AND THE CHEMICAL POTENTIAL OF THE SYSTEM

The total Helmholtz free energy, ΔA, of the network-solvent system may be obtained by the sum of two components, the Helmholtz free energy of mixing, ΔA_m, of polymer chains with solvent and the elastic free energy, ΔA_{el}, as

$$\Delta A = \Delta A_m + \Delta A_{el} \qquad (9)$$

The additivity of the mixing and elastic free energies stated by Eq. (9) has previously been shown, by experiments, to be valid for non-ionic systems. The additivity may be assumed to hold also for systems where the fraction of structural units bearing ionized substituents on network chains is very small.

The free energy of mixing of polymer chains with solvent may satisfactorily be represented by the Flory-Huggins theory according to which,

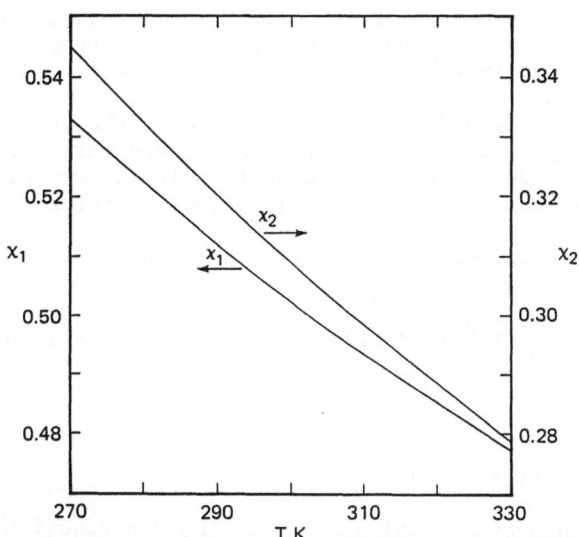

Figure 1. Values of X_1 (left ordinate) and X_2 (right ordinate)
calculated as a function of temperature for the
system cyclohexane-polystyrene. Calculations are
performed according to the equation-of-state formulation for polymer solutions and is described in
References 3 and 6. According to the theory, X_1
and X_2 are functions of the thermal expansivities,
characteristic volumes, pressures and temperatures
of the components, of the ratio of surface areas
for a segment of the polymer and solvent, of the
exchange free energy and entropy for formation of
a contact between a solute segment and a solvent
molecule, and of temperature, pressure and degree
of ionization.

$$\Delta A_m = kT(n_1 \ln(v_1) + n_2 \ln(v_2) + X n_1 v_2) \tag{10}$$

where n_1 and n_2 are the numbers of solvent and polymer molecules, respectively, v_1 is the volume fraction of solvent, k is the Boltzmann constant, T is the temperature and X is the parameter representing the interaction between the solvent and polymer. In general X depends on the composition as well as on the temperature. It may be expressed by the first two terms of the Taylor series expansion, as

$$X = X_1 + X_2 v_2 \tag{11}$$

where X_1 and X_2 are functions of temperature[3]. For various synthetic polymers X_1 and X_2 have been determined experimentally, over wide ranges of temperature[3,4]. The dependence of X_1 and X_2 on temperature for the system cyclohexane-polystyrene[6] is shown in Figure 1.

The parameters X_1 and X_2 play major role in the volume changes of network-solvent systems. Although they have been carefully and precisely determined for a wide variety of synthetic polymer-solvent systems, experimental data is scarce for biological polymeric systems.

The elastic free energy of the network is obtained according to statistical theory[6,8] as

$$\Delta A_{el} = (3/2)kT\{\xi(\lambda^2-1) + \mu(1+\lambda^2/\kappa)B - \mu\ln[(1+B)(1+\lambda^2 B/\kappa)]\} + \Delta A_i \tag{12}$$

where

$$B = (\lambda^2 - 1)/(1 + \lambda^2/\kappa)^2$$

and λ is as defined by Eq. (8) and ΔA_i is the contribution of ionizable groups on chains to network potential (see sequel). It is seen from Eq. (12) that the elastic behavior is determined in terms of the set of parameters ξ, μ, κ, and i. and the linear dilation ratio, λ.

The elastic free energy given by Eq. (12) reduces to the neo-Hookean form if $\kappa = 0$. For nonzero values of κ, and for the general three dimensional form of the deformation gradient tensor, λ, Eq. (12) has been shown to be in satisfactory agreement with experiments of diverse nature[9].

The excess chemical potential, $\Delta\mu_1$, for the solvent in the network-solvent system relative to the surroundings may be written as[1]

$$\Delta\mu_1 = RT \frac{\partial \Delta A}{\partial n_1} = RT\left(\frac{\partial \Delta A_m}{\partial n_1} + \frac{\partial \Delta A_{el}}{\partial n_1}\right) \tag{13}$$

where n_1 is the number of solvent molecules. Substituting from Eqs. (10) and (12), after performing the differentiation, one obtains

$$\Delta\mu_1 = RT\{\ln(1 - v_2) + v_2 + X_1 v_2^2 + X_2 v_2^3$$
$$+ (V_1/N_A V^0 \lambda)[\xi + \mu K(\lambda)] - i v_2/x_c\} \tag{14}$$

where

$$K(\lambda) = B[\dot{B}(1 + B)^{-1} + (\lambda/\kappa)^2(B + \lambda^2\dot{B})(1 + \lambda^2 B/\kappa)^{-1}]$$

with

$$\dot{B} = \partial B / \partial \lambda^2$$

$$= B \ (\lambda^2 - 1)^{-1} - 2(\lambda^2 + \kappa)^{-1}$$

The term $-iv_2/x_c$ in Eq. (14) is obtained by differentiation[1] of the term ΔA_i of Eq. (12) with respect to n_1.

At equilibrium of the network-solvent system with the surrounding pure solvent, Eq. (14) equates to zero (the activity of the pure solvent being unity). Solution of the resulting equation then leads to the determination of the equilibrium degree of swelling, $\lambda^3 = (V/V^o) = v_2^o/v_2$. The degree of equilibrium swelling is thus a function of the two parameters χ_1 and χ_2. Changing the thermodynamic conditions of the surrounding pure solvent affects χ_1 and χ_2. For a given set of χ_1, χ_2, the degree of equilibrium swelling of the network may be obtained by solving Eq. (14).

CRITICAL PHENOMENA AND PHASE TRANSITIONS IN NETWORK-SOLVENT SYSTEMS

Equation (14), when equated to zero, is in general a single valued function of v_2. For certain values of χ_1 and χ_2, however, critical conditions may obtain, at which the first and second derivatives of the chemical potential, taken with respect to composition, vanish. Thus,

$$\Delta\mu_1 = 0$$

$$\partial\Delta\mu_1/\partial v_2 = 0 \tag{15}$$

$$\partial^2\Delta\mu_1/\partial v_2^2 = 0$$

These three equations may be solved for the three unknowns, v_2 (or, equivalently, λ), χ_1 and χ_2. The solution set thus gives the values of the swelling ratio and χ_1 and χ_2 required for critical conditions. Inasmuch as Eqs. (15) are nonlinear in v_2, they may conveniently be solved by numerical techniques. Results of such calculations are shown in Figure 2 for $p = 1.5$. (This value of p being the commonly encountered one for synthetic polymer-solvent systems.) Each curve describes critical conditions for a given number of ions on a network chain. Each point on a curve represents the χ_1, χ_2 values required for critical conditions for a chain length x_c. Various chain lengths are indicated by points on the curves. Point A is the limit point for a network with $x_c = \infty$. Point B, for example, shows that for a network with chain length of $x_c = 300$, on which there are 5 ionic groups, the values of χ_1 and χ_2 necessary for critical conditions to exist are 0.59 and 0.17, respectively. In a network with these values of the parameters, critical conditions hold and the compressibility and fluctuations become infinitely large.

The regions exterior to the convex domain delineated by the critical curves given for several values of i in Figure 2 may be referred to as the triphasic regions. In the triphasic region, the system may coexist in three phases at equilibrium; the pure solvent, a highly dilute and a concentrated phase. The conditions for triphasic equilibrium are

$$\Delta\mu_1(v_2) = 0$$

$$\Delta\mu_1(v_2') = 0 \tag{16}$$

$$\int_{v_2}^{v_2'} \Delta\mu_1 v_2^{-2} dv_2 = 0$$

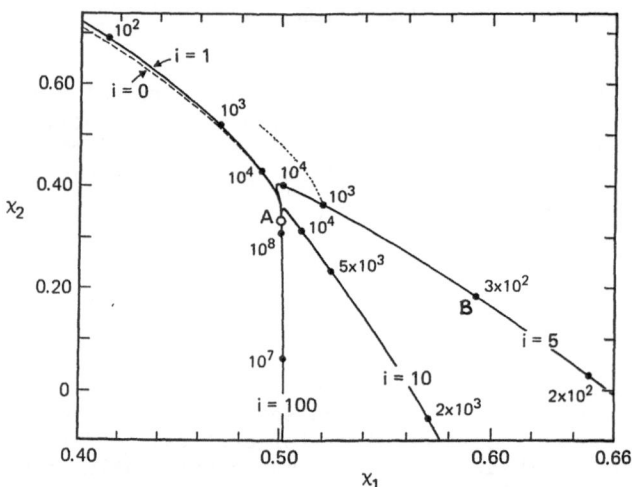

Figure 2.   Relationship between χ_1 and χ_2 required for critical behaviour
in ionic networks. Point A is the limiting point for a network
with $x_c = \infty$. Values of i and of the representative chain lengths
x_c are indicated on each curve. The dashed curve refers to the
nonionic network. The dotted curve indicates the values of χ_1
and χ_2 for triphasic equilibrium for a network with $x_c = 10^3$
and i= 1.5. All calculations are for P=1.5.

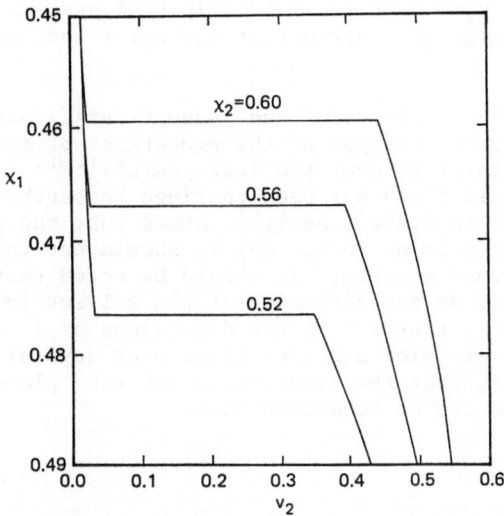

Figure 3.   Illustrative calculations of phase equilibria
for a network with $v_2^0 = 0.035$.

where v_2 and v_2' are polymer concentrations in the concentrated and dilute phases, respectively. Numerical solution of Eqs. (16) leads to values of v_2 and v_2' in terms of the parameters X_1 and X_2. In the X_1-X_2 plane of Figure 2, the solution set corresponds to a triphasic curve, as shown, for example, by the dotted line. A point on this curve corresponds to a first order phase change. As the point (X_1,X_2) of the system moves through the triphasic curve, the network discontinuously and reversibly changes from a highly swollen to a collapsed state, or vice versa.

Illustrative calculations for such a phase change are shown in Figure 3 for a network with $p = 1.5$, $x_c = 10^3$ and cross-linked at $v_2^0 = 0.035$. For simplicity, X_2 is assumed to be constant, and X_1 is varied. The upper curve is for $X_2 = 0.60$. For values of X_1 larger than 0.459, the network is in a highly swollen state. Polymer concentration in the system is less than 3%. When X_1 becomes larger than 0.459, the network suddenly and discontinuously shrinks, expelling a large volume of solvent to its surroundings. The volume fraction of polymer increases to more than 40% as a result of this process. With increasing X_1, the network further expells solvent, in a gradual and continuous manner. The remaining two curves are obtained similarly for $X_2 = 0.56$ and 0.52. In real systems, X_2 may not be constant. This however does not effect the nature of the resulting transition.

Occurrence of critical phenomena and phase transitions in networks upon temperature changes or changes of the properties of solvent has been of great experimental interest over the last years[6,10,11]. Although the experimental work in this field has been confined to synthetic polymeric systems, the present thermodynamic analysis shows that the phenomenon is universal, and that there is no reason why it should not take place in biological network-solvent systems. It should be noted that the finite transition is controlled by the diffusion of the solvent in and out of the network. In networks that are of the dimensions of 1 cm., or larger the transition may be very slow and take place over several hours or days. Analysis shows, however, that the transitions may take place within microseconds if the networks are of submicron size.

CONCLUSION

In the above paragraphs, a molecular theory is presented according to which first order transitions are predicted in network-solvent systems. Necessary conditions for the occurence of such a transition is formulated in terms of the interaction parameter, X, reflecting the quality of the solvent in which the system is immersed. That such transitions are physically possible have recently been shown by the detailed experiments[10,11]. All the present experiments in which the networks are shown to undergo transitions are of synthetic polymeric molecules, however. The X parameter for biological polymer-solvent systems are not well characterized. However, it may confidently be stated from the examination of Figure 2 that the necessary X_1 and X_2 values for transitions in a biological network-solvent system falls within experimentally accessible values, these values being in the vicinity of 1/2 for X_1 and 1/3 for X_2.

ACKNOWLEDGEMENT

This study has been partly supported by the Research Fund of Bogazici University.

REFERENCES

1. P.J. Flory, "Principles of Polymer Chemistry", Cornell University Press, Ithaca, New York (1953).
2. P.J. Flory and Y. Tatara, J. Polym. Sci. Polym. Phys. Ed., 13:683 (1975).
3. P.J. Flory, Disc. Faraday Soc., 49:7 (1970).
4. B.E. Eichinger and P.J. Flory, Trans. Faraday Soc., 64:2035 (1968).
5. P.J. Flory, Proc. R. Soc. Rond. A, 351:351 (1976).
6. B. Erman and P.J. Flory, to appear in Macromolecules, Sept. 1986.
7. F. Harrary, "Graph Theory", Reading, Mass., U.S.A.: Addison-Wesley (1971).
8. P.J. Flory and B. Erman, Macromolecules, 15:800 (1982).
9. B. Erman and P.J. Flory, Macromolecules, 15:806 (1982).
10. T. Tanaka, Phys. Rev. Lett., 40:820(1978).
11. M. Ilavsky, Polymer, 22:1687 (1981); 15:782 (1982).

THE ORIGIN AND ACTION OF THE CONTRACTILE RING

Thomas E. Schroeder

Friday Harbor Laboratories, University of Washington
620 University Road
Friday Harbor, Washington USA 98250

INTRODUCTION

In animal cells (and a few selected plant cells) cell division is
concluded by formation of a cleavage furrow that constricts the entire
cell, ultimately pinching it into two parts (Figure 1). Cell cleavage
by constriction is one of two general strategies of cytokinesis. In
most plant cells a phragmoplast directs a restructuring process that
begins inside the cell, resulting in an outward-growing wall. Each
mechanism of cytokinesis is coordinated with mitosis (or meiosis), but
the actual mechanisms of cytokinesis are known to be quite separate from
the mechanisms that causes chromosome separation.

This chapter examines the nature and origin of contractile ring,
the cytoskeletal specialization that is widely accepted to be the
structure that causes the cytokinetic constriction known as the cleavage
furrow. At the present stage of knowledge there is a fair amount of
specific data about the structure, composition and behavior of the
contractile ring, some of them quite enigmatic, but substantially less
is precisely known about its origin. Accordingly, the exploration of
origin of the contractile ring includes searching for clues among the
cytological events in the mitotic apparatus and cell cortex that precede
the onset of cleavage. Many of the unreferenced details mentioned here
may be found in earlier reviews (Wolpert, 1960; Rappaport, 1971;
Schroeder, 1975; Arnold, 1976; Beams and Kessel, 1976; Hiramoto, 1981;
Sisken, 1981; Conrad and Rappaport, 1981; Schroeder, 1981a).

General points-of-view in this chapter include the following
propositions: that cell cleavage is a complex biological phenomenon
involving interrelations of morphological, biochemical, physiological,
and biomechanical processes; that the mechanism of cleavage is best
elucidated by observation and experimentation of the widest possible
variety of cells; that it is difficult to distinguish which details of
a cleaving cell might reflect a universal mechanism common to all cell
types and which details betray redundancies and idiosynchrases in the
mechanism; and, where concrete facts are unavailable, that induction
and conjecture based upon demonstrable facts are more useful guides to
reality than formalized theorizing or model-building based on inanimate
materials.

The contractile ring exhibits a structural organization that
is intimately suited to its putative mechanical action, which is to
exert active tensile forces against the cell membrane of the cleavage
furrow. Once established in the right part of the cell at the right
time, it continues to exist and function until the cell is completely
constricted. Much of the organization and behavior of the contractile
ring suggests that its physiological action is, at least superficially,
closely related to muscle contractility. On the other hand, the analogy
with muscle should not be accepted too literally, for some of the
contractile ring's most interesting and enigmatic properties remain
unacknowledged and poorly understood. For example, the muscle analogy
fails to elucidate: how the contractile ring can be so short-lived,
seeming to assemble and disassemble in 10 minutes; how it contracts down
to nothing; how unilateral furrows elongate; or how the multiple,
intersecting furrows of the cellularization process in insect embryos
progress into the cell.

The existence of the contractile ring was specifically predicted
many years before it was first concretely described about 20 years ago.
That early prediction (see Marsland and Landau, 1954; Wolpert, 1961)
asserted that the cytoplasm beneath a cleavage furrow, when explored in
detail, would prove to be physiologically and structurally specialized
for contraction. The proposition grew out of an empirical tradition of
carefully observing and manipulating living cells (mostly sea urchin
eggs), and it provided a reasonable and realistic explanation of the
cleavage mechanism, even though it lacked specific supporting data. It

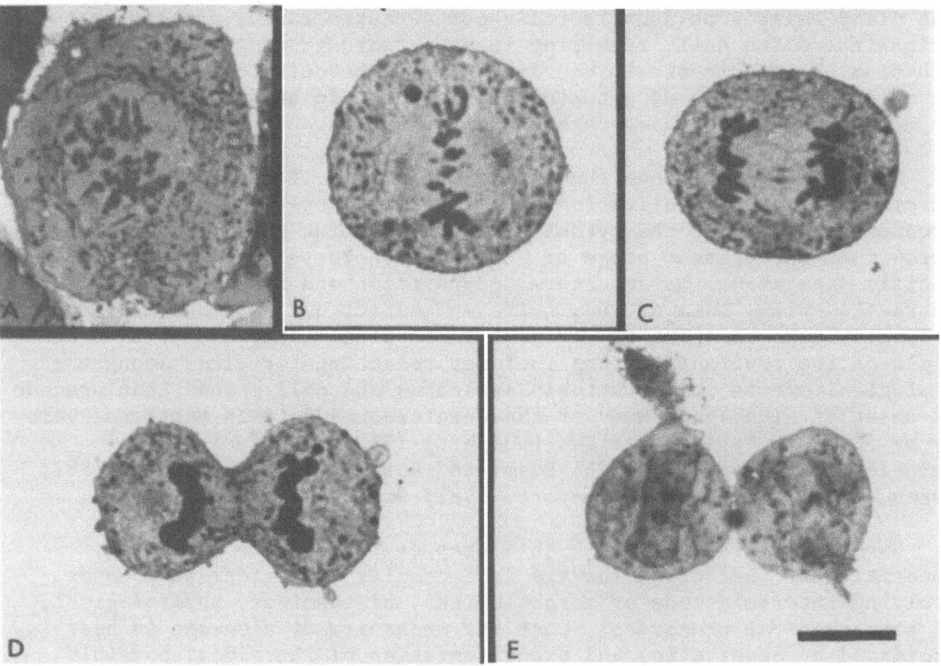

Figure 1. Photomicrographs of mammalian (HeLa) tissue culture cells at various stages of division: A, prophase; B, metaphase; C, end of anaphase;
D, mid-cleavage; E, end of cleavage. The cleavage furrow (B) constricts
the cell. A midbody connects the pair of daughter cells (E). The contractile ring cannot be seen lining the bottom of the cleavage in these 1 um
Epon sections. Bar = 10 um.

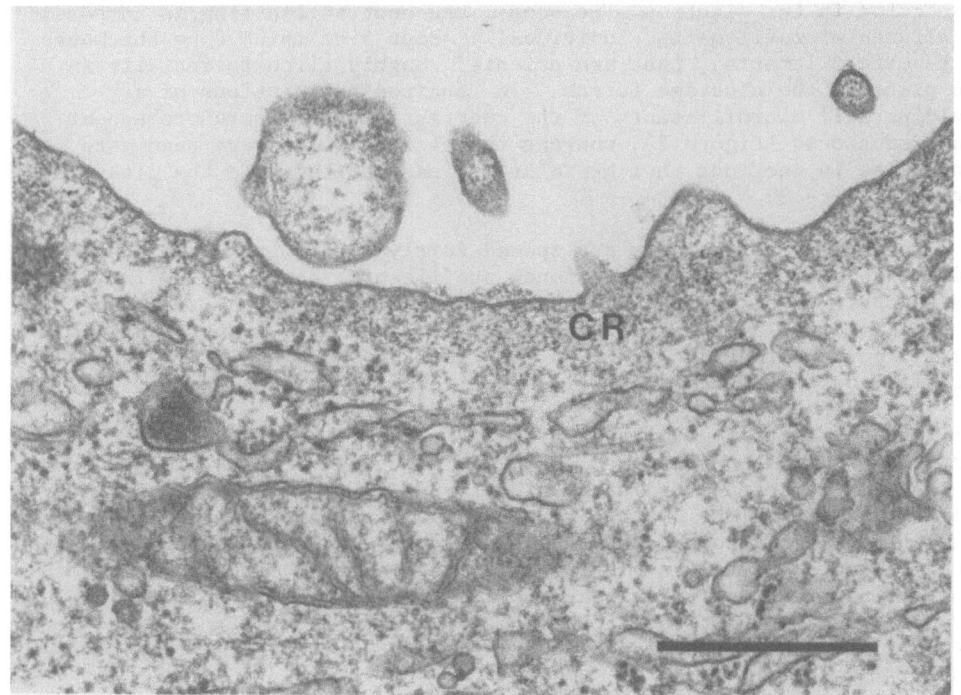

Figure 2. Electron micrograph through the bottom of a cleavage furrow of a HeLa cell at mid-cleavage. The contractile ring appears as a layer of cross-sectioned microfilaments (CR) immediately beneath the membrane. Bar = 0.5 μm.

successfully competed with a variety of vague and implausible theories in which "membrane" growth was often proposed as the active agent of cleavage. Subsequent advances in microscopy, biochemical detection, and cellular biomechanics have largely fulfilled the predictions of the contractile ring theory and have thus elevated the contractile ring to the level of an accepted reality.

Electron microscopic observations of well-preserved cells consistently reveal that the cytoplasm beneath a cleavage furrow is indeed structurally specialized in a way not detectable by ordinary light microscopy. Directly beneath the plasma membrane of the cleavage furrow there is a unique, thin "dense layer" that does not occur elsewhere (Figure 2). This "dense layer" appears suprisingly similar in organization in a wide variety of cell types, from tiny amoebae to very large amphibian eggs and algal cells, in addition to the more familiar tissue culture cells and sea urchin eggs. It is consistently and precisely confined to the bottom, concave portion of a cleavage furrow, regardless of the stage of cleavage, and does not extend up the walls of the cleavage furrow. This singular feature has lent pursuasive morphological support to the idea that this material is actually responsible for causing the concavity and, thus, the furrow.

The "dense layer", now called the contractile ring, is characteristically a band or ring about 0.1 μm thick and 10 μm wide underlying the plasma membrane of the cleavage furrow (Figure 2). Remarkably, these generalized dimensions occur in most cells regardless of size or stage of cleavage (an observation discussed further below), with only occasional exceptions such as algae whose contractile rings seem to be narrower (Pickett-Heaps et al., 1978; Scott, 1986). On close

inspection in the electron microscope, the contractile ring is composed
of aligned microfilaments, individually about 5 nm thick (the thickness
of F-actin filaments), that are oriented roughly circumferentially in
the plane of the cleavage furrow. In longitudinal sections of a
dividing cell microfilaments of the contractile ring therefore appear
cross-sectioned (Figure 2), whereas the microfilaments are seen more
lengthwise in sections that graze or are perpendicular to the cleavage
furrow (Figure 3).

Because microfilaments are spaced fairly uniformly about 15 nm
apart throughout the cross-sectional profile of a contractile ring,
there appear to be about 5000 filaments per cross-section. The length
of these microfilaments is undetermined, however they are probably at
least 1 um long, certainly not so long that any single microfilament
actually courses around the full cell circumference.

The entire array of microfilaments appears to be morphologically
very cohesive and homogeneous; it excludes cytoplasmic particles,
inclusions, membranes, and other fiber types. Unlike bundles of
microfilaments in less dynamic structures such as microvilli, there is
generally no evidence of cross-linking or higher degrees of organization
within this loosely aligned array of microfilaments, although in one
case grazing sections through a contractile ring have revealed an
irregular banding pattern (Sanger and Sanger, 1980) and in another
oriented macromolecules of myosin have been detected (Yumura and Fukui,
1985).

The contractile ring is closely applied to the plasma membrane,
which would seem to be necessary for its action to be translated into
cell deformation. Its microfilaments approach the membrane laterally,
at least as close as 10 nm, but obvious intervening specializations have
not yet been discovered. (Many people anticipate that the contractile
ring will be shown to attach to the plasma membrane by specializations
equivalent to adhesion plaques which ordinarily mediate the linkage of
stress fibers to extracellular substrates of adherent cells. This idea
seems naive since stress fibers are noted primarily for their stability,
immobility, and non-contractility, contrary to the evident dynamism of
the contractile ring). Neither morphological sub-membrane densities or
adhesion plaque molecules such as vinculin, spectrin, or talin have so
far been described; on the other hand, alpha-actinin has been shown to
be concentrated in the cleavage furrow (Fujiwara et al., 1979; Nunnally
et al., 1980; Mabuchi et al., 1985), as has an unidentified membrane
glycoprotein (Rogalski and Singer, 1985).

The contractile ring is too thin to be seen by ordinary light
microscopy (Figure 1). Even the most sensitive polarization microscopy,
designed to detect aligned materials in cells, cannot unambiguously
reveal it because it does not exhibit a high enough degree of structural
order. Thus, until recent successes with light microscopic
immunocytochemistry for localizing particular biochemical constituents,
there have been few other options besides electron microscopy for
analyzing the contractile ring. Since ultrastructural preservation of
delicate cytoplasmic structures is notoriously capricious, it is
gratifying not to be obliged to rely upon this single method for
exploring contractile ring organization and composition.

Gradually, biochemical information is gathering about the
contractile ring (see review by Mabuchi, 1986), but it often lacks the
required level of quantitative and organizational precision. Direct
biochemical study of has been frustrated by the inability to isolate the
contractile ring. Bulk isolation by cell fractionation has not been

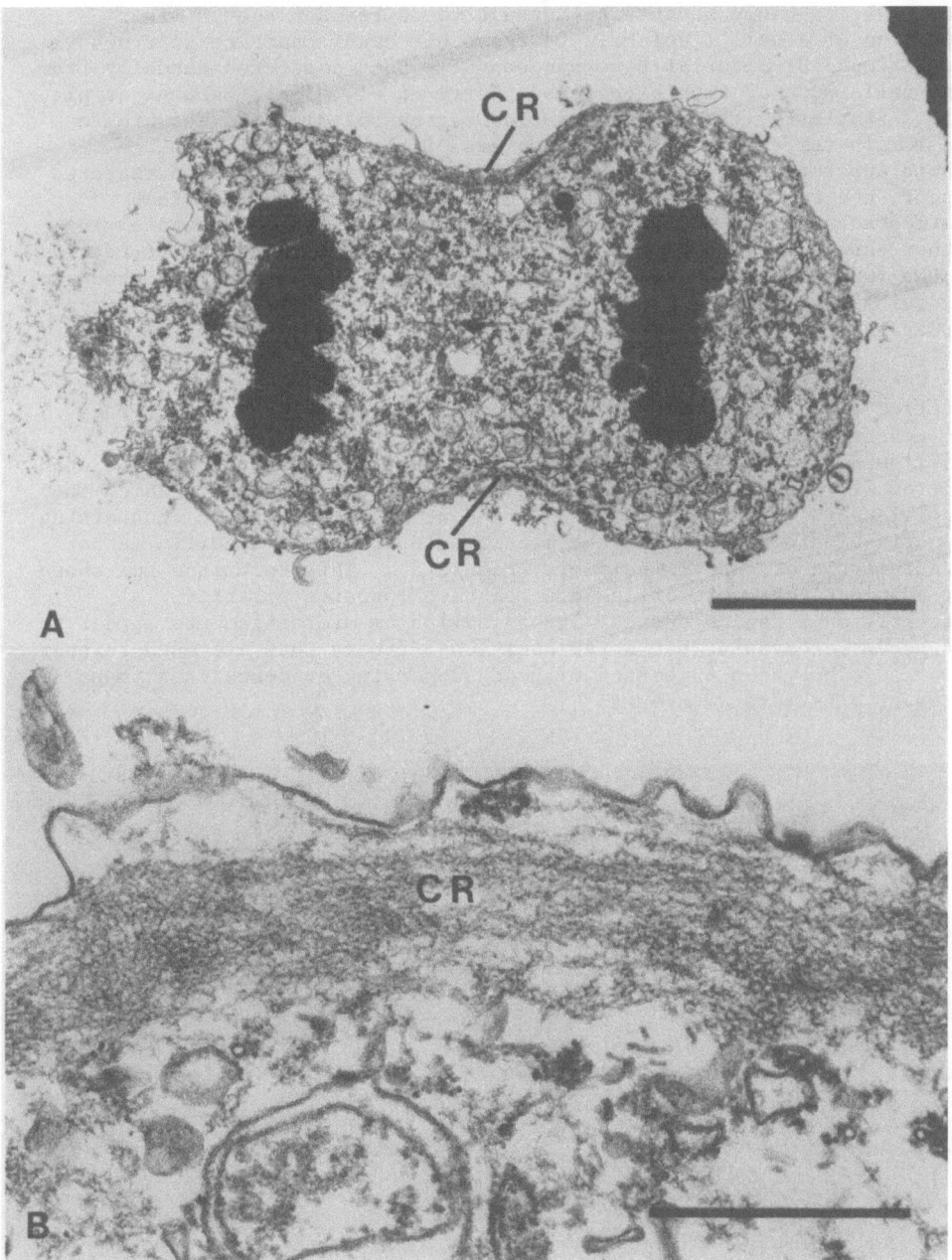

Figure 3. Electron micrographs of HeLa cells that have
been extracted in glycerol and irrigated with heavy meromyosin
to "decorate" the actin filaments. A, low-power view of an
entire cell at midcleavage; the contractile ring (CR) is
evident. B, section perpendicular to the plane of (A) through a
contractile ring (CR), showing its "decorated" microfilaments.
Bars = 5 μm (A) and 0.5 μm (B).

successful, probably because the structure represents such a minor proportion of a cell's volume. On the other hand, contractile rings plus surrounding material have successfully been dissected manually from large cells such as amphibian eggs (Perry et al., 1971; Mabuchi et al., 1985). Similarly, the cleavage furrow regions of dividing sea urchin eggs remain remarkably intact during mass-isolations of late-stage mitotic apparatuses. Although the yield and purity of these structures is poor, the fact that any isolated cleavage furrow survives such treatments testifies to the physical integrity of the contractile ring. Another experimental approach is the semi-isolation technique of peeling off the furrow cortex containing the contractile ring, recently shown to be feasible by Yonemura and Kinoshita (1986). Analysis of these kinds of preparations deserves to be exploited more extensively in the future.

When contractile ring microfilaments were first observed, their ultratructural appearance and dimensions strongly suggested that they were constructed of F-actin polymers, a point that has since been established without question. Initially, this demonstration relied upon electron microscopy (Perry et al., 1971; Schroeder, 1973), in which the microfilaments were shown to bind exogenous heavy meromyosin (containing the actin-binding portion of the myosin molecule) in a specific manner characteristic of F-actin polymers (Figure 3). This technique has shown that adjacent actin microfilaments can have opposite polarity, apparently on a random basis. Indeed, all 5 nm microfilaments appear to bind heavy meromyosin, but a few additional fibers that are much thinner do not; it has been suggested without supporting evidence that these represent oligomers of myosin.

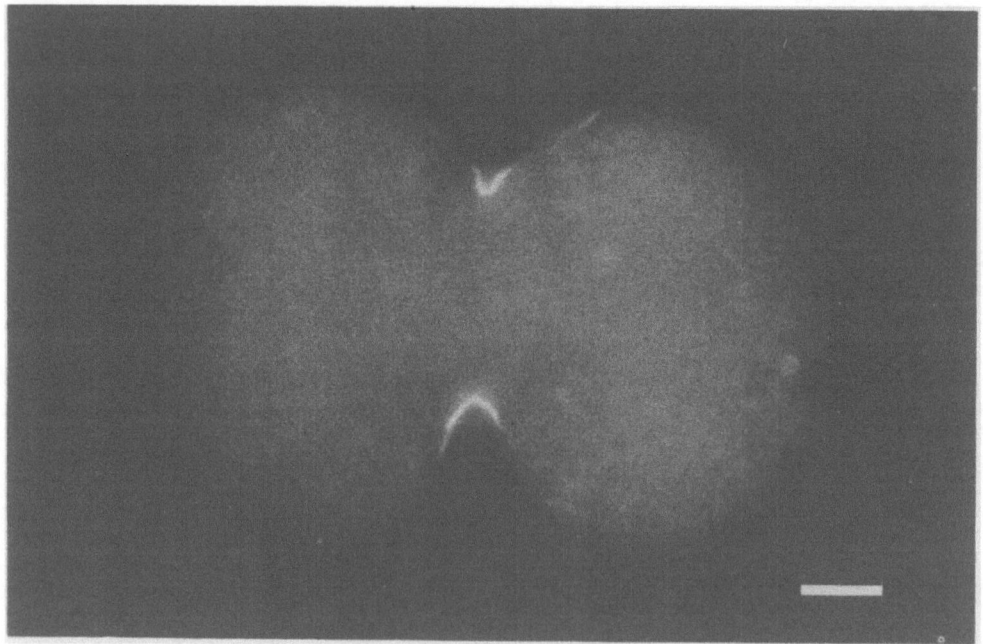

Figure 4. Immunofluorescence image of isolated blastomere of a sea urchin embryo (mesomere-mesomere) at midcleavage after staining with antimyosin. The brightly staining regions at the bottom of the cleavage furrow, containing the contractile ring, are rich in myosin. Bar = 5 μm.

The presence of actin in the contractile ring has also been visualized in the light microscope by staining cells with actin-specific markers, such as fluorochrome-conjugated actin-binding proteins (including actin, tropomyosin, alpha-actinin, and heavy meromyosin), antiactin antibodies, or phallotoxins that are specific for F-actin. These latter stains are very effective for revealing somewhat thicker bundles of actin, but they often fail to reveal the contractile ring, probably because it is so thin. When successful these methods reveal the width and overall configuration of the contractile ring in whole-mounted cells but they exaggerate its thickness.

By itself, of course, actin does not contract, as illustrated by many kinds of immotile and non-contractile cell structures that are caused by arrays of actin, for example microvilli or acrosomal processes. By analogy with muscle, it has long been suspected that the contractile ring contained either myosin or some related mechanoenzyme. Thanks to indirect immunocytochemistry using antimyosin antibody, myosin has now been co-localized with actin in the contractile rings of several kinds of tissue culture cells (Aubin, 1981), Dictyostelium amoebae (Yumura and Fukui, 1985), and sea urchin blastomeres (Schroeder, 1987). The results (Figure 4) confirm the dimensions and overall configurations obtained for actin, but the finer details of how the actin and myosin are spatially distributed are generally not resolved (except in the case of Dictyostelium amoebae prepared by a special technique where there is evidence that contractile ring myosin is concentrated into discrete rods aligned in the plane of cleavage).

PHYSIOLOGY OF CONTRACTION

The essential conclusion of many kinds of mechanical studies, usually involving sea urchin eggs, is that a cell divides by constriction because the tensile forces at its cleavage furrow exceed the forces of resistance elsewhere in the cell. These tensile forces have been quantitatively determined by direct measurement and by various calculations derived from the cell geometry, and it is generally inferred that they are directly due to the action of the contractile ring residing beneath the cleavage furrow.

In a qualitative sense, the functional importance of the contractile ring was first demonstrated by micromanipulations in which furrowing activity regressed when the base of the cleavage furrow was disrupted. Once contractile ring microfilaments were discovered, their particular functional significance was demonstrated by the consistent effect of the inhibitor cytochalasin in arresting or reversing the furrow as well as obliterating the microfilaments (Schroeder, 1970; 1972; 1978). Indeed, those findings first suggested that cytochalasin might act specifically against microfilaments in general, which is now known to be true because of its ability to disassemble actin polymers. The more refined idea that dynamic assembly-disassembly of actin is important for contractile ring function has been further supported by the inhibitory effect of phalloidin (Cande et al., 1981; Hamaguchi and Mabuchi, 1982), a substance that promotes actin polymerization.

The specific role of myosin in cleavage has been tested by injecting substances that specifically inhibit its function. When injected into starfish blastomeres, antimyosin antibodies (depending on their particular activities as functional blockers) can rapidly inhibit cleavage without affecting mitosis (Mabuchi and Okuno, 1977; Kiehart et al., 1982). Inactivated heavy meromyosin injected into cleaving frog eggs similarly inhibits cleavage (Meeusen et al., 1980). Thus, it has been reasoned that myosin is important to the functioning contractile ring but not to the functioning mitotic apparatus.

On the basis of the actomyosin model of contractile ring function, the putative role of calcium of ions in activation has been explored with particular interest. While it is known that extreme depression of the intracellular calcium concentration by injecting a buffered chelator prevents cleavage, there is very little evidence that free calcium levels actually increase during cleavage (Schantz, 1985; Poenie et al., 1985). On the contrary, one study of changing calcium levels during the cell cycle indicates that cleavage is accompanied by a cell-wide decrease in calcium concentration (Yoshimoto et al., 1985). Clearly, the role of calcium requires further investigation. An alternate possibility that contractile ring activity is regulated by phosphorylation of a myosin light-chain has not been addressed experimentally, although the issue has been raised (Ezzell et al., 1983).

The intracellular conditions required for contractile ring action have proven difficult to specify from studies of whole cells, probably because the important reactions are locally confined to such a small organelle in a relatively large cell. These conditions, however, have been successfully studied in permeabilized cell "models" in which cells are lysed mid-cleavage in detergents or other extraction media and then constriction is reactivated by optimizing the media conditions in terms of ions and nucleotides. Furrowing ring activity have recently been reexamined in such "models" prepared from tissue culture cells (Cande et al., 1981) and sea urchin eggs (Yoshimoto and Hiramoto, 1985). Furrowing can be reactivated (or maintained) only in cells that had already begun to cleave; furrows cannot be initiated if they had not yet developed at the time of lysis. Thus, while these studies help to elucidated some of the requirements for ATP and various ions, they do not comment on the mechanism by which a contractile ring is established, except to substantiate that there is no organized precursor of the contractile ring before the onset of cleavage.

DIMENSIONAL CHANGES

With strange uniformity, contractile rings consistently seem to be about 0.1 μm thick and 10 μm wide (the same dimension as the concavity at the bottom of the furrow), regardless of cell type, cell size or stage of cleavage. That is, neither the thickness or width of the contractile ring appear to scale according to the dimensions of the cell in which it occurs. This was originally noted in the pioneering electron microscope studies of marine invertebrate eggs and has since been extended to include protozoans, various somatic cells of mammals and birds, tiny slime-mold amoebae, and large amphibian and fish eggs. (Algae represent the single known exception, as mentioned earlier). There is presently no clear understanding why either of these dimensions should be so consistently uniform in contractile rings of such diverse cells, so it seems especially significant that this assertion has been corroborated in sea urchin eggs by three totally different techniques: in sections viewed in the electron microscope (Schroeder, 1972); in whole cells stained with antimyosin (Schroeder, 1987); and in shear-isolated furrow fragments stained for F-actin (Yonemura and Kinoshita, 1986).

In addition to the comparative aspect, it is even more curious that neither the thickness or the width of a contractile ring seems to change appreciably according to the stage of cleavage, at least in sea urchin eggs. Dimensional changes seem to be restricted to changes in length (i.e. circumference in this case). By systematically comparing cells of all stages, Schroeder (1972) found that the contractile ring is full-sized when it abruptly appears at the beginning of cleavage, then somehow shrinks in volume exclusively by reducing its circumference, and finally disappears abruptly at the end of cleavage. The volume is thus greatest at the beginning of cleavage and reduces to nothing at the end. Unfortunately for efforts to understand this peculiar behavior, there is no evidence whatsoever as to how the volume is reduced. Electron microscopy has failed to provide evidence of intermediate states, for example of contractile ring microfilaments in the process of disassembly.

Conversely, at the onset of cleavage one might expect a contractile ring to gradually acccumulate, but this has not yet been observed. In sea urchin eggs and tissue culture cells that cleave "equilaterally", the definitive contractile ring seems to arise simultaneously, and inexplicably, with the first signs of cell constriction. From the outset, it is typically about 0.1 μm thick and 10 μm wide, dimensions that it then maintains throughout the cleavage process. So long as such a cell is still spherical and unconstricted, a contractile ring cannot be anticipated as an accumulation of microfilaments (ultrastructurally) or actin or myosin (immunocytochemically), even as briefly as one minute before furrowing is anticipated on the basis of mitotic activity. Moments later, using the exact same methods, a cell with even the shallowest furrow exhibit a fully-formed contractile ring. Results from electron microscopy and antimyosin staining agree on these points.

The simultaneity of the appearance of the structured contractile ring with a visible furrow is dramatically reinforced by ultrastructural work on cells that cleave "unilaterally", such as some partilcarly large cells (cephalopod, fish, and amphibian eggs) and even some smaller ones (coelenterate eggs). In these eggs cleavage begins as a short crease at the animal pole, not as a constriction encircling the entire cell. In such cases, the contractile ring is accordingly a "contractile band" of microfilaments with definite ends which coincide with the extremities of the crease-like furrow. As cleavage progresses, the furrow and the contractile ring elongate simultaneously by a kind of self-propagation. Immediately ahead of the defined contractile ring, in a region where the contractile ring will momentarily appear, no organized precursor has actually been found (Arnold, 1969). In these cases, the cleavage furrow deepens as the overall length of the contractile ring increases.

When considered mechanistically, the behavior of the contractile ring exhibits dimensional changes and behaviors that are difficult to explain. Certainly they are not elucidated by a simple model of muscle-like contractility; shortening of a myofibril in a muscle is limited to a fraction of its starting length and is compensated by an increase in the cross-sectional area so that its volume is conserved. Superimposed on its so-called contractile behavior, the contractile ring is obviously characterized by a high degree of dynamic assembly-disassembly, a topic which is regrettably poorly understood. Indeed, the dynamics of this structure contrasts so sharply with that of myofibrillar actomyosin since a contractile ring can even be set up in differentiated smooth or cardiac muscle without affecting the myosibrils (Schroeder, 1976).

SURFACE INVOLVEMENT

Since cell cleavage divides a cell of fixed total volume into two daughter cells, the overall surface area must increase. In response to this truism, there has been a long-standing interest in the detailed interrelationship between proposed mechanisms of area increase and cleavage itself. Is membrane production required for cleavage to proceed? Must membrane production be specifically localized? Do demands for additional surface area also signify that the sub-surface cortex expands? How is overall expansion (of the surface or the cortex) compatible with the putative contraction that occurs at the cleavage furrow? These and other aspects of the "new surface" problem have been explored in many ways (see reviews by Dan, 1960; Bluemink and de Laat, 1977: Schroeder, 1981a), and it now seems that each kind of cell finds its own unique solution. Comparisons suggest that the production of the needed new surface is an accommodation and corollary of cleavage but not a causative factor or even a rate-limiting pre-requisite.

Attempts to ascertain the involvement of the cell surface during cleavage is complicated by the demonstrable fact that many cell types are covered with microvilli. In sea urchin eggs microvilli account for a redundancy of surface area of about 400%, some of which could easily be "unfolded" or "pulled out" in order to provide the needed area. (If this happens, and it may, it has not been detected so far). On the other hand, in smaller cells where microvilli are proportionately fewer and the area redundancy is less, microvilli are apparently heavily consumed; nevertheless, there appears to be no uniform pattern to this utilization, as in some types microvilli disappear at the equator and in others they disappear at the poles.

Strict localization of membrane production has recently been worked out with amphibian eggs (Byers and Armstrong, 1986). In these large, rapidly dividing cells there are few microvilli to exploit; instead, new plasma membrane is assembled during the middle and late stages of each division along the lateral walls of the furrow, regions that will later become the surfaces between adjacent blastomeres. In these same regions, "new" cortex is also elaborated, lacking the pigment inclusions characteristic of the "old" cortex. Areas that do not obviously participate in new surface production (either as membrane or cortex) include the shallow furrow of early stages, the convex shoulders of the later furrow, the embryo's outer aspect, and the 10 μm wide strip of surface at the leading edge of the furrow that is associated with the contractile ring. (This latter observation suggests that the plasma membrane, to which the contractile ring is originally attached, remains associated during cleavage).

A related pattern of new surface production occurs in dividing sea urchin embryos but with an important difference. New surface, lacking both microvilli and pigment inclusions, also appears on the aspects of adjacent blastomeres that face one another, but in sea urchin embryos it appears only after cleavage is complete, that is between divisions. In this case, the timing suggests that new surface is not required for the actual cleavage process, perhaps because microvilli can temporarily donate surface, but it is nevertheless eventually restored. At the next division, a cleavage furrow forms equally well beneath "old" as well as "new" surface, proving that local microvilli are not necessary for a successful furrow.

CORTICAL SOURCE OF MICROFILAMENTS

Describing the contractile ring once is known to exist is relatively straightforward, even if one admits that the techniques for detecting it are imperfect. Discovering its origin, however, is severely impeded by the fact that the contractile ring has no well-organized precursor. Even though one knows where and when the contractile ring will appear, concrete morphological clues to its origin are lacking. To fill in for the lack of evidence, the origin of the

contractile ring will be speculative, in which ad hoc mechanisms are
invented to suit whichever subsets of data happen to be popular.

Typically, as explored in sea urchin eggs and tissue culture cells,
the visible "dense layer" of aligned microfilaments exists only in
conjunction with a clear cleavage furrow; even as little as one minute
before a furrow appears (anticipated on the basis of mitotic activity),
the equatorial cortex, where the contractile ring will appear, is not
ultrastructurally or cytochemically distinctive. Moments later,
however, the identical experimental methods reveal that a cell with even
the shallowest furrows already possesses a fully-formed contractile
ring. From the outset, such a contractile ring has the typical cross-
sectional profile 0.1 μm thick and 10 μm wide, dimensions that it
maintains throughout the cleavage process. Incongruously, the
definitively organized contractile ring seems to arise precisely
simultaneously as the onset of constriction, its putative function.

One periodically recurring suggestion concerning the origin of the
contractile ring is that its microfilaments arise from core
microfilaments of microvilli. The transition has not actually been
observed, and a priori it seems hardly credible that a bundle of
microfilaments that is heavily cross-linked for the purpose of
structural stability and rigidity could readily convert to a contractile
state. Indeed, the idea is not supported by the facts: microvilli and
their microfilaments are not consumed during division in sea urchin
eggs, where the amount of actin in the initial contractile ring would
require the depletion of about 50% of all microvilli cores (Schroeder,
1981a); furthermore, furrows (and presumably contractile rings) form
perfectly well on surfaces that lack microvilli, including "exovates"
squeezed out of eggs (Rappaport, 1976; 1983) and the "new surfaces" of
sea urchin blastomeres mentioned above (Schroeder, 1986). In addition
to refuting the idea that the contractile ring is derived from
microvilli cores, these latter observations also rule out any role of
microvilli cores as essential anchoring agents for the contractile ring
(Begg et al., 1983).

It seems more plausible that the structure of the contractile ring
arises from microfilament precursors already present in the sub-surface
cortex before cleavage begins. The idea of a cortical reservoir of
available actin is supported by the consistent finding that isolated
cortices from cells generally contain more actin than what is required
for their known actin-containing structures. It is furthermore known
that most cell types can undergo episodes of cortical contractility,
both natural and inducible, that involve microfilaments. Presumably,
such activities reveal that the cortex is structurally pre-conditioned
to respond to a variety of stimuli and to engage in the organizational
changes necessary for contraction.

Examples of natural, localized contractions that are mediated by
cortical microfilaments include: the traveling waves of shallow
constrictions exhibited by amphibian eggs after fertilization and before
each cleavage (Hara, 1971; Hara et al., 1980; Sawai, 1979; Yoneda, etal,
1982); the intricate deformations of the Tubifex eggs during meiotic
and mitotic divisions (Shimizu, 1975); persitaltic constrictions that
sweep along barnacle eggs (Lewis, et al., 1973; Schroeder, 1975); and
the fixed polar lobe constrictions of annelid and molluscan eggs
(Conrad and Williams, 1974). Transitory arrays of microfilaments at the
sites of most of these constrictions have been described
ultrastructurally.

The prodigious ability of the amphibian eggs cortex to respond rapidly to experimental stimuli by contracting is well-known. Cortical contractions are readily inducible by such treatments as wounding, calcium ion injection, and ionophore application (see reviews by Schroeder, 1975; Ezzell et al., 1983), and these contractions have also been correlated with the localized appearance of extensive mats of actin microfilaments that cannot be seen in uncontracted cortices.

These examples serve to suggest that the contractile ring could also originate, as in the above contractions, from cytoskeletal materials already present within the cortex in response to a particular stimulus correlated to the end of mitosis. The unactivated cortical cytoskeleton could exist in a variety of structural and physiological states of readiness, according to the specific cell type, requiring either assembly, recruitment of formed components, or activation before becoming functionally contractile. Conceivably, the stimulus could differ somewhat from cell to cell. Perhaps such distinctions could explain how a unilateral cleavage furrow self-propagates in some cells but not in others, why real cleavage furrows constrict to completion whereas other constrictions relax, why cortical microfilaments have been seen in some cells prior to cleavage (Usui and Yoneda, 1982) and not others, and why endoplasmic myosin may be seen to migrate to the cortex in some cells but not others (Yumura and Fukui, 1985). The pre-conditions and stimuli for contractile ring formation and function could be somewhat different in each example. If such variations actually exist, it will be difficult indeed to specify universal principles of cytoskeletal reorganization that account for the origin of the contractile ring as a morphological entity.

AUTONOMOUS CORTICAL CONTRACTILITY

One implication of the ability of some cells to undergo natural, localized contractions in addition to cleavage furrows may be that the cytoskeleton is non-homogeneously distributed in the cortex, either structurally or in terms of activity. Another implication, more relevant to the origin of the contractile ring, is that the cortical cytoskeleton is capable of responding to stimuli quite unrelated to those involved in cell division. Lending credence to this latter possibility are experiments with Tubifex eggs (Shimizu, 1979; 1981) and snail eggs (Conrad, 1973) in which characteristic, localized contractions occur despite obliteration of the mitotic apparatus by inhibitors or be enucleation. Although normally correlated with cell division, these deformations are obviously pre-localized and independently stimulated.

Cells do not have to deform when they undergo contractions, so long as all parts contract equally and synchronously. Recent biomechanical studies of sea urchin and starfish eggs (Yoneda et al., 1978; Yoneda and Yamamoto, 1985) have shown that such spherical cells experience periods of cortical contractility that are not triggered by mitosis. Indeed, they exhibit complete cycles of contraction and relaxation even when mitosis is completely inhibited by drugs or enucleation (Yoneda and Schroeder, 1984; Yoneda and Yamamoto, 1985; Shinagawa, 1986). Since the cells remain spherical throughout these changes, these autonomous cycles of mechanical change are interpreted as being isometric and global in extent (meaning that they occur throughout all parts of the cortex simultaneously). When mitosis is inhibited, the period of these persisting contraction-relaxation cycles is noticeably prolonged compared to the division cycle of intact cells (Figure 5).

Proof that such global contractions are really capable of doing work (i.e. are actual contractions, not merely passive stiffness

changes) has been demonstrated in various ways, including by showing
that the ability of global contractions to lift static weights
(Schroeder, 1981b). Such global contractions are presumably mediated by
microfilaments since they are inhibited by cytochalasin operating by
an actomyosin-like mechanism. Morphological evidence for a build-up of
cortical microfilaments during cortical contraction has been provided by
Usui and Yoneda (1982).

The mechanical changes in the cortex bear a particular temporal
relationship with cell division that is easily resolved at lower culture
temperatures and tends to be obscured at higher temperatures, when cells
develop more rapidly. The contraction phase is similar in
rate in normal or colchicine-inhibited eggs (Schroeder, 1981b; Yoneda
and Schroeder, 1984) and coincides with the early stages of mitosis. In
the normal cell, therefore, cleavage begins after a peak of cortical
tension has already been attained globally; subsequently, as cleavage
proceeds overall tension then declines. (This noteworthy event seems
to contradict the demonstrable fact that cleavage occurs by furrow
contractility, but the point is that the the onset of cleavage occurs at
high global tension whereas actual furrowing proceeds by localized
contractility). In contrast to the global development of tension, the
decay of tension (i.e. relaxation) during cleavage of normal cells is
much more rapid than the autonomous decay of tension in mitotically
inhibited cells (Figure 5). The more gradual rate of relaxation of
inhibited cells may account for the prolonged period of contraction-
relaxation and raises the possibility that cortical relaxation is
accelerated by the presence of the mitotic apparatus in a normal cell.

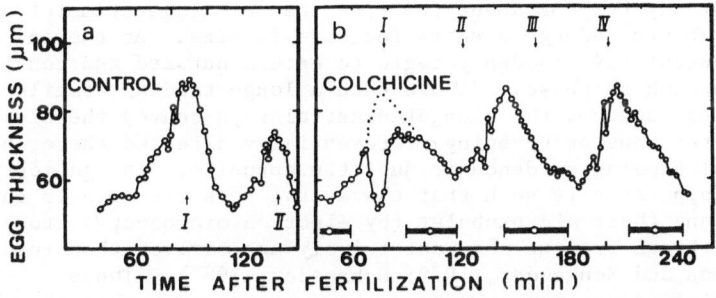

Figure 5. Cycles of cortical contraction and relaxation in
sand dollar eggs probed with a standard force of compression,
expressed as resulting egg thickness (a thicker egg exhibits
higher cortical tension). In (a) an intact embryo passes
through the first two divisions (arrows indicate the onset of
furrowing). In (b) a similar embryo treated with colchicine to
inhibit mitosis and thus cleavage still exhibits both global
contraction and global relaxation; the cycle period is longer
than in the controls, and in particular the relaxation phase is
longer. The dotted line disguises an unexplained, but
consistent artifact of the first contraction phase. (Modified
after Yoneda and Schroeder, 1984).

THE CLEAVAGE STIMULUS vs STRUCTURE OF THE MITOTIC APPARATUS

In the normal coordination between mitosis and cytokinesis, the cleavage furrow forms after the chromosomes have been separated by the anaphase movements of the mitotic apparatus. The location of the furrow is determined by the localtion of the mitotic apparatus. It is now well-accepted that the mitotic apparatus is instrumental in stimulating the process of cleavage, even though it is the contractile ring that generates the actual forces. Extensive micromanipulation studies, typically in sea urchin eggs, demonstrate that the cleavage stimulus is imparted to the cortex by a pair of nearby mitotic asters at a specific time (see review by Rappaport, 1986). The cleavage stimulus differentiates the previously uniform cortex into a zone of active contractility (the cleavage furrow) and non-furrow zones that are relatively more yielding (ordinarily referred to as the "poles"). The location of the pair of asters establishes the location of the cleavage furrow, hence eccentric asters establish lop-sided furrows characteristic of unilateral or unequal divisions. The timing of the cleavage stimulus is generally considered to be around mitotic anaphase in animal cells (Hamaguchi, 1975; Hiramoto, 1971; Rappaport, 1981), but it is noteworthy that in diatoms the cleavage furrow is determined during prophase, before the mitotic apparatus is even established (Wordeman et al., 1986).

Discovering the specific nature of the cleavage stimulus is one of the great remaining frontiers in cell division research. It is generally assumed that elucidating the cleavage stimulus will also explain much about the origin of the contractile ring. The elegant experiments by Rappaport (1971; 1973; 1982; 1985), in which the spatial and temporal relationships between the mitotic apparatus and cortex are rearranged, are usually interpreted to mean that the mitotic asters effect a positive change in the equatorial cortex resulting in localized furrow formation; they thus imply that the stimulus initiates assembly of a contractile ring.

In sea urchin eggs, the structure of the mitotic apparatus is intimately linked to the effectiveness of the cleavage stimulus, whatever that may be. At about the time that cleavage stimulation occurs, the asters undergo a sharp increase in size. At the end of metaphase, astral rays suddenly begin to extend outward and continue extending through anaphase. It has been a long-standing, facile assumption that somehow the elongated astral rays convey the cleavage stimulus to the equator by being preferentially directed there, but the morphological finding evidence is just the opposite. The geometry of the mitotic apparatus is such that the astral rays (as seen in the light microscope) and their microtubules (by electron microscopy) project into the polar cortical regions more extensively and earlier than into the equator (Asnes and Schroeder, 1979; Schroeder, 1987). These morphological findings are so pursuasive as to suggest that the absence of astral rays at the equator might even correlate with the aster-induced stimulation of equatorial furrowing. For such a mechanism of "reverse" cleavage stimulation to apply, it would seem to require a fundamental reappraisal of the entire concept of cleavage initiation.

Selective physical contact between astral rays and the cortex has been inferred from various morphological changes that occur in sea urchin eggs (Schroeder, 1985). Particularly interesting is a shape-change called "oblation" which occurs naturally in some species at the end of anaphase as an immediate precursor to furrowing; just before cleavage begins, the equator momentarily bulges outward (the pole-to-pole diameter decreases) and then, a few minutes later, the equator constricts. Although not yet proven, an interpretation of this event is that it is caused by an internal shortening change in the mitotic apparatus (perhaps the disintegrating spindle) that is translated into a shape-change by physical contact between the asters and the polar cortex. It is the last point that is important here: that at the critical time of the putative cleavage stimulus the mitotic asters may be in close contact with the poles.

MODELS OF CLEAVAGE INITIATION

Two classes of mechanical models have recently emerged to explain the intiation of cleavage, that is the origin of the initial force imbalance between the equatorial cortex and the polar cortex. The result of this imbalance is the onset of constriction in a previously spherical cell. Although confined to qualitative mechanical events, the models address the question of the geometry and timing of the cleavage stimulus without dealing specifically with either ultrastrucural organization or biochemistry; only incidentally do they comment upon the origin of the contractile ring and then only as a manifestation of relatively elevated contractility at the equator.

One model asserts that contractile activity is specifically imparted to the equatorial cortex by the cleavage stimulus. Since this stimulus emanates from the mitotic asters, the initiation of differential contractility at the equator can be said to result from "aster-mediated stimulation of equatorial contractility". This scheme does not acknowledge, explain or depend upon autonomous contraction-relaxation phenomena and is inconsistent with the evidence that the onset of cleavage is associated with a decrease, rather than an increase, in cortical tension. Nevertheless, the idea is popular despite the obvious morphological detail that the mitotic asters are closer to the poles than to the equator where the model claims they are

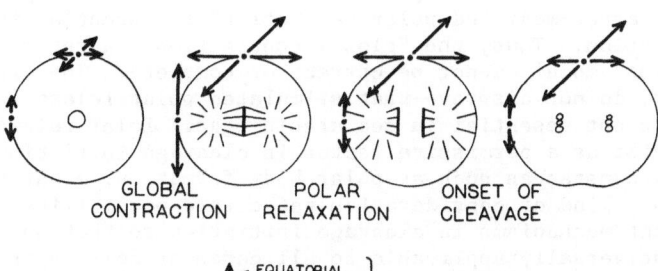

Figure 6. Diagram of the "global contraction-polar relaxation" model of cleavage initiation. Two separate steps are shown: cortical contraction that is global and autonomous of the mitotic apparatus and a relaxation at the poles induced by the nearby mitotic asters. The equator continues to exhibit unchanged contraction. (From Schroeder, 1981b).

expected to produce a positive change. On the other hand, this model is abundantly supported by a large number of experiments based upon micromanipulation in which establishment of a furrow was only susceptible to interference by alterations performed in the region between the mitotic apparatus and the equatorial cortex (see Rappaport and Rappaport, 1983).

The second mechanical model operates from the premise that autonomous, global contraction of the cortex is an indispensable precursor to cleavage. It asserts that force-generation develops autonoumously and globally; then, as the result from aster-mediated changes of the poles, this global contraction is mechanically channeled into the equator. The morphological correlate of this channeling is the rearrangement of the structural elements responsible for force-production (the contractile microfilaments) into the contractile ring. As originally proposed (Schroeder, 1981b), this two-step "global contraction – aster-mediated polar relaxation" model postulated that the cleavage stimulus changes the poles during (or after) the global contraction by causing them to relax (or cease contracting)(Figure 6). By this means, the equatorial cortex would continue to contract as the exclusive, regional vestige of the autonomous contraction. Beyond the acceptance of the global contraction as vital to the cleavage process, the essential feature of this scheme is that the cleavage stimulus is an inhibitory influence of the asters upon the polar cortex.

This model is supported by the morphology of the normal sea urchin egg in which the asters at late anaphase are clearly closer to the polar cortex than to the equator, as well as by the "oblation" phenomenon (as interpreted above). Concrete ultrastructural evidence that a global array of microfilaments indeed coalesces into the contractile ring, as conceptualized recently by Bray et al. (1986) and Mabuchi 1986), has been provided by the work of Usui and Yoneda (1982). Furthermore, the ability of an aster to stimulate a cortical relaxation and loss of microfilaments has been indicated in the case of polar body formation in the Tubifex egg (Shimizu, 1983). On the other hand, the idea of aster-mediated polar relaxation in this model has distinctly not been supported by the results of numerous micromanipulation experiments; in many variations, interference between and aster and any non-equatorial cortex never prevented the onset of furrowing (Rappaport and Rappaport, 1983; 1984; 1985).

So persuasive are these experiments that they convincingly refute the idea that "aster-mediated polar relaxation" is essential for cleavage initiation. Thus, the "global contraction – aster-mediated polar relaxation" model cannot be correct or complete. The experimental tests, however, do not disprove that stimulated polar relaxation occurs, only that it is not essential in sea urchin eggs. Polar relaxation could still exist as a permissive factor in cleavage initiation, which, in selected circumstances such as polar body formation, might still be essential. This kind of consideration reflects a possibility that there may be redundant mechanisms in cleavage initiation so that no single mechanism is universally applicable to all cases or cell types. Regardless, the proposition that the global cortical contraction contributes to the mechanism of cleavage initiation is not refuted or weakened (but neither is it proved). The challenge, then, is to reconcile disparate aspects of these models into a more acceptable scheme of cleavage initiation, one which preserves the postulated relevance of the global contraction and the (frequently) essential aster-mediated change at the equator.

Toward this end, this discussion concludes with a modified model of cleavage initiation in which global contraction is retained as the essential source of cortical contractile force and in which a positive, aster-mediated change of the equatorial cortex is also admitted. The scheme is inherently ambiguous in order to accommodate the possibility of redundant mechanisms by which regions of the cortex become regionally differentiated in terms of contractile capability.

It is relevant to recall that, in addition to an autonomous contraction, the cortex of mitotically inhibited cells also undergoes an autonomous relaxation, interestingly at a rate somewhat slower than control cells (Figure 5). In order to exploit the autonomous contractile properties of the cortex in cleavage initiation, it seems theorectically possible that differential manipulation of the <u>rates of relaxation</u> could result in the relative prolongation of force-generating capability by the equator while poles relatively relax, as pictured in Figure 7.

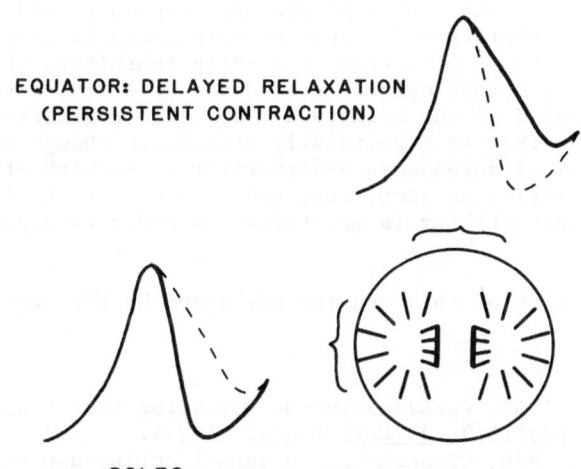

EQUATOR: DELAYED RELAXATION
(PERSISTENT CONTRACTION)

POLES:
ADVANCED RELAXATION

Figure 7. Diagram of the "global contraction-differential relaxation" model of cleavage initiation in which cortical relaxation is relatively delayed at the equator and advanced at the poles because of unequal interaction with mitotic asters or the a specific "protective" signal. This temporal force imbalance drives the initiation of cleavage.

This model could be called the "global contraction – differential relaxation" mechanism of cleavage initiation. According to it (Figure

7), it is not essential that the poles are actively stimulated to relax (although this would be permitted), rather they are simply allowed to relax on the schedule determined by the autonomous cycle of cortical contraction-relaxation (as in the control cell in Figure 5a). In contrast, the equatorial cortex must be stimulated (by an unknown mechanism) to retain its contractility or literally to relax at a slower-than-normal rate (as in the mitotically inhibited cell of Figure 5b). Cleavage is then initiated by a differentiation of cortical regions in terms of their rates of relaxation in which the equator exerts a greater tendency to contract and the poles a greater tendency to yield. By this mechanism the necessary imbalance of cortical force-generating capability could commence at any time during the development of autonomous contraction or of autonomous relaxation (for example, by culturing at different temperatures to change the development rate) without changing the outcome; as an example, Figure 7 illustrates differentiation beginning at the end of the contraction phase, that is the peak of cortical tension.

To satisfy this theoretical scheme, the precise nature of the aster-mediated cleavage stimulus remains unspecified. Asters could mediate changes at either (or both) the equator and the poles, depending upon circumstances or cell types, with equivalent results (namely the enhancement of relative contractility at the equator). The localized absence of astral rays from the equatorial cortex could be the morphological correlate that allows this region to undergo "delayed relaxation"; interestingly, delayed relaxation is shown in eggs whose mitotic apparatus is prevented by mitotic inhibitors (Figure 5). Conversely, the morphological interaction of astral rays with the polar cortex be related to the relatively "advanced relaxation" postulated for this region, either by a positively stimulated change or by attenuating the peak level of autonomous contractility. As with other models of cleavage initiation as such, this scheme does not obviously elucidate how furrow contractility is maintained in order to achieve complete cell constriction.

(Preparation of this chapter was aided by NIH grant HD/GM 20306).

REFERENCES

Arnold, J.W., 1969, Cleavage furrow formation in a telolecithal egg (Loligo peallii), J. Cell Biol., 41:894.
Arnold, J.W., 1976, Cytokinesis in animal cells: new answers to old questions, in: The Cell Surface in Animal Embryogenesis and Development," G. Poste and G.L. Nicolson, eds., Elsevier/North-Holland, Amsterdam.
Asnes, C.F., and Schroeder, T.E., 1979, Cell cleavage: ultrastructural evidence against equatorial stimulation by aster microtubules, Exptl. Cell Res., 122:327.
Aubin, J.E., 1981, Immunofluorescence studies of cytoskeletal proteins during cell division, in: "Mitosis/Cytokinesis," A.M. Zimmerman and A. Forer, eds., Academic Press, New York.
Beams, H.W., and Kessel, R.G., 1976, Cytokinesis: a comparative study of cytoplasmic division in animal cells, Amer. Sci. 64:279.
Begg, D.A., Salmon, E.D., and Inoue, S., 1983, Changes in the structural organization of actin in the sea urchin egg cortex in response to hydrostatic pressure, J. Cell Biol., 97:1795.
Bluemink, J.G., and de Laat, S.W., 1977, in: "The Synthesis, Assembly and Turnover of Cell Surface Components," G. Poste and G.L. Nicolson, eds., Elsevier/North-Holland, Amsterdam.

Bray, D., Heath, J., and Moss, D., 1986, The membrane-associated 'cortex' of animal cells: its structural and mechanical properties, J. Cell Sci. Suppl., 4:71.

Byers, T.J., and Armstrong, P.B., 1986, Membrane protein redistribution during Xenopus first cleavage, J. Cell Biol., 102:2176.

Cande, W.Z., McDonald, K., and Meeusen, R.L., 1981, A permeabilized cell model for studying cell division: a comparison of anaphase chromosome movement and cleavage furrow constriction in lysed PtK1 cells, J. Cell Biol., 88:618.

Conrad, G.W., 1973, Control of polar lobe formation in fertilized eggs of Ilyanassa obsoleta Stimson, Amer. Zool., 13:980.

Conrad, G.W., and Rappaport, R., 1981, Mechanisms of cytokinesis in animal cells. in: "Mitosis/Cytokinesis", A.M. Zimmerman and A. Forer, eds., Academic Press, New York.

Conrad, G.W., and Williams, D.C., 1974, Polar lobe formation and cytokinesis in fertilized eggs of Ilyanassa obsoleta, Dev. Biol. 36:363.

Dan, K., 1960, Cytoembryology of echinoderms and amphibia, Int. Rev. Cytol., 9:321.

Ezzell, R.M., Brothers, A.J., and Cande, W.Z., 1983, Phosphorylation-dependent contraction of actomyosin gels from amphibian eggs, Nature, 306:620.

Ezzell, R.M., Cande, W.Z., and Brothers, A.J., 1985, Ca2+-ionophore-induced microvilli and cortical contractions in Xenopus eggs. Evidence for involvement of actomyosin, Roux's Arch. Dev. Biol. 194:140.

Fujiwara, K., Porter, M.E., and Pollard, T.D., 1979, Alpha-actinin localization in the cleavage furrow during cytokinesis, J. Cell Biol., 79:268.

Fujiwara, K., and Pollard, T.D., 1976, Fluorescent antibody localization of myosin in the cytoplasm, cleavage furrow, and mitotic spindle of human cell, J. Cell Biol., 71:848.

Hamaguchi, Y., 1975, Microinjection of colchicine into sea urchin eggs, Dev. Growth Differ., 17:111.

Hamaguchi, Y., and Mabuchi, I., 1982, Effects of phalloidin microinjection and localization of fluorescein-labeled phalloidin in living sand dollar eggs, Cell Motil., 2:103.

Hara, K., 1971, Cinematographic observations of 'surface contraction waves' (SCW) during early cleavage of axolotl eggs, W. Roux Arch. Entwicklunsmech. Organ., 167:183.

Hara, K., Tydeman, P., and Kirschner, M., 1980, A cytoplasmic clock with the same period as the division cycle in Xenopus laevis eggs, Proc. Nat. Acad. USA, 77:462.

Hiramoto, Y., 1971, Analysis of the cleavage stimulus by means of micromanipulation of sea urchin eggs, Exptl. Cell Res., 68:291.

Hiramoto, Y., 1981, Mechanical properties of dividing cells, in: "Mitosis/Cytokinesis", A.M. Zimmerman and A. Forer, eds., Academic Press, New York.

Kiehart, D., Mabuchi, I., and Inoue, S., 1982, Evidence that myosin does not contribute to force production in chromosome movement, J. Cell Biol., 94:165.

Lewis, C.A., Chia, F.-S., and Schroeder, T.E., 1973, Peristaltic constrictions in fertilized barnacle eggs (Pollicipes polymerus), Experientia, 29:1533.

Mabuchi, I., 1986, Biochemical aspects of cytokinesis, Int. Rev. Cytol., 101:175.

Mabuchi, I., Tsukita, Sh., Tsukita, Sa., and Sawai, T., 1985, Properties of the cleavage furrow isolated from newt eggs, Zool. Sci., 2:949.

Mabuchi, I., and Okuno, M., 1977, The effect of myosin antibody on the division of starfish blastomeres, J. Cell Biol., 74:251.

Marsland, D., and Landau, J., 1954, The mechanisms of cytokinesis: temperature-pressure studies on the cortical gel system in various marine eggs, J. Exptl. Zool., 125:507.

Meeusen, R.L., Bennett, J., and Cande, W.Z., 1980, Effect of microinjected N-ethylmaleimide-modified heavy meromyosin on cell division in amphibian eggs, J. Cell Biol., 86:858.

Nunnally, M.H., D'Angelo, J.M., and Craig, S.W., 1980, Filamin concentration in cleavage furrow and midbody region: frequency of occurrence compared with that of alpha-actinin and myosin, J. Cell Biol., 87:219.

Ohtsubo, M., and Hiramoto, Y., 1985, Regional differences in mechanical properties of the cell surface in dividing echinoderm eggs, Develop. Growth Differ., 27:371.

Perry, M.M., John, A.H., and Thomas, N.S.T., 1971, Actin-like filaments in the cleavage furrow of newt eggs, Exptl. Cell Res., 65:249.

Pickett-Heaps, J.D., Tippit, D.M., and Andreozzi, J.A., 1982, Cell division in the pennate diatom Pinnularia. II, Later stages of mitosis, Biol. Cell., 33:79.

Poenie, M., Alderton, J., Tsien, R.Y., and Steinhardt, R.A., 1985, Changes of free calcium levels with stages of the cell division cycle, Nature, 315:147.

Rappaport, R., 1971, Cytokinesis in animal cells, Int. Rev. Cytol., 31:169.

Rappaport, R., 1973, On the rate of movement of the cleavage stimulus in sand dollar eggs, J. Exptl. Zool., 183:115.

Rappaport, R., 1976, Furrowing in altered cell surfaces, J. Exptl. Zool., 195:271.

Rappaport, R., 1981, Cytokinesis: cleavage furrow establishment in cylindrical sand dollar eggs, J. Exptl. Zool., 217:365.

Rappaport, R., 1982, Cytokinesis: the effect of initial distance between mitotic apparatus and surface on the rate of subsequent cleavage furrow progress, J. Exptl. Zool., 221:399.

Rappaport, R., 1983, Cytokinesis: furrowing activoty in nuclrated endoplasmic fragments of fertilized sand dollar eggs, J. Exptl. Zool., 227:247.

Rappaport, R., 1985, Repeated furrow formation from a single mitotic apparatus in cylindrical sand dollar eggs, J. Exptl. Zool., 234:167.

Rappaport, R., 1986, Mitotic aparatus-surface interaction and cell division, Int. J. Invert. Reprod. Dev., 9:263.

Rappaport, R., and Rappaport, B., 1983, Cytokinesis: effects of blocks between the mitotic apparatus and the surface on furrow establishment in flattened echinoderm eggs, J. Exptl. Zool., 227:213.

Rappaport, R., and Rappaport, B.N., 1984, Division of constricted and urethane-treated sand dollar eggs: a test of the polar stimulation hypothesis, J. Exptl. Zool., 231:81.

Rappaport, R., and Rappaport, B.N., 1985, Surface contractile activity associated with isolated asters in cylindrical sand dollar eggs, J. Exptl. Zool., 235:217.

Rogalski, A.A., and Singer, S.J., 1985, An integral glycoprotein associated with the membrane attachment sites of actin microfilaments, J. Cell. Biol., 101:785.

Sanger, J.M, and Sanger, J.W., 1980, Banding and polarity of actin filaments in interphase and cleaving cells, J. Cell Biol., 86:568.

Sawai, T., 1979, Cyclic changes in the cortical layer of non-nucleated fragments of the newt's egg, J. Embryol. Exp. Morph., 51:183.

Schantz, A.R., 1985, Cytosolic free calcium-ion concentration in cleaving embryonic cells of Oryzias latipes measured with calcium-selective electrodes, J. Cell Biol., 100:947.

Schroeder, T.E., 1970, The contractile ring. I. Fine structure of dividing mammalian (HeLa) cells and the effects of cytochalasin B, Z. Zellforsch., 109:431.

Schroeder, T.E., 1972, The contractile ring. II. Determining its brief existence, volumetric changes, and vital role in cleaving Arbacia eggs, J. Cell Biol., 53:419.

Schroeder, T.E., 1973, Actin in dividing cells: contractile ring filaments bind heavy meromyosin, Proc. Nat. Acad. Sci. USA, 70:1688.

Schroeder, T.E., 1975, Dynamics of the contractile ring, in: "Molecules and Cell Movement", S. Inoue and R.E. Stephens, eds., Raven Press, New York.

Schroeder, T.E., 1976, Actin in dividing cells: evidence for its role in cleavage but not mitosis, in: "Cell Motility", R. Goldman, T. Pollard, and J. Rosenbaum, eds., Cold Spring Harbor Laboratory.

Schroeder, T.E., 1978, Cytochalasin B, cytokinesis, and the contractile ring, in: Cytochalasins - Biochemical and Cell Biological Aspects", S.W. Tanenbaum, ed., Elsevier/NorthHolland, Amsterdam.

Schroedr, T.E., 1981a, Interrelations between the cell surface and the cytoskeleton in cleaving sea urchin eggs, In "Cytoskeletal Elements and Plasma Membrane Organization", G. Poste and G.L. Nicolson, eds., Elsevier/North Holland, Amsterdam.

Schroeder, T.E., 1981b, The origin of cleavage forces in dividing eggs, Exptl. Cell Res., 134:231.

Schroeder, T.E., 1985, Physical interactions between asters and the cortex in echinoderm eggs, in: "The Cellular and Molecular Biology of Invertebrate Development", R.H. Sawyer and R.M. Showman, eds., Univ. S. Carolina Press, Columbia.

Schroeder, T.E., 1986, The egg cortex in early development of sea urchins and starfish, in: "Developmental Biology, Vol. 2", L. Browder, ed., Plenum, New York.

Schroeder, T.E., in preparation.

Scott, J.. 1986, Ultrastructure of cell division in the unicellular red alga Flintiella sanguinaria, Can. J. Bot., 64:516.

Shimizu, T., 1975, Occurrence of microfilaments in the Tubifex eggs undergoing the deformation movement, J. Fac. Sci. Hokkaido Univ. (Ser. 4, Zool.), 20:1.

Shimizu, T., 1979, Surface contractile activity of the Tubifex egg: its relationship to the meiotic apparatus functions, J. Exptl. Zool., 208:361.

Shimizu, T., 1981, Cyclic changes in shape of a non-nucleate egg fragment of Tubifex, Dev. Growth Differ., 23:101.

Shimizu, T., 1983, Organization of actin filaments during polar body formation in eggs of Tubifex, Eur. J. Cell Biol., 30:74.

Shinagawa, A., 1986, Close correlation of the periodicities of the cleavage cycle and cytoplasmoc cycle in early newt embryos, Dev. Growth Differ., 28:251.

Sisken, J.E., 1981, Inhibitors and stimulators in the study of cytokinesis, in: "Mitosis/Cytokinesis", A.M. Zimmerman and S. Forer, eds., Academic Press, New York.

Usui, N., and Yoneda, M., 1982, Ultrastructural basis of the tension increase in sea-urchin eggs prior to cytokinesis, Dev. Growth Differ., 24:453.

Wolpert, L., 1960, The mechanics and mechanism of cleavage, Int. Rev. Cytol., 10:163.

Wordeman, L., McDonald, K.L., and Cande, W.Z., 1986, The distribution of cytoplasmic microtubules throughout the cell cycle of the centric diatom Stephanopyxis turris: their role in nuclear migration and positioning the mitotic spindle during cytokinesis, J. Cell Biol., 102:1688.

Yoneda, M., Ikeda, M., and Washitani, S., 1978, Periodic change in the tension at the surface of activated non-nucleate fragments of sea-urchin eggs, Dev. Growth Differ., 20:329.

Yoneda, M., Kobayakawa, Y., Kubota, H.K., and Sakai, M., 1982, Surface contraction waves in amphibian eggs, J. Cell Sci., 54:35.

Yoneda, M., and Schroeder, T.E., 1984, Cell cycle timing in colchicine-treated sea urchin eggs: persistent coordination between the nuclear cycles and the rhythm of cortical stiffness, J. Exptl. Zool., 231:367.

Yoneda, M., and Yamamoto, K., 1985, Periodicity of cytoplasmic cycle in non-nucleate fragments of sea urchin and starfish eggs, Dev. Growth Differ., 27:385.

Yonemura, S., and Kinoshita, S., 1986, Actin filament organization in the sand dollar egg cortex, Develop. Biol., 115:171.

Yoshimoto, Y., and Hiramoto, Y., 1985, Cleavage in a saponin model of the sea urchin egg, Cell Struct. Funct., 10:29.

Yoshimoto, Y., Iwamatsu, T., and Hiramoto, Y., 1985, Cyclic change in intracellular free calcium levels associated with cleavage cycles in echinoderms and medaka eggs, Biomed. Res., 6:387.

Yumura, S., and Fukui, Y., 1985, Reversible cyclic AMP-dependent change in distribution of myosin thick filaments in Dictyostelium, Nature, 314:194.

LARGE DEFORMATION THEORY OF MEMBRANES AS APPLIED TO CELL DIVISION AND POSSIBILITY OF VARIOUS STRAIN ENERGY DENSITY FUNCTIONS

Murat Dikmen

Boğaziçi University
Istanbul

INTRODUCTION

The purpose of this paper is to make some purely mechanical comments on the phenomenon of cytokinesis, viewed as the division of a membrane filled with a liquid. Thus, except explanatory arguments in connection with the working hypotheses adopted in the next section and a few closing remarks, we shall not dwelve on the biological aspects of the phenomenon. These aspects can be found in the now extensive literature and they have been indeed thourcughly discussed with authority in other lectures of this workshop, and need not be reviewed here again.

Mechanical interpretations have been given in the last two decades by various authors to the cleavage of animal cells. However the first paper using the theory of large deformations of membranes and nonlinear elasticity is due to Pujara and Lardner (1979). They considered a spherical membrane enclosing an incompressible fluid. They admitted the theory of equatorial constriction, according to which cell division is caused by forces acting along a contractile ring. They used constitutive equations of the type of Mooney-Rivlin material and STZC material (Skalak et al, 1973), the latter also with a modified version to take account of the wrinkling effect in the case of a negative principal stress. They concluded that the STZC material provides with a good description of the elastic properties of the cell membrane. Some of the results of this paper were discussed by Akkaş (1980 a) and Pujara and Lardner (1980) in their reply. This discussion drew attention to the fact that for a theory to be in concordance with experimental findings (Hiramoto, 1968), it should be able to describe the decrease of the internal pressure after an increase which accurs in the early stages of cleavage. This observation lead Akkaş (1980 b) to the adoption of a "stage-dependent" coefficient of elasticity so that "the numerical results on the intracellular and the ring force will agree with the corresponding experimental results". Akkaş also considers an initially spherical elastic membrane, dividing by contraction of an equatorial ring. In contrast to the assumptions made by Pujara and Lardner (1979), he assumes that the membrane is initially slightly inflated and that the volume enclosed by the cell membrane may change during cleavage. He adopts the constitutive equations of the Mooney-Rivlin (or, more specifically, Neo-Hookean) material. Later, Akkaş (1981) developed a viscoelastic model, but concluded that this model did not result in any conclusion which was different from or not arrived at by his previous elastic model.

Considering the ultrastructure of animal cells, Akkaş and Engin (1981) discussed a mechanism that may give an acceptable explanation for the change of the elastic properties of the membrane during cytokinesis.

As it shall be discussed in the last section of this paper, the question of deciding in favour of a definite constitutive equation remains open at present. It seems therefore worth trying to see how far one can go in the mechanical study of cytokinesis <u>without</u> a specific constitutive law. This is what shall be attempted here. The mathematical and mechanical background required is rather elementary. The strict minimum of the needed differential geometry of plane curves and surfaces of revolution is rapidly reviewed. The solution of a simple differential equation is shown in detail. The values of elliptic integrals introduced by this solution are available in standard tables. The equations of equilibrium are rederived here. All of the numerical computations needed in this work could be made by hand, but a non-programmable pocket calculator was most conveniently used.

HYPOTHESES

We shall first list and discuss a number of consistent hypotheses. They correspond to the image we make of a cell as a mass of plasma enclosed in a membrane, but they are not evident. They may be considered however as convenient working hypotheses.

H_o: <u>The "cell membrane" behaves mechanically like a "membrane"</u>.
This is the fundamental hypothesis. It implies the existence of an outer layer structurally independent of the contained material. Strictly speaking, this cortical layer consists of a lipid bilayer and underlying cytoskeletal components. A well-defined internal boundary does not exist. Hence the thickness of the cortical layer also is not well-defined. Nevertheless, Pujara and Lardner (1979) and Akkaş (1980 b) base their investigations on this fundamental hypothesis.

A rigorous discussion of the mechanical concept of membrane on the basis of general shell theory has been given by Dikmen (1982). This concept implies that the membrane be very thin so that in mathematical modelling it can be assimilated to a surface, and perfectly flexible. The actual thickness of the membrane does not enter the equations of equilibrium. The thickness of the membrane appears in the constitutive equation when this latter is obtained from the three-dimensional continuum model of the material.

H_1: <u>At the onset of cleavage, the cell membrane has a spherical shape</u>.
This hypothesis relies on the observation that prior to division the cell attains a roughly spherical shape. It may therefore be considered as a convenient approximation providing with an appreciable simplification The surface representing the membrane at the onset of cell division is then, according to the hypothesis, a sphere. Actually, the cell membrane exhibits processes called microvilli and also depressions. Thus, even in terms of mathematical modelling, the representative surface of the cell at any stage (spherical at the onset of the cleavage) should be considered as an average over a less regular representative surface. This remark is important from the membrane-theoretical point of wiev (Dikmen, 1970 and 1982) and we shall come back to this remark in H_2 and H_5.

H_2: <u>The membrane is initially slightly inflated</u>.
There is an important theorem of differantial geometry, stating that isometric closed convex surfaces are congruent. But if an

arbitrarily small portion is cut off, the remaining surface ceases
to be the only surface with a given metric. There are, nevertheless,
congruence theorems for isometric convex caps also. For more details,
we refer to Dikmen (1982). In particular, these congruence theorems
will retain their validity in the cases of complete spheres and
spherical caps. But as pointed out above, neither the cell at the
onset of cleavage is everywhere locally convex, nor do the two lobes
of the dividing cell constitute everywhere locally convex caps.

Besides these purely geometric considerations which suggest that
there should be some internal pressure in excess of the pressure
of the external medium, biological observations also show that there
is an initial stretch of the cell surface. Akkaş (1980 b and 1981)
assumed initial inflation.

H_3: <u>Cleavage is caused by contraction of an "equatorial ring".</u>
Among the many theories advanced with the purpose of explaining the
mechanism of cytokinesis, the theory of equatorial constriction
finds currently wide acceptance. Since the initial surface in the
process of cleavage has been assumed to be a sphere (cf.H_1), any
great circle should be equally liable to act as the contractile
equatorial ring. Biologically, however, the location of the contractile ring is determined by the position of the mitotic spindle.

H_4: <u>The constitutive law describing the behaviour of the membrane is</u>
<u>time-independent. The material is assumed to be inert.</u>
According to H_8, the division of cell membrane will be considered
as a quasi-static process, i.e. as a succession of independent
states of equilibrium. It is therefore consistent to admit only
those constitutive laws which are time-independent. In this respect,
elasticity does not have to be postulated at the outset. Material
properties do not affect the equilibrium conditions for membranes.
It will be therefore sufficient to assume that possible heterogeneity or anisotropy of the material will be such as to preserve the
rotational symmetry of the dividing cell, throughout the process
of cleavage (see remark at the end of this section). The last part
of the statement excludes chemical interaction or loss of material
in the cortical layer. We also do not consider the material properties peculiar to the narrow furrow region where the contraction
ring is situated.

H_5: <u>The membrane cannot sustain compression stresses.</u>
This is an assumption which may be thought of either as a part of
the specification of material properties and hence relate to the
constitutive law (cf.H_4) also, or as a geometric ability to deform
in absence of tensile forces (cf.H_2).

H_6: <u>The intracellular pressure is uniform, but may change with the</u>
<u>deformation of the membrane.</u>
Even with H_7, the intracellular pressure will change as tensile
forces in the membrane change.

H_7: <u>The volume enclosed by the cell membrane remains nearly constant</u>
<u>during cleavage.</u>
Invariance of the volume was assumed by Pujara and Lardner (1979).
Akkaş (1980 b) admitted possibility of change of volume during the
process of cleavage. In his viscoelastic model (Akkaş, 1981), he
reverted to the constant volume hypothesis, observing that "the
numerical results presented in the previous work showed that the
constant and changing volume conditions both yield qualitatively
the same results, and the quantitative difference between the re-

sults is less than 2 % ". Our assumption, that the volume remains "nearly" constant permits to neglect variations of volume in calculations, while small compressibility and/or exchange of matter in small amounts with the external medium are not physically excluded.

H_8: <u>The division of the cell membrane may be considered as a quasi-static process.</u>
I.e. the process of division can be considered as a succession of independent states of equilibrium. Hereby, dynamic effects are neglected and the time variable eliminated. This hypothesis is justified by the fact that cell division is a rather slow process and does therefore not involve appreciable accelerations.

It is clear that as a consequence of H_1, H_3 and H_6, the membrane will retain rotational symmetry about an axis orthogonal to the plane of the equatorial ring.

ON THE GEOMETRY OF SURFACES OF REVOLUTION

Let us start with a plane curve which exhibits symmetry with respect to a straight line, and take this line as the z-axis while we denote by r the distance of a generic point of the curve to the same line (Fig.1). The position of any point of the curve is then determined by giving z as a function $z(r)$ of r. The radius of curvature of the curve is given by

$$R_1 = \frac{(1 + z')^{3/2}}{z''} \tag{1}$$

where primes stand for differentiation with respect to r. For later use, we call ϕ the angle made by the z-axis and the normal to the curve, oriented according to this order.

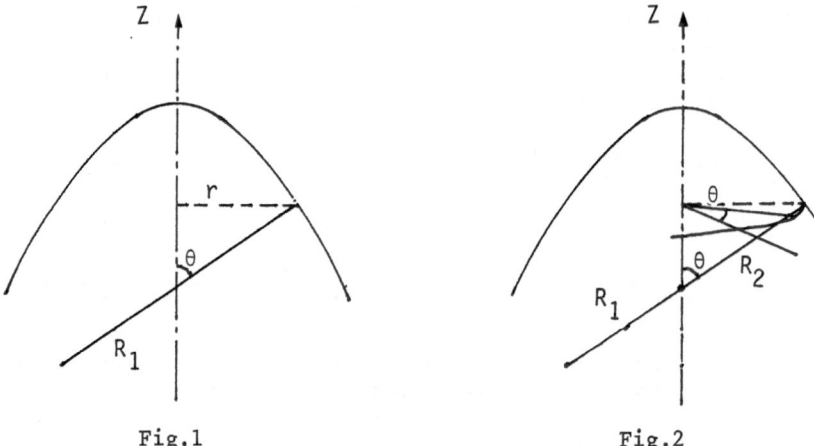

Fig.1   Fig.2

A surface of revolution is generated by the plane curve when its plane is made to rotate about the z-axis. The various positions of the curve are then called the meridians of the surface, and the circles described by each point of the curve are called its parallels. The z-axis is the axis of revolution (Fig.2). The meridians and parallels are mutually orthogonal everywhere on the surface. One of the principal radii of curvature of the surface is R_1, given by eqn. (1), while the second of these is given by (Fig.2)

$$R_2 = \frac{r}{\sin\phi} \tag{2}$$

EQUATIONS OF EQUILIBRIUM

Instead of taking the general equations of equilibrium of membranes, to then simplify them in the case of membrane of revolution under uniform internal pressure, we proceed directly rederiving the equations for this simple case. We consider therefore the equilibrium of a membrane element, cut out by two meridians and two parallels, each of the pairs corresponding to the infinitesimal angle increment $d\phi$ and $d\theta$, respectively (Fig.3). Because of the symmetry of membrane and loading, no shear forces (in the tangent plane) are acting, while there are meridional forces N_ϕ and N_θ, normal to the edges (per unit of length of the edges). p is internal pressure.

By projection along the normal to the surface, we have

$$N_\theta \, d\theta \, R_1 \, d\phi \, \sin\phi + N_\phi \, d\phi \, r \, d\theta = p \, r \, d\theta \, R_1 \, d\phi$$

Dividing by rR_1 and taking into account (2), we find

$$\frac{N_\phi}{R_1} + \frac{N_\theta}{R_2} = p \tag{3}$$

By projection along the meridian (Fig.4), we have

$$(\pi r^2)p = N_\phi \, (2\pi r) \, \sin\phi$$

or

$$N_\phi = \frac{1}{2} \, p \, \frac{r}{\sin\phi}$$

or, by virtue of (2),

$$N_\phi = \frac{1}{2} \, pR_2 \tag{4}$$

Combining (3) and (4), we obtain the set

$$\left.\begin{aligned} N_\phi &= \frac{1}{2} \, R_2 \, p \\[2mm] N_\theta &= \frac{2R_1 - R_2}{2R_1} \, R_2 \, p \end{aligned}\right\} \tag{5}$$

Thus for given shape and pressure, the membrane forces are expressed explicitly, without intermediary constitutive equation.

THE "BUBBLE" MODEL

Since the macro-description of the cell at the onset of cleavage is given by a sphere (cf.H_1), the membrane equations (5) yield

$$N_\phi = N_\theta = \frac{1}{2} R p \tag{6}$$

where

$$R = R_2 = R_1 \tag{7}$$

is the radius of the sphere.

The mechanical meaning of (6) is that the membrane is under tension, equal in all directions and of the same amount everywhere. Conversely, if the tension is to be equal in all directions, the principal radii of curvature must be equal at any point, and it is known from geometry that such a surface can only be a sphere (if not a plane).

Now let us consider a case of a dividing cell, in which the two lobes which are going to become the daughter cells remain spherical throughout the process. I.e., membrane (tensile) forces are everywhere the same. In other words, the lobes behave like soap "bubbles" (Fig.5).

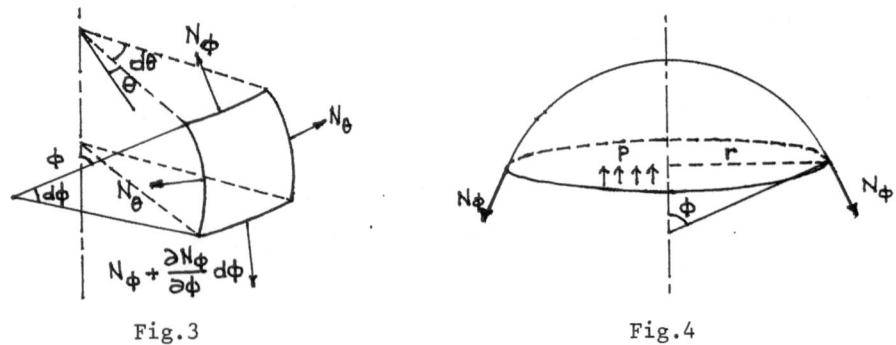

Fig.3   Fig.4

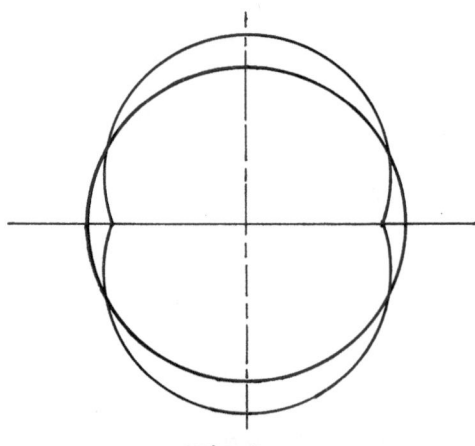

Fig.5

At this stage, it is useful to list the elementary formulas for the volume V and the area A of the complete sphere with radius R

$$V = \frac{4\pi}{3} R^3$$

$$A = 4\pi R^2$$

as well as the volume V_H and area the A_H of the segment of sphere of radius R and height H

$$V_H = \frac{1}{6} \pi H (3 r_f^2 + H^2)$$

$$A_H = \pi (r_f^2 + H^2)$$

Here, r_f stands for the furrow radius (Fig.6). If, in view of H_7, we equate half of the volume of the complete sphere to the total volume of a segment, we have

$$\frac{V}{2} = \frac{1}{2} \frac{4\pi}{3} R^3 = \frac{1}{6} \pi H (3 r_f^2 + H^2) = V_H$$

and therefore

$$H^3 + (3 r_f^2) H - (4R^3) = 0 \tag{8}$$

which is an equation of third degree for H, when r_f and R are given. Here

$$0 \le r_f \le R$$

With the furrow ratio f defined by

$$Rf \equiv r_f$$

$$(0 \le f \le 1)$$

eqn. (8) becomes

$$H^3 + (3R^2 f^2) H - (4R^3) = 0 \tag{9}$$

It is easy to see that this equation has only one real root $\bar{H}$ (and two complex conjugate roots) which is given by the Cardano formula:

$$\bar{H} = R \sqrt[3]{2+\sqrt{f^6+4}} + R \sqrt[3]{2-\sqrt{f^6+4}} \tag{10}$$

$\bar{H}$ is called the polar length.
Setting

$$\ell \equiv \frac{\bar{H}}{R} \tag{11}$$

for a non-dimensional measure of the polar length, eqn. (10)

$$\ell = \sqrt[3]{2+\sqrt{f^6+4}} + \sqrt[3]{2-\sqrt{f^6+4}} \tag{12}$$

The variation of ℓ as a function of f is shown in Fig.7.

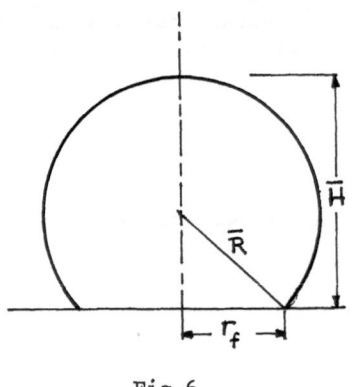

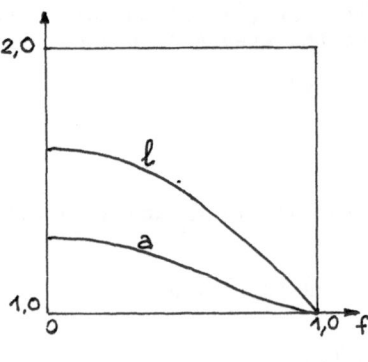

Fig.6　　　　　　　　　　　　　　Fig.7

Under these circumstances, the total area of the dividing cell membrane is given by

$$2A_{\bar{H}} = 2\pi R^2 \ (f^2 + \ell^2)$$

or, in non-dimensional form, by dividing by A,

$$a \equiv \frac{2A_{\bar{H}}}{A} = \frac{f^2 + \ell^2}{2} \tag{13}$$

The variation of a as a function of f is also shown in Fig.7.
We find here the increase by about 26 percent observed in cytokinesis. We shall comment later on the polar length.

It is easy to calculate the ring force F acting in the plane of the furrow ring along the unit length of the boundary of one of the lobes (Fig.8). From the equilibrium of membrane forces and ring forces at the boundary, and a geometric relation which is obvious from Fig.8, we obtain

$$F = \sqrt{(\frac{p\bar{R}}{2})^2 - (\frac{pRf}{2})^2} = \frac{p}{2}\sqrt{\bar{R}^2 - R^2 f^2} = \frac{p}{2}\sqrt{(\frac{R^2 f^2}{2H} + \frac{H}{2})^2 - R^2 f^2} = [\frac{1}{2}(\ell - \frac{f^2}{\ell})](\frac{pR}{2}) \tag{14}$$

F is directed inward of the furrow ring.
For f=1, F is equal to zero, and for f=0 the ring force becomes equal to the membrane force in the daughter cell, i.e.

$$F = \frac{1}{2} \ (\frac{pR}{2}) \quad \ell = \frac{p\bar{R}}{2}$$

in this particular stage, as it should be.
Again for one of the lobes, the annular tension is given by

$$[\frac{1}{4} (\ell - \frac{f^2}{\ell}) \ f \ ] \ (pR^2) \tag{15}$$

and the axial ring force by

$$[\ f^2] \ p \ (\pi R^2) \tag{16}$$

The quantities in square brackets in (14),(15) and (16) are plotted in Fig.9 against f, to show the variation of ring force, annular tension and axial ring force, respectively, in different stages of cleavage.

238

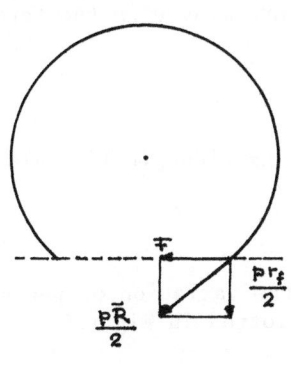

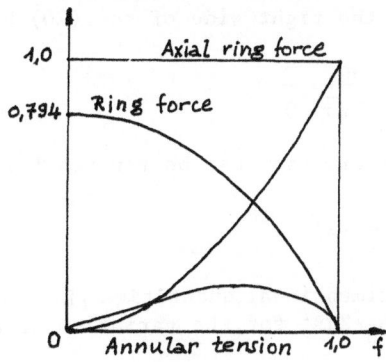

Fig.8   Fig.9

According to H_6, the pressure may change during the process, which is considered to be quasi-static in conformity with H_8. Hence the principle of virtual work applies:

$$-2\pi \, r_f \, F \, \delta r_f + p \, \delta V_{\bar{H}} + \alpha \, \delta A_{\bar{H}} = 0 \tag{17}$$

where α is a (dimensional) surface tension coefficient.

It can be readily verified that $\delta V_{\bar{H}} = 0$, as it should be in view of H_7. The second term in (17) therefore drops. For the remaining variation $\delta A_{\bar{H}}$, we note that

$$\delta \bar{H} = R \, [(2-\sqrt{f^6+4})^{2/3} - (2+\sqrt{f^6+4})^{2/3}] \, \frac{f}{\sqrt{f^6+4}} \, \delta f \tag{18}$$

and find

$$\delta A_{\bar{H}} = \{-\sqrt{f^6+4} \, -f^2 \, [(2-\sqrt{f^6+4})^{1/3}-(2-\sqrt{f^6+4})^{1/3}]\} \, \frac{f}{\sqrt{f^6+4}} \, \delta f \tag{19}$$

While the term containing expicitly p, i.e. the second term, in eqn.(17) disappears, p enters again via F in the first term [cf.eqn.(14)]. Thus p can be found from eqn.(17)

$$p = \frac{4\alpha}{R} \cdot \frac{-(\sqrt{f^6+4}) \, [(2+\sqrt{f^6+4})^{1/3}+(2-\sqrt{f^6+4})^{1/3}]-f^2 \, [(2-\sqrt{f^6+4})^{2/3}-(2+\sqrt{f^6+4})^{2/3}]}{(\sqrt{f^6+4})\{3 \, f^2-[(2-\sqrt{f^6+4})^{2/3}+(2+\sqrt{f^6+4})^{2/3}]\}} \tag{20}$$

Obviously, p is explicitly a function of f but also a function of the values taken by α. On the other hand, α is a function of strain, which is again related to f, i.e. the stage of cleavage.

α is a constitutive function. So far, p cannot be determined unless α is known.

It is interesting to look at the values taken by the right side of eqn. (20) in the particular cases in which f=0 and f=1, respectively. For f=0, we find

$$p = \frac{\alpha}{R} \, (4)^{2/3} \tag{21}$$

For f=1, the right side of eqn.(20) becomes indeterminate in the form

$$p = \frac{4\alpha}{R\sqrt{5}} \frac{0}{0}$$

This indeterminacy can be removed however by using L'Hospital's rule to find

$$p = \frac{\alpha}{R} \quad (2)$$

The non-dimensional quantities pR/α, $pR/2\alpha$ for the variation of pressure, as well as $A_H/2\pi R^2$ for the variation of area are plotted in Fig.(10).

THE "BAG" MODEL

Here we invoke the hypothesis H_5 and ask for a surface of revolution representative of a membrane under uniform pressure in which the hoop forces vanish everywhere, i.e. $N_\theta=0$. In this case, eqn.(5)$_2$ yields immediately

$$R_1 = \frac{R_2}{2}$$

Then, using eqns.(1) and (2), we have the ordinary differential equation of second order

$$\frac{z' (1+z'^2)}{z''} = \frac{1}{2} r \quad (22)$$

for the meridian of the surface.

Setting

$$w \equiv z' \quad (23)$$

eqn.(22) can be rewritten in the form

$$\frac{w'}{w(1+w^2)} = \frac{2}{r} \quad (24)$$

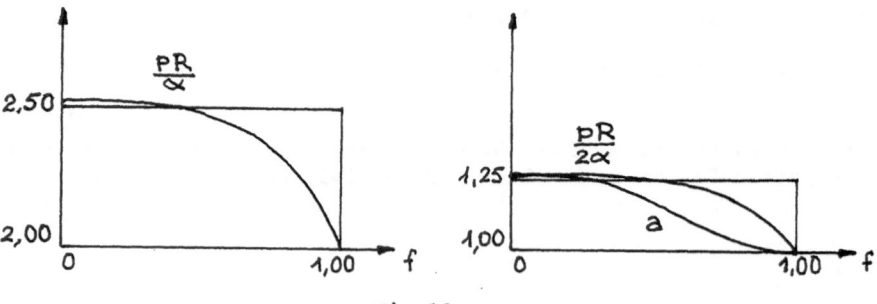

Fig.10

or

$$\frac{1}{2} d \left[\ln\left(\frac{w^2}{1+w^2}\right)\right] = \frac{2}{r} dr$$

which can be immediately integrated to give

$$\frac{w^2}{1+w^2} = \left(\frac{r^2}{C}\right)^2 \qquad \text{(C: arbitrary positive constant)} \qquad (25)$$

(25) in conjunction with (23) gives

$$\frac{dz}{dr} = \pm \frac{\frac{r^2}{C}}{\sqrt{1-\left(\frac{r^2}{C}\right)^2}} \qquad (26)$$

with

$$\left.\begin{array}{l} \dfrac{r}{\sqrt{C}} \equiv \xi \\[12pt] \dfrac{z}{\sqrt{C}} \equiv \zeta \end{array}\right\} \qquad (27)$$

eqn. (26) takes the non-dimensional form

$$\frac{d\zeta}{d\xi} = \pm \frac{\xi^2}{\sqrt{1-\xi^4}} \qquad (28)$$

In order to have a differential equation with real coefficients, there must be

$$\frac{r^2}{C} \leq 1$$

or

$$\xi^2 \leq 1$$

ζ as a function of ξ is obtained by quadrature from (28):

$$\zeta = \pm \int \frac{\xi^2}{\sqrt{1-\xi^4}} d\xi \qquad (29)$$

This implies the use of elliptic integrals of the first and second kinds. The standard form of an elliptic integral of the first kind is

$$F(\psi,k) = \int_0^\psi \frac{d\theta}{\sqrt{1-k^2 \sin^2\theta}} \qquad (30)$$

and the standard form of an elliptic integral of second kind is

$$E(\psi,k) = \int_0^\psi \sqrt{1-k^2 \sin^2\theta} \; d\theta \qquad (31)$$

241

The values taken by these integrals, for different values of the modulus k and of the variable limit ψ are tabulated.

It is easy to see, that since $-1 \leq \xi \leq 1$,

$$\zeta(\psi) = \pm \frac{1}{\sqrt{2}} \{2 \, E \, (\psi, \frac{1}{\sqrt{2}}) - F \, (\psi, \frac{1}{\sqrt{2}})\} \tag{32}$$

The closed convex curve representing this solution is shown in Fig.11.

As it can be seen, this curve is very flat in the vicinity of the points $(0,1)$ and $(0,-1)$, and it can be verified that the curvature vanishes at these points, i.e. the radius of curvature becomes infinite.

If we let this curve rotate about the ζ-axis, we obtain a closed convex surface. A family of such surfaces may then be obtained from (32) and (27), by assigning various values to the positive constant C.

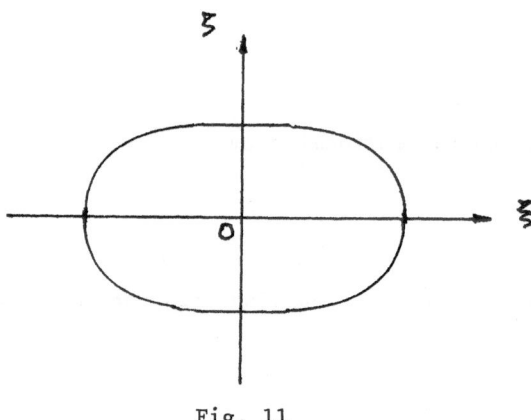

Fig. 11

Volumes enclosed by these surfaces, between the equatorial plane and a parallel plane which passes through the point $(0,z)$ are given by

$$\hat{V} = \pi \int_0^z r^2 \, dz = \pi \, C^{3/2} \int_1^\xi \frac{\xi^4}{\sqrt{1-\xi^4}} \, d\xi$$

Introducing again the elliptic integrals of first kind,

$$\hat{V} \, (\psi) = \frac{\pi}{3} \, C^{3/2} \{ \frac{1}{\sqrt{2}} \, F \, (\psi, \frac{1}{\sqrt{2}}) + [\cos\theta \, \sqrt{1-\cos^4\theta}]_0^\psi \} \tag{33}$$

The area of the portion of surface between the two parallel planes specified as above, is obtained by elementary integration:

$$\hat{A} = \pi \, C \, \text{arc sin } \xi^2 \tag{34}$$

Let us now assume that an initially spherical shell, represented by
a unit sphere, i.e. with radius equal to one unit of length, divides into
two lobes which finally have the shape of the surface described above. Each
of these daughter cells will have a volume equal to one half of the volume
of the sphere (cf. H_7):

$$2 \, \hat{V} \left(\frac{\pi}{2} \right) = 2 \, \frac{\pi}{3} \, C^{3/2} \{ \frac{1}{\sqrt{2}} \, F \left(\frac{\pi}{2} , \frac{1}{\sqrt{2}} \right) \} = \frac{1}{2} \, \frac{4\pi}{3}$$

It can be read from the tables for elliptic integrals of the first kind that

$$F(\frac{\pi}{2} , \frac{1}{\sqrt{2}}) = 1,311$$

Hence

$$C^{1/2} = 0,914$$

This value gives also the equatorial radius of the daughter cells. With this
value of $C^{1/2}$, the total area of each of these cells can be calculated to
be equal to 1,312. The corresponding value of the polar length is 1,097.

This model of membrane shall be referred to here as the "bag" model.

COMPARISON OF "BUBBLE" AND "BAG" MODELS

First of all, we point out that both models are statically and kine-
matically possible.

It is easy to see that each of models corresponds to an extreme case
concerning the mechanical behaviour of the cell membrane. In fact, if one
considers the non-dimensionalized polar lengths for f=0, one sees that the
value is 1,587 for the "bubble", while it is just 1,092 for the "bag". An
envolope of experimental results obtained by Hiramoto and By Yoneda and Dan
can be found in the papers by Akkaş (1980 b and 1981). For f=0, this envelope
presents lower and upper limits approximately equal to 1,4 and 1,5, respectively. So, the "bubble" gives too high values, and the "bag" gives too low
values.

At this point, we draw attention to the fact that we did not use as
yet explicite the hypothesis H_2. Akkaş (1980 b) assumes that the spherical
cell is initially inflated by 5% of its radius. What he calls "initial radius" and uses in calculating the non-dimensionalized polar lengths is this
inflated radius.

Returning to the comparison of the two models, we observe that the
increase by about 25% in the total surface area, found experimentally, is
best approximated by the "bubble", while this increase is about 31% in the
case of the "bag".This suggests that the actual deformation of the cell
membrane should be closer to the bubble model, at least in the early stages
of cleavage. Indeed, a sudden jump from the finite radius of the initial
sphere to the infinite radius of curvature of the bag meridian at the distal
vertices of the daughter cells is unrealistic. Another important feature of
the "bag" is that at the vertices the membrane forces become infinite. This
also is physically inadmissible. Thus it is reasonable to think that the
cleavage, beginning first by costriction of the equatorial ring, progresses
gradually so that, the distal part of the initial half sphere retains an

almost spherical shape, while the region in the vicinity of the equator will assume more and more the shape of the "bag", like a closing draw-string bag. This modelling is adequate also in taking account of the wrinkles appearing near the equatorial ring. In fact the "bag" model, with vanishing hoop forces and viewed as a macro-description as already indicated, allows the existence of wrinkles.

These considerations lead to the idea of a hybrid representation of the shape of the lobes of a dividing cell. We proceed now to a brief account of two of such representations.

HYBRID MODELLING

A hybrid model can be obtained by taking a portion of the bag surface and completing it by adjoining a portion of a surface of revolution (about the z-axis), such that the two portions form a convex closed surface, satisfying some conditions. Namely:

i) The two portions must have the same curvature R_1 at the joint. This implies that the function representative of the meridian curve as well as the first and second order derivatives be continuous.

ii) The volume of the closed surface thus formed must be equal to half of the volume of the initial sphere.

Two possible cases are shown in Fig.12. Fig.12 a corresponds to a case in which the portion of the bag surface used includes the equatorial zone of this latter. The meridian of the complementing surface may be specified by a polynomial approximation of the form

$$z = a_o + a_2 x^2 + a_4 x^4 + a_6 x^6 \qquad (35)$$

This model is convenient whenever wrinkling occurs also at considerable distance from the equatorial ring of the dividing cell. On the contrary, if wrinkling is to remain within a narrow zone near this ring, then the hybrid model shown in Fig.12 b is advantageous. The corresponding approximation to the meridian curve can be chosen as

$$x^2 = d_o + d_1 y + d_2 y^2 + d_3 y^3 \qquad (36)$$

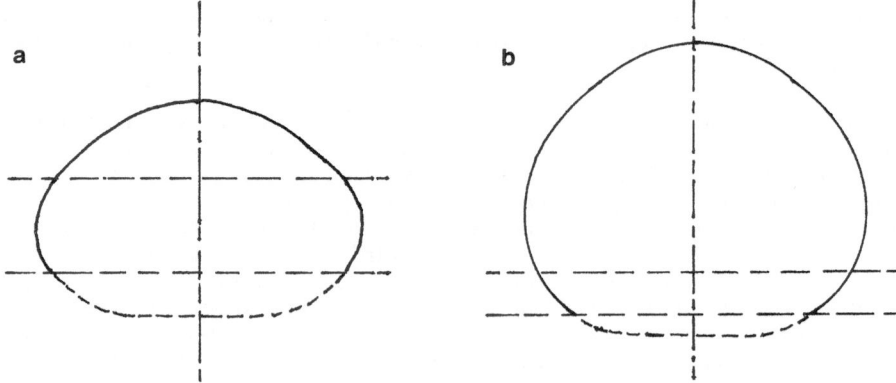

Fig. 12

The sets of coefficients $\{a_0, a_2, a_4, a_6\}$ and $\{d_0, d_1, d_2, d_3\}$ can be adequately determined by somewhat lengthy but straightforward calculations. We refrain from reproducing here all details of the computations and the numerical results obtained thereby. But we bring to attention that, within the understanding of this and the preceding section, the shape of the dividing cell can be calculated to find results which very satisfactorily approximate the geometric results obtained by Akkaş (1980 b).

DISCUSSION

As it is seen, the deformation of a dividing cell is subject to limitations which do not depend on a specific constitutive law of the membrane material. The geometric properties are therefore not very likely to give decisive criteria for the appropriate choice of such a law. On the other hand, the objection raised by Akkaş (1980 a and b) in relation with the decreasing internal pressure in the late stages of the cleavage cannot be settled unless a more refined study of the microstructure of the membrane and its mechanical properties is made.

REFERENCES

Akkaş,N.,1980 a, Letter to the editor, J.Biomechanics, 13:459.
Akkaş,N.,1980 b, On the biomechanics of cytokinesis in animal cells, J.Biomechanics, 13:977.
Akkaş,N.,1981, A viscoelastic model for cytokinesis in animal cells, J.Biomechanics, 14:621.
Akkaş,N.,and Engin,A.E.,1981, Ultrastructure of animal cells and its role during cytokinesis from a structural mechanics viewpoint, in: Proc. Symp. Mechanical Behavior of Structured Media, A.P.S. Selvadurai, ed., Elsevier, Amsterdam.
Dikmen,M., 1982, Theory of Thin Elastic Shells, Pitman, London.
Hiramoto,Y., 1968, The mechanics and mechanism of cleavage in the sea urchin egg, Symp.Soc. Exp. Biol., 22:311.
Pujara,P., and Lardner,T.J., 1979, A model for cell division, J.Biomechanics, 12:293.
Pujara,P., and Lardner,T.J., 1980, Reply to letter by Prof. Akkaş on a "Model for cell division". J.Biomechanics, 13:460.
Skalak,R., Tözeren,A., Zarda,R.P., and Chien.S., 1973, Strain energy function of red blood cell membrane, Biophys,J., 13:245.

ELASTIC MODELS OF CYTOKINESIS

Thomas J. Lardner

Department of Civil Engineering
University of Massachusetts
Amherst, Massachusetts 01003

INTRODUCTION

Cytokinesis, the division of the cytoplasm and the plasma membrane of cells, is a fundamental process in cell physiology. The process in some cells, for example the spherical cells of fertilized sea urchin eggs, can be thought of as the division of a sphere containing cytoplasm into two equal spheres with no change in cell volume [1]. The process is driven by a contractile ring or furrow region near, or attached to, the cell membrane and localized near the cell equator of the dividing cell. The mechanism responsible for the initiation of furrowing and furrow development is apparently a biochemical interaction of the internal cell structures with structures near the cell membrane [2,3].

A complete theory to describe the biomechanics of cell division should provide, first, an explanation of the initial alignment of the cell membrane contractile microfilaments into the equatorial contractile ring, and second, a description of the relation between the force in the contractile ring and the deformation of the cell. Figure 1 shows a simple sequence of mechanical events during cell division. Fundamental to the process is the signal to start the formation of the fibers into the contractile ring, Fig. 1(a). Once formed, the contractile fibers need a stimulus to continue to contract, Fig. 1(b), and finally a stimulus to cause the fibers to break down, Fig. 1(c). The precise driving forces or stimulii for these different steps in cell division are only now becoming fully understood; many different aspects of the process of cytokinesis will be discussed in detail in this volume. It is appropriate to keep in mind a comment by Rappaport [4] in regard to work on understanding

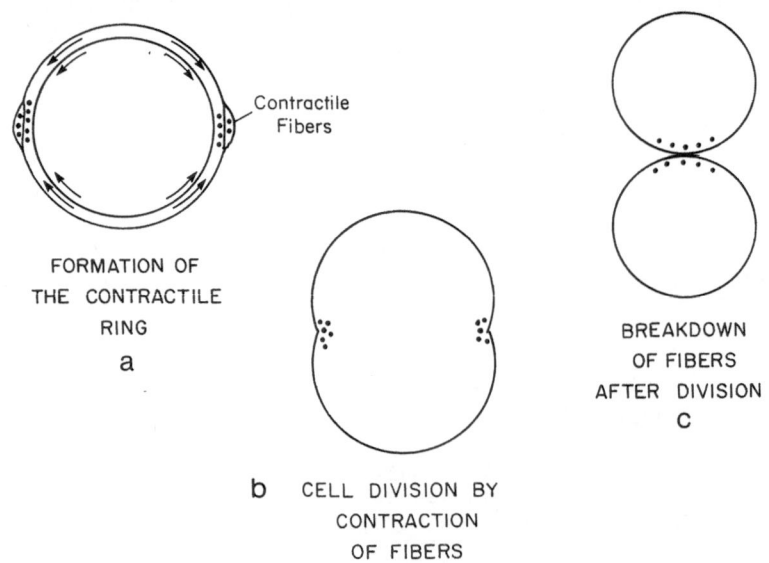

FORMATION OF
THE CONTRACTILE
RING

a

Contractile
Fibers

b   CELL DIVISION BY
CONTRACTION
OF FIBERS

BREAKDOWN
OF FIBERS
AFTER DIVISION

c

Fig. 1.   Sequence of mechanical events during cell division.

cytokinesis.   "...Progress has often been hampered by a kind of
overintellectualization and that the systematic linking together of
observations and measurements with a series of minimal hypothetical
mechanisms that invite experimental analysis has sometimes been
difficult."

The literature on the physiology of cell division of sea urchin eggs
and other species is enormous; references [1-15] are only representative,
and additional references can be found in this volume.   The fertilized egg
of the sea urchin is often chosen for a study of cell division because it
is a convenient species with which to work.   A review of the physical
properties of sea urchin eggs can be found in [9].   The properties and
deformations of biological tissues and cell membranes, especially for red
blood cell membranes, have also been the subject of many investigations in
the biomechanics literature; see [17-25] as representative of some of
these investigations.

In this paper we will discuss and review different aspects of elastic
models for cytokinesis which attempt to relate the overall deformation of
the cell and the cell membrane to the force in the furrow.   We do not

consider the mechanism of the initial formation of the contractile ring, for example, as shown in Fig. 1(a). We will present results for "passive" models that do not contain the active contractile mechanisms driving the constricting furrow. However, we will discuss briefly possible constitutive relations for membrane behavior that do incorporate active contractile mechanisms. The contractile models are preliminary and additional work is needed to include information on the magnitude of the force during furrowing. Theories for active stresses exist in fluid mechanics [16], and these theories might provide a means for modelling the initial stages of contractile ring formation.

The selection of the appropriate description of the elastic properties of the cell membrane is the most important part of any model of cell division. The deformation pattern of the membrane depends on the constitutive relation chosen for the characterization of the membrane material. We will discuss a number of potential membrane constitutive relations below.

DISCUSSION

As we have noted, numerous investigations exist in the physiology and cell literature describing the problem of cell division; see [1-15] to name a few, and the references contained in this volume. However, there have been only a few investigators, to our knowledge, who have considered the problem from a mechanics point of view [26-35].

The first investigation of a model for cell division is due to Prothero and Rockafellar in 1967 [26]; this paper seems to have been overlooked in more recent investigations of the mechanics of cytokinesis.

Prothero and Rockafellar attempt to describe through the use of geometrical modelling the observed changes in cell shape and the movement of surface markers during cell division. The assumptions in their model are those assumptions which are basic in later models. We quote from their introduction:

"One possibly attractive approach to characterizing the shape of a cell during cleavage is to assign mechanical properties to the cell membrane and to impose some specified mechanism for constricting the cell while treating the cell contents as a fluid. Thus, for example, one might plausibly treat the membrane as a thin elastic sheet and assume that constriction is due to contractile fibers around the cleavage furrow ... Nevertheless, it seems unlikely that the cell membrane is purely elastic, and rather more likely that it has mechanical properties possibly intermediate between, or an admixture of, those of an elastic material and a monolayer. Furthermore, an important constraint in the problem, namely that of constant volume, may be difficult to incorporate in a purely mechanical model. Accordingly, we eschew the strictly mechanical approach and adopt instead more general constraints which may be consistent with, but not confined to, a model of the elastic type alluded to above."

Basically, Prothero and Rockafellar investigated the geometric and mechanical consequences of the division with constant area strain of a sphere into two daughter spheres. What is important in their contribution is the clear description of the required ingredients of any model of cell division, as the above brief extract from their paper shows.

Greenspan [27] presented a theory for the division of spherical cells based on a fluid mechanics model. He introduced the concept of an effective surface tension to approximate the tension in the membrane of the cell, and then postulated that the magnitude of the effective surface tension depends on the local surface-concentration of tension elements. One may consider the tension elements to be representative of the mechanical behavior of the filaments of the cytoplasmic matrix or of the behavior of an as yet undiscovered tensile component of the membrane. Before the onset of cleavage, the cell is in equilibrium and the tension elements are assumed to be uniformly distributed over the surface. The tension of the surface membrane is therefore constant and is counterbalanced by the internal pressure. When equilibrium is disturbed, the surface-concentration of tension elements is altered and a relative difference in tension between the equator and the pole is established. This difference in tensions causes the membrane to deform inward at the equator and to draw the surface toward it to create a furrow. The resultant shear stresses set the protoplasm streaming in the same direction. This motion tends to increase the concentration of the tension elements near the furrow thereby increasing the contractile forces still further which in a self-perpetuating manner drives the furrow deeper. In essence, Greenspan's theory predicts that the dynamical process of division is unstable and, once triggered, develops rapidly without further stimuli. It also explains how the initial alignment of the surface bound tension elements into an equatorial belt can occur from fluid circulation of the cytoplasm; see Fig. 1(a).

In a later experimental study on aspects of cell division [28], Greenspan showed how a variable surface tension can divide an oil droplet and how instability of a fluid-filled thin elastic membrane can be triggered by surface distributions of elastic bands.

Doerner [29] in a model of the process of cell division introduced the assumption that at the beginning of cell division the shape of the cell changes from a sphere to a short central cylinder with ellipsoidal ends. The uniform internal pressure (or "prime mover for the system" as stated by Doerner) creates a discontinuity between the radial displacement of the cylinder and that of the ellipsoidal ends. This requires shearing forces and moments to be uniformly distributed along the circumference where the sections join and to be of magnitudes such as to eliminate this discontinuity. Doerner's analysis suggests that the shear forces will decrease in magnitude as the difference between the two displacements gradually diminishes. He does not report, however, any result that can be compared with experimental data.

A more recent paper by Catalano and Eilbeck [30] investigates the more general problem of a series of cell divisions or embryonic development. The approach is to consider the membrane of the cell possessing a surface cleavage field which in turn is related to an internal field. The internal field determines the cleavage plane, and after cleavage a new surface field will be present related to the old surface field by the cleavage deformation in the membrane. In this way a membrane surface field determines a cleavage which in turn modifies the membrane surface field leading to a subsequent cleavage. By this process a differentiation phenomenon of the original spherical cell shape can occur. Reference [30] shows a number of results of multiple cell divisions for the case of two dimensional cells.

The more recent models employing the mechanics of elastic and viscoelastic membranes are those of Akkas [31-33], and Pujara and Lardner [34-35]. In these investigations, the assumptions of the different models are almost the same, differing slightly in one or two assumptions; we will consider elastic behavior of membranes only in this paper. These investigations begin with the following assumptions [32]:

   a) At the onset of cell division, the shape of the cell membrane is a sphere;

   b) The membrane is slightly inflated initially [32], or the membrane is initially stress-free with zero initial pressure in the cytoplasm [34];

   c) Cell division is caused by the contraction of a ring about the equator

   d) The constitutive relations describing the behavior of the membrane material are nonlinear. The selection of the membrane description is what separates these investigations. In a later paper, Akkas [33] models the cell membrane of the spherical cell as a non-linear viscoelastic material. The features of a Mooney-Rivlin model of non-linear elasticity and the standard solid of linear viscoelasticity are combined. The results of the viscoelastic model appear to be in good agreement with many of the experimental results available in the literature and with the results of the earlier elastic models [32,34] as far as the comparison of the geometric variables of the problem are concerned. The results on the stiffness related parameters, such as the intracellular pressure and the equatorial ring force, however, do not agree with the corresponding experimental results. According to Akkas [33], this means that the incorporation of viscoelastic constitutive models do not resolve the differences between the predicted values and the experimental values. In fact, Akkas concludes that his viscoelastic model does not result in any conclusion which does not hold also in the elastic models. He does allow, however, the value of the stiffness coefficient to vary during division in both his elastic model [32] and his viscoelastic model [33]. The elastic model used in [32] will be discussed later;

e) The viscous properties of the cell cytoplasm are neglected, and it is assumed that the internal pressure is uniform;

f) The cell volume is a constant value [34] or is allowed to change [32] during cell division;

g) Cell division is a quasi-static problem of large deformation of an elastic membrane stretched over a changing shape.

In [32,34], the elastic problem is formulated as a deformation problem in nonlinear membrane elasticity, i.e., an examination of the deformation pattern of an initially spherical membrane for prescribed displacement at the equator subject to the condition that the enclosed volume is constant or changes slightly. The deformation pattern will obviously depend upon the constitutive relations chosen to describe the behavior of the membrane material. Analyses of the deformation of elastic membranes are numerous [36-45] and many investigations have applied the results to biological problems, e.g., the deformation of red blood cells [20-25] and the compression of sea urchin eggs [19].

In this paper we will discuss the well known Mooney-Rivlin material [36] which has been used extensively for the description of rubber-elasticity since the time of Mooney [37-38], and the strain energy function originally proposed by Skalak, Tozeren, Zarda and Chien (STZC material; [17,18,21]) for the description of the membrane of a red blood cell.

We find that the results for a modified form of the STZC material agree well with certain of the experimental values for cell division [34], as do results for a Mooney material [32] if the sphere is initially inflated, and therefore either material description may be applicable for the properties of the dividing sea urchin egg cell membrane outside the region of the contractile ring.

252

FORMULATION

Equilibrium and Geometry.

The undeformed configuration of an axisymmetric membrane is given in
the form

$$z = z(r) \qquad\qquad (1)$$

The coordinates (ρ, η) define the deformed position of the point initially at
(r,z) in the undeformed membrane, Fig. 2.

Along the membrane surface, an infinitesimal arc length ds is deformed
to the infinitesimal arc length dS. The principal stretch ratios in the
meridian and the circumferential directions denoted by λ_1 and λ_2 respectively are related to the variables ρ, η, z and r by

$$\lambda_1 = \frac{dS}{ds} = \left(\frac{\rho'^2 + \eta'^2}{1 + z'^2}\right)^{1/2} \quad ; \quad \lambda_2 = (\rho/r) \qquad\qquad (2)$$

where the primes denote differentiation with respect to the independent
variable r.

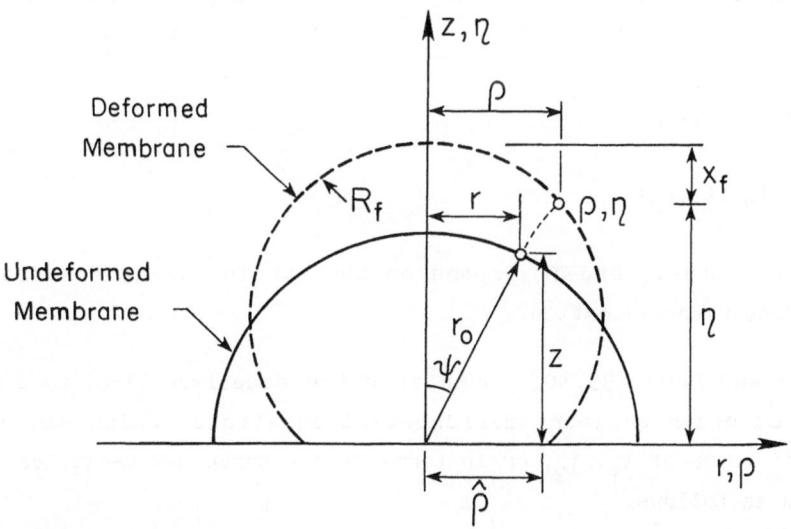

Fig. 2.   Geometry of Deformation of a Hemispherical Membrane by Equatorial
Constriction.

The basic equilibrium equations for membranes of revolution in the meridian-tangent and normal directions are, respectively,

$$\frac{dT_1}{d\rho} + \frac{1}{\rho} (T_1 - T_2) = P_t$$

$$\kappa_1 T_1 + \kappa_2 T_2 = P_n \qquad (3)$$

where T_1 and T_2 are the stress resultants with dimensions of force per unit edge length of the membrane; κ_1, κ_2 are the principal curvatures; and P_n, P_t are the external loads per unit area in the normal and the tangential directions of the deformed membrane, respectively.

The principal curvatures can be defined by the relations

$$\kappa_1 = \frac{d\theta}{dS}$$

$$\kappa_2 = \frac{\sin \theta}{\rho} \qquad (4)$$

where θ is the angle measured from the positive axis of symmetry to the outward normal of the deformed surface.

From the geometry of the deformed surface we can write

$$\frac{d\eta}{d\rho} = -\tan \theta \qquad (5)$$

The principal stress resultants per unit length of the deformed surface are

$$T_1 = T_1 (\lambda_1, \lambda_2)$$

$$T_2 = T_2 (\lambda_1, \lambda_2) \qquad (6)$$

where functions T_1 and T_2 depend on the strain energy function of the material under consideration.

Yang and Feng [39,40] formulated the equations (1-6) as a system of three first order ordinary differential equations, which can either be written in terms of λ_1, λ_2, or in terms of the three new variables u, v, and w defined as follows:

$$u(r) = \left(\rho'^2 + \eta'^2\right)^{1/2}$$

$$v(r) = \frac{\rho}{r} = \lambda_2 \qquad\qquad (7)$$

$$w(r) = \rho'$$

The equations (1-7) can be reduced by simple manipulations to the following first order system; see Yang and Feng [39] for further details.

$$\frac{du}{dr} = \frac{uz'z''}{\left(1+z'^2\right)} + \left(1 + z'^2\right)^{1/2} \left[-\frac{w}{vr}\frac{f_3}{f_1} + \frac{wP_t}{f_1} - \left(\frac{w-v}{r}\right)\frac{f_2}{f_1}\right]$$

$$\frac{dv}{dr} = \frac{w-v}{r} \qquad\qquad (8)$$

$$\frac{dw}{dr} = \frac{w}{u}u' - u\left(u^2 - w^2\right)^{1/2}\left[\frac{P_n}{T_1} - \frac{\left(u^2-w^2\right)^{1/2}}{uvr}\frac{T_2}{T_1}\right]$$

where

$$f_1 = \frac{\partial T_1}{\partial \lambda_1} \; ; \; f_2 = \frac{\partial T_1}{\partial \lambda_2} \quad \text{and} \quad f_3 = T_1 - T_2 \qquad\qquad (9)$$

For the case of a spherical membrane with $P_t = 0$, the equations can be written in the following form:

$$\lambda_1' = \left(\frac{\delta \cos\psi - w \sin\psi}{\sin^2\psi}\right)\left(\frac{f_2}{f_1}\right) - \frac{w}{\delta}\left(\frac{f_3}{f_1}\right) \quad ; \quad \delta' = w$$

$$\qquad\qquad (10)$$

$$w' = \frac{\lambda_1'w}{\lambda_1} + \frac{\left(\lambda_1^2 - w^2\right)}{\delta}\left(\frac{T_2}{T_1}\right) - \frac{\lambda_1\left(\lambda_1^2 - w^2\right)^{1/2}P_n r_o}{T_1}$$

where

$$f_1 = \frac{\partial T_1}{\partial \lambda_1} \; ; \; f_2 = \frac{\partial T_1}{\partial \lambda_2} \; ; \; f_3 = T_1 - T_2$$

and where primes indicate differentiation with respect to ψ, the angular position on the undeformed sphere, and,

$$\delta = \lambda_2 \sin\psi \qquad\qquad (11)$$

255

In order to describe the process of cell division of a spherical cell, we examine the deformation pattern of an initially spherical membrane with a prescribed displacement at the equator subject to the condition that the enclosed volume is constant.

The governing equations are given by equations (10). The constant volume condition can be written as:

$$3 \int_{0}^{\frac{\pi}{2}} \delta^2 (\lambda_1^2 - w^2)^{1/2} d\psi = 2 \tag{12}$$

The boundary conditions are as follows:

at the pole ($\psi = 0$),

$$\lambda_1 = \lambda_2 = \omega = \lambda_o \; ; \; \delta = \lambda_2 \sin\psi = 0, \text{ and}$$

at the equator ($\psi = \pi/2$),

$$\delta = \lambda_2 \sin\psi = \hat{\lambda}_2 \tag{13}$$

where $\hat{\lambda}_2$ is the specified value of the circumferential stretch-ratio at the equator that determines the stage of deformation.

The numerical integration method can be started by expanding the three dependent variables λ_1, δ and ω in a power series about $\psi=0$, e.g.,

$$\lambda_1 = \lambda_o + \alpha_{11}\psi^2 + \alpha_{12}\psi^4 + \ldots \tag{14}$$

and retaining only the leading terms of order ψ^2 in the expansions. The α_{ij} are constant coefficients to be determined by substituting (14) into (10) and comparing the coefficients of ψ^2 on both sides of the equations.

The numerical scheme can be outlined as follows:

1. Assume a value for λ_o and P_n and integrate equations (10) with the initial conditions given by the expansion of the variables (around $\psi = 0$) until ψ equals $\pi/2$.

2. Check to see if equation (13) is satisfied. If not, assume a new value for P_n and repeat procedure 1 until $\delta = \hat{\lambda}_2$ at $\psi = \pi/2$.

3. With this combination of λ_o and P_n, the volume condition, equation (12), should be checked. If this requirement is not met, λ_o should be changed and the procedures repeated.

The formulation and method of solution used by Akkas [32] is essentially similar to the formulation above except for the inclusion of the initial deformation of the sphere and the use of different variables. We note that in this approach and that of Akkas, the model is a passive model

256

in that the specification of $\hat{\lambda}_2$ at the equator ($\psi=\pi/2$) gives rise to the deformation of the membrane.

Constitutive Relations

The deformation pattern of the membrane depends on the constitutive relation chosen for the characterization of the membrane material. The constitutive relations selected here are those of the well-known Mooney-Rivlin material [36] and the relations proposed by Skalak et al. [21] for the red blood cell membrane. The Mooney-Rivlin material description has been used extensively by several investigators [37-45] for the characterization of the behavior of a rubber material since its inception. Hiramoto [46] and Yoneda [47], for example, have performed experiments on rubber balls as large scale models of biological cells in order to investigate theories on the deformation of cells under load. Comparisons of theoretical results for the Mooney-Rivlin material with the results obtained by Hiramoto [46] and Yoneda [47] from experiments on rubber balls can be found in [19].

The constitutive equation we use for representing the behavior of the membrane of a sea urchin egg cell is that of a STZC material (Skalak et al., [21]) which was proposed for the description of the deformation of the membrane of a red cell. We will use a modification of the STZC material for the modeling of cell division. The modification is convenient in order to account for the negative membrane tensions found in the calculations for the STZC material. The behavior of the modified STZC material will be discussed and compared with the available results from experiments on dividing sea urchin eggs. Akkas [32] has made similar comparisons using the Mooney-Rivlin material description.

The strain energy function for a Mooney-Rivlin material [36] has the form

$$W = C_1(\tilde{I}_1 - 3) + C_2(\tilde{I}_2 - 3)$$
$$= C_1[(\tilde{I}_1 - 3) + \tilde{\Gamma}(\tilde{I}_2 - 3)] \tag{15}$$

where C_1 and C_2 are the material constants with the dimensions of stress; $\tilde{I}_1$, $\tilde{I}_2$ are the stretch-ratio invariants given by

$$\tilde{I}_1 = \lambda^2_1 + \lambda^2_2 + \frac{1}{\lambda^2_1\lambda^2_2}$$
$$\tilde{I}_2 = \frac{1}{\lambda^2_1} + \frac{1}{\lambda^2_2} + \lambda^2_1\lambda^2_2 \tag{16}$$

and

$$\tilde{\Gamma} = \frac{C_2}{C_1}$$

With this strain energy function the relations between the stress resultants T_1, T_2 and the principal stretch ratios λ_1, λ_2 become

$$T_1 = 2hC_1 \left(\frac{\lambda_1}{\lambda_2} - \frac{1}{\lambda_1^3 \lambda_2^3} \right) (1 + \tilde{\Gamma} \lambda_2^2)$$

$$ \tag{17} $$

$$T_2 = 2hC_1 \left(\frac{\lambda_2}{\lambda_1} - \frac{1}{\lambda_1^3 \lambda_2^3} \right) (1 + \tilde{\Gamma} \lambda_1^2)$$

where h is the initial thickness of the membrane.

The STZC material [21] is defined by a strain energy function of the form

$$W = \frac{B}{4} \left[\frac{1}{2} I_1^2 + I_1 - I_2 \right] + \frac{C}{8} I_2^2 \tag{18}$$

where B and C are membrane material properties, assumed to be constants; I_1, I_2 are the stretch-ratio invariants defined as:

$$I_1 = \lambda_1^2 + \lambda_2^2 - 2$$

$$ \tag{19} $$

$$I_2 = \lambda_1^2 \lambda_2^2 - 1$$

In this case the tensions per unit length in the deformed position T_1, T_2 are related to the principal stretch-ratios λ_1, λ_2 by

$$T_1 = \frac{C}{2} \frac{\lambda_1}{\lambda_1} \left[\Gamma (\lambda_1^2 - 1) + \lambda_2^2 I_2 \right]$$

$$ \tag{20} $$

$$T_2 = \frac{C}{2} \frac{\lambda_2}{\lambda_1} \left[\Gamma (\lambda_2^2 - 1) + \lambda_1^2 I_2 \right]$$

where $\Gamma = \frac{B}{C}$

See [17,18,21,48] for a further discussion of the nature of the strain energy function (18), its application to cell membrane elasticity and for results for the stress resultants in uniaxial tension and isotropic tension.

258

Finally, the strain energy function for a modified STZC material is defined as

$$W = \frac{B}{4} \left(\frac{1}{2} I_1^2 + I_1 - I_2 \right) + \frac{C}{8} f(\psi) I_2^2 \qquad (21)$$

where B, C, I_1 and I_2 carry the same meaning as in equations (19,20) and the function $f(\psi)$ is defined by

$$f(\psi) = 1 - \frac{1}{\pi} \left[\tan^{-1} (\psi - \psi_0) + \frac{\pi}{2} \right] \qquad (22)$$

where ψ_0 corresponds to an angular position which is to be selected. With this strain energy function, the stress-strain relations become

$$T_1 = \frac{C}{2} \frac{\lambda_1}{\lambda_2} \left[\Gamma(\lambda_1^2 - 1) + f(\psi)\lambda_2^2 I_2 \right]$$

$$\qquad (23)$$

$$T_2 = \frac{C}{2} \frac{\lambda_2}{\lambda_1} \left[\Gamma(\lambda_2^2 - 1) + f(\psi)\lambda_1^2 I_2 \right]$$

The expressions for f_1, f_2, and f_3 in equation (10) for different constitutive relations can be obtained from the equations (17,20,23).

It is clear from the discussions above of the constitutive relations, that the membrane material has no active components corresponding to the presence of contractile filaments as needed in a model of cytokinesis. Attempts have been made to add another term in an ad-hoc fashion to the expression for T_2 to account for a contractile force; however, no firm conclusions could be drawn from numerical investigations with this additional term.

An approach to include a contractile effect that may be reasonable is to assume that the λ_2 stretch is measured relative to an intermediate stress free contracted state which arises from the initial stress free state upon contraction in the λ_2 direction. In this case,

$$\lambda_2 = \frac{\ell_2}{\ell_{20}^*} = \frac{\ell_2}{\ell_2^*} \frac{\ell_2^*}{\ell_{20}^*} = \tilde{\lambda}_2 \exp(-\int \hat{\alpha} dc) \qquad (24)$$

where λ_2 is the total stretch ratio measured from the initial stress free state, ℓ_{20}^*; ℓ_2^* is the length of a line element in a stress free state after contraction of the line element ℓ_{20}^*, where the contraction is proportional to the length,

$$d\ell_2 = -\hat{\alpha}\ell_2$$

$$\ell_2^* = \ell_{20}^*\exp(-\int\hat{\alpha}dc)$$

(25)

where c is the local value of the contraction and $\hat{\alpha}$ is a positive contraction function. It follows that the total deformation of an element of the membrane can be considered as made up of two steps: the first step is a stress-free contraction of line elements parallel to the two direction, followed by the second step arising from stresses in the membrane due to loading or constraint.

Arguments of this sort suggest we modify the constitutive relations for the Mooney-Rivlin material to the form

$$T_1 = 2h_oC_1\ C(c)\left(\frac{\lambda_1 a}{\lambda_2} - \frac{a^3}{\lambda_1^3\lambda_2^3}\right)\left(1 + \Gamma a^{-2}\lambda_2^2\right)$$

(26)

$$T_2 = 2h_oC_1 C(c)\left(\frac{\lambda_2}{a\lambda_1} - \frac{a^3}{\lambda_1^3\lambda_2^3}\right)\left(1 + \Gamma\lambda_1^2\right)$$

where

$$a = \exp(-\int\hat{\alpha}dc)$$

and C(c) is a function of the local contraction value c, and is equal to unity when c=0. Partial justification for this form of the constitutive relations follows by considering (by analogy) the contraction of a line element as driven by a decrease in temperature, e.g., as in thermal stress problems. Constitutive relations can be derived for temperature effects in membranes following [53,54] and the form of the relations obtained are the same as Eqs. (26); we present a brief derivation of these constitutive relations in the Appendix.

By analogy again, we replace λ_2 in the STZC constitutive relations by (λ_2/a) to obtain

$$T_1 = \frac{C}{2}\frac{\lambda_1}{a\lambda_2}\left[\Gamma(\lambda_1^2 - 1) + f(\psi)(\lambda_2/a)^2 I_2\right]$$

(27)

$$T_2 = \frac{C}{2}\frac{\lambda_1}{a\lambda_2}\left[\Gamma((\lambda_2/a)^2 - 1) + f(\psi)\lambda_1^2 I_2\right]$$

$$I_2 = (\lambda_1\lambda_2/a)^2 - 1$$

where the coefficient C is now dependent on the value of c. With these constitutive relations containing contractile effects, the boundary condition of prescribing the constriction $\hat{\lambda}_2$ at the equator is not needed. Instead c and $\hat{\alpha}$ are prescribed as a function of position on the undeformed membrane and the resulting form of the deformed membrane is found. (This method of formulation is equivalent to the formulation of a thermal stress problem in which the temperature is prescribed initially.) If c and $\hat{\alpha}$ are functions which are approximately zero except near the equator, we have a model of the contractile ring and we can investigate the effect of different distributions and values of c on the deformed shape.

However, the constitutive relations are still deficient in that the contraction c must be prescribed as a function of the stage of division. However, it is interesting to note that the above arguments give a variable coefficient C in the constitutive relations which is to be function of the stage of division. The variability of membrane stiffness was postulated by Akkas [32] as a means to improve comparisons between theoretical predictions and the experimental results; we will discuss this point later.

RESULTS

The results we will present are for the constitutive relations (17,20,23). Since little information is available on the magnitude of contractile forces, calculations using the constitutive relations with contractile elements does not appear warranted.

The equilibrium equations (10) were numerically integrated for the two constitutive relations (17,20) with the appropriate boundary condition of a prescribed deflection at the equator, i.e., for various stages of cell division (15%, 30%, 50%, 70%, and 90%). Equilibrium along the circumference where ρ is a maximum checked numerically to within 1% and the constancy of volume condition checked to within 1.5% in the numerical calculations.

Akkas [32] carried out similar calculations for the case when $C_2 = 0$ in Eqs. (17) corresponding to a neo-Hookean material. The cell membrane in his case was assumed to be slightly stretched at the onset of cleavage and the radius of the initial configuration is 1.05 times the radius of the reference configuration.

We will compare the results of the different numerical calculations with the results obtained from experiments on dividing sea urchin eggs. Data are available for the deformed profiles, areal change, principal stretch ratios λ_1, [1] and for estimates of the force in the furrow [47-50] at different stages of cleavage.

Figure 3 shows the variation of the polar deflection $(\bar{\eta}/r_o)$ for different stages of division for $\Gamma = 0$, 0.25 and for $\bar{\Gamma} = 0$, 0.2 and 2.0. Also shown are the results obtained by Akkas [32].

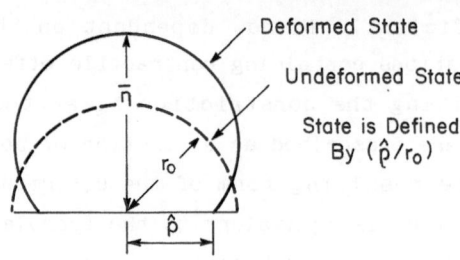

Deformed State

Undeformed State

State is Defined
By $(\hat{\rho}/r_0)$

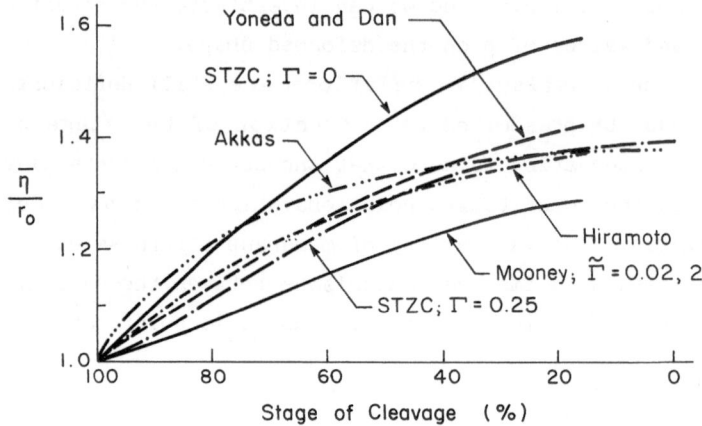

Fig. 3. Polar deflection versus stage of cleavage curves for STZC
material, $\Gamma = 0$, 0.25; Mooney material, $\tilde{\Gamma} = 0$, 0.2, 2; Mooney
material, Akkas [32]. Experimental results of Hiramoto (Table 2,
[1]) and of Yoneda and Dan (Fig. 5, [47]).

     The results obtained by Akkas, with an initial inflation of the
membrane for the Mooney material, $\tilde{\Gamma} = 0$, are in approximate agreement with
the experimental values. The effect of the initial inflation on the results
can be seen by comparing the curve for the Mooney material, $\tilde{\Gamma} = 0$ with the
curve of Akkas. Also shown in this figure are the results obtained by
Hiramoto (Table 2; [1]) and Yoneda and Dan (Figure 5, [47]) from the
experiments. Agreement between theory and experiment is good for the STZC
material when Γ is equal to 0.25. It should be noted that the results for
the Mooney material are insensitive to the values of $\tilde{\Gamma}$.

For the case with $\Gamma = 0$ for the STZC material, the equilibrium equations simplify with $T_1 = T_2 = T$, a constant and they can be integrated to show that the deformed profiles are spherical with uniform areal strain; see [34] for the details of this derivation and for expressions for the stretch-ratio λ_o, the stress resultants and the internal pressure. Another interesting feature of the STZC material with $\Gamma = 0$ is that the same spherical deformed shape of a membrane of this material can be reached not only by radial constriction at the equator of the membrane, but also by the inflation of a flat circular or a spherical cap membrane. The expressions for the pressures and areal stretches for each of these cases are presented in [19].

Since the values of $(\bar{\eta}/r_o)$ with $\Gamma = 0.25$ for the STZC material are in good agreement to the two sets of experimental values, we conclude that the areal strain is nonuniform. This is confirmed by Hiramoto's [1] observation that there appears to be a nonuniform stretching of the cell membrane at all stages of cell division. In the calculations that follow, we select $\Gamma = 0.25$; we will also compare our results with those of Akkas [32] for the case of $\tilde{\Gamma} = 0$ with initial inflation.

The results from the calculations for both a STZC material (with $\Gamma = 0.25$) and a Mooney material (with $\tilde{\Gamma} = 0$), indicate that in a small region near the equator the stress resultant T_2 is negative. The values of ψ for which T_2 becomes zero are plotted in Figure 4. For the STZC material, the dimensions of the region in which $T_2 \leq 0$ are approximately equal to the physical dimensions of the contractile ring in a dividing sea urchin egg, whereas for a Mooney material this region is considerably larger (about twice the width of the region for a STZC material). Wrinkling will occur in the region with a negative tension [51,52]. Beams and Kessel [15] report the presence of small folds at right angles to the furrow in dividing cells which could be a manifestation of wrinkling possibly due to negative tensions.

The existence of a region in which both T_2 was negative and T_1 was large suggested a modification to the constitutive equations in order to have T_2 approximately zero in a region near the furrow. The intent of the modification was to cause a slight change in the constitutive relation over most of the sphere with the greatest change occuring over the region near $\psi = \pi/2$. To do this, the constitutive relations for the STZC material, equation (20), were modified. The material constant C is replaced by $Cf(\psi)$ where $f(\psi)$ is given by equation (22). In equation (22), ψ_o represents the angular position of a point on the undeformed sphere which upon deformation separates the negative T_2 region from the positive T_2 region according to the STZC material description. The value of ψ_o for various stages is ob-

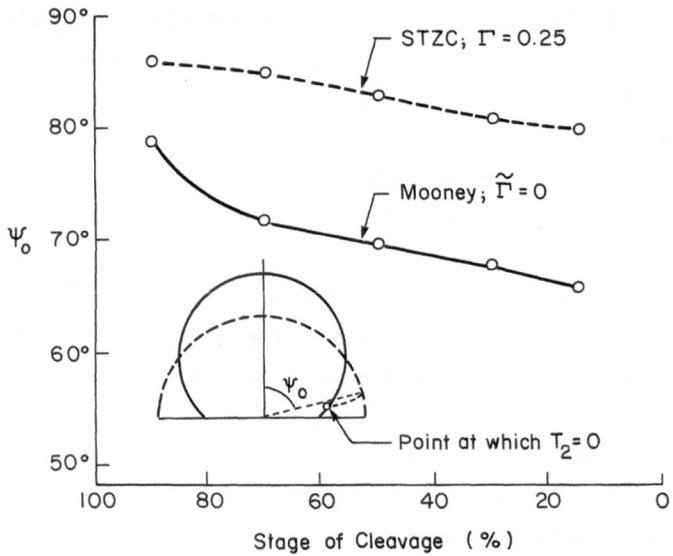

Fig. 4. Variation of the polar angle ψ_0 at which T_2 becomes zero with stage of cleavage for $\Gamma = 0.25$, STZC material and $\tilde{\Gamma} = 0$, Mooney material.

tained from calculations for the STZC material (Figure 4). The values of $(\bar{\eta}/r_0)$ versus stage of cleavage for this modified set of constitutive relations with $\Gamma = 0.25$ are almost identical to the results for STZC material with $\Gamma = 0.25$ and are not shown in Figure 3.

The variation of T_1, T_2 with ψ for both the STZC and the modified STZC material, is shown in Figure 5 for a 30% stage of cleavage. The value of the circumferential stress resultant T_2 in the furrow region is negative and approximately equal to zero for all stages of division with the modified constitutive relation. A different approach was used by Akkas [32] in that in his numerical procedure whenever the stress resultant T_2 was negative it was set equal to zero in the computations.

The comparison of the deformed profiles for the modified STZC material with the experimental values of Hiramoto (Table 2, [1]) is shown in Figure 6. Agreement with the experimental results is good for all stages.

Figure 7 shows the corresponding results from Akkas [32] in which the motion of points on the meridian is also shown. Of interest is the large stretching of the membrane near the equator during the later stages of division.

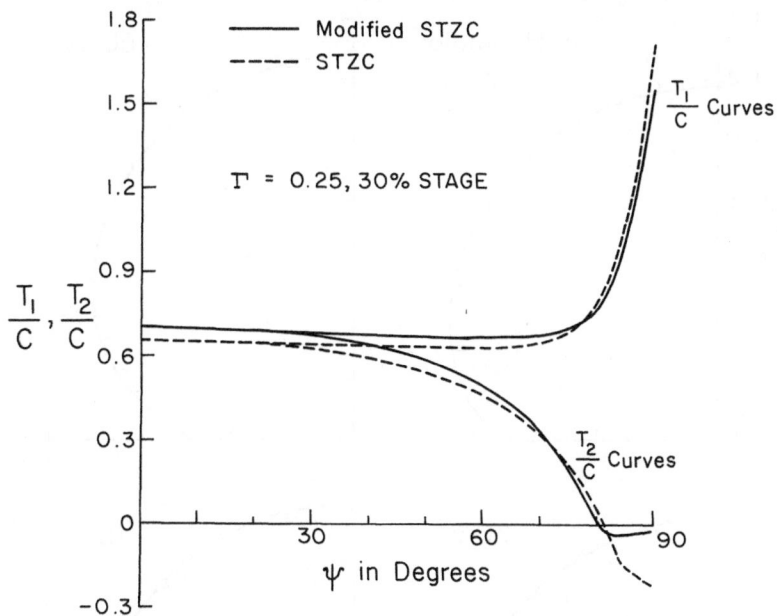

Fig. 5. Variation of stress resultants (T_1/C), (T_2/C) with the polar
angle ψ for $\Gamma = 0.25$, STZC and modified STZC materials for a 30%
stage of cleavage.

The values of λ_1 at the pole and at the equator obtained from the
calculations are plotted in Figure 8. Hiramoto (Figure 5, [1]) measured
approximately the linear change of segments of the meridional circumference
of deformed sea urchin eggs during cleavage. This change is an average
measure of the principal stretch ratio λ_1. The λ_1 values for the polar and
furrow regions are compared with the calculated values in Figure 8.

Figure 9 shows the variation of the ratio of the total surface area of
the deformed stage to the area of the undeformed membrane, (A_f/A_o), with
stages of cleavage as determined from our analysis and as calculated by
Hiramoto (Figure 7A, [1]) from his experiments. These results agree well
with Hiramoto's values.

In the numerical formulation, we select a starting value of the stretch
ratio at the pole and march the solution forward until the solution
satisfies the boundary condition of a prescribed $\hat{\lambda}_2$ with constant volume.

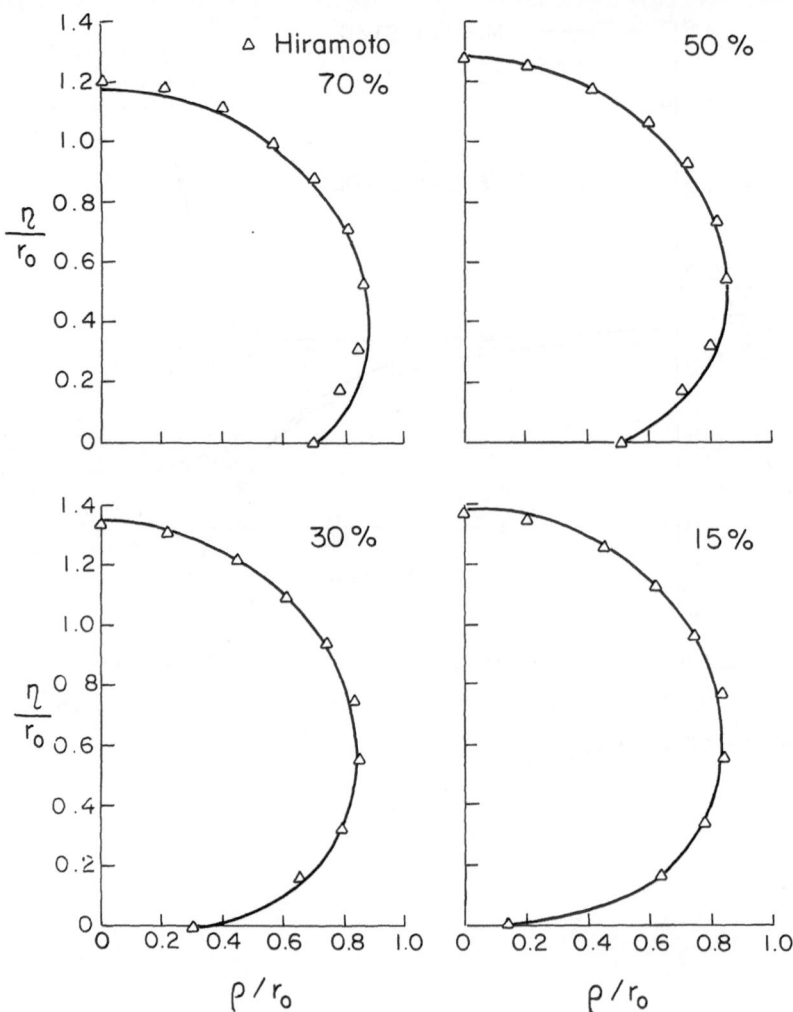

Fig. 6. Profiles of the deformed membrane for 70%, 50%, 30%, and 15%
stages for $\Gamma = 0.25$, modified STZC material. Experimental
profiles of Hiramoto (Table 2 and Fig. 4, [1]).

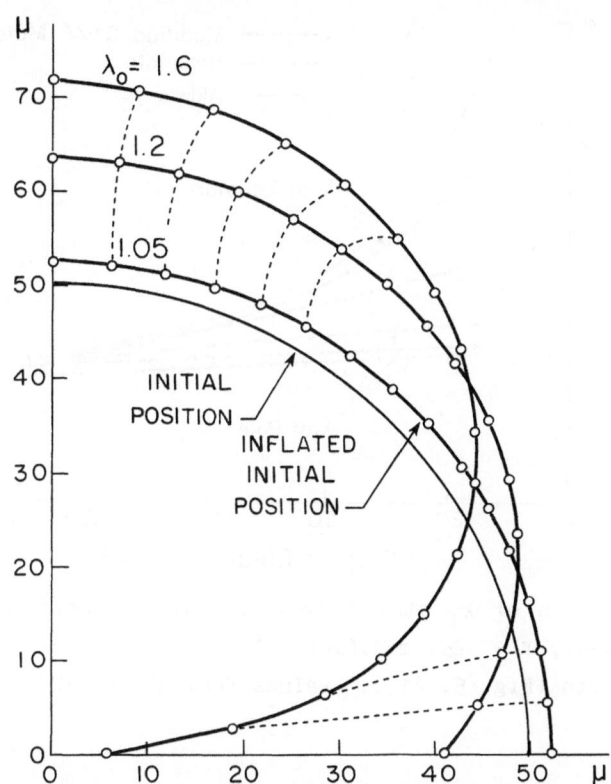

Fig. 7. Profiles of the deformed membrane for different stages for a Mooney material with initial inflation of five percent, Akkas [32].

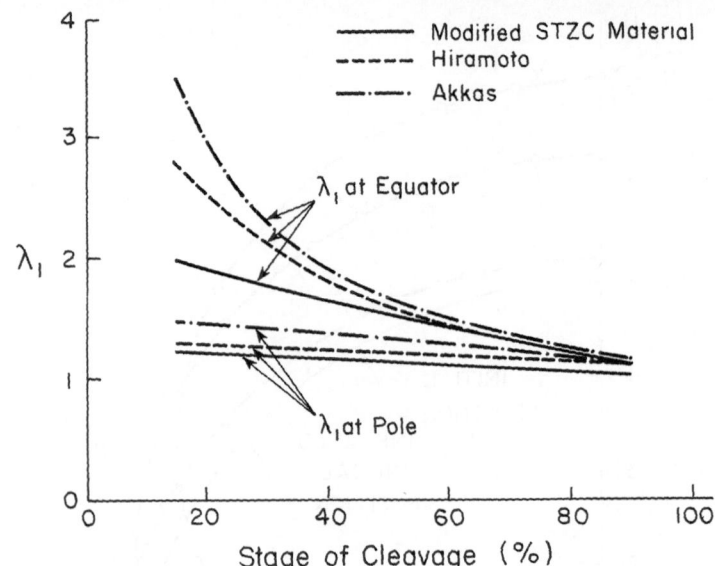

Fig. 8. Variation of λ_1 at the pole and at the equator with stage of cleavage, Γ = .25, modified STZC material. Calculated values of Hiramoto (Fig. 5, [1]). Values from Akkas [32].

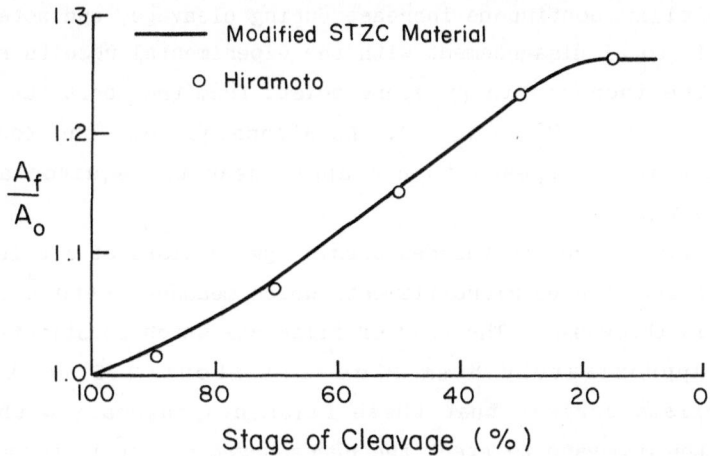

Fig. 9. Areal ratio, (A_f/A_o), versus stage of cleavage curves for $\Gamma =$
0.25, modified STZC material. Values calculated from experiment
by Hiramoto (Fig. 7A, [1]).

In doing this calculation, we adjust the value of the internal normal pressure p_n. As the membrane stretches over the deformed shape, the internal pressure will increase. The results from our numerical calculations for the pressure for the STZC material with $\Gamma = 0.25$ are shown in Fig. 10. We note that the value of p when $pr_o/C = 1$ and $C = 2$ dyne/cm is approximately 400 dyne/cm^2 which is a reasonable value. However, the monotonic increase in pressure with stage of division is a defect in this model. Akkas [31,32]

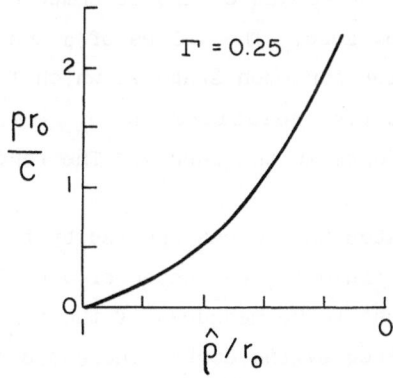

Fig. 10. Nondimensional pressure (pr_o/C) versus furrow radius ($\hat{\rho}/r_o$),
modified STZC material.

obtains a similar continuous increase during cleavage, and notes that this increase is in total disagreement with the experimental results reported by Hiramoto. The increase in pressure arises from the constitutive relation giving large values of T_1 as $\lambda_2 \to 0$. As a consequence, the model has the shortcoming that it appears to not apply near the equator at the later stages of cleavage.

The cleavage furrow in the sea urchin egg consists of a uniform ring of circumferentially aligned microfilaments which measure 35-60 $\overset{\circ}{A}$ in diameter and 0.1 µm in thickness. The band of filaments which constitute the furrow region is approximately 8 µm wide and about 0.2 µm thick [3]. Experimentalists suggest that these filaments interact with each other and/or with the membrane to exert the necessary contractile forces to cause division. We will estimate here the value of the force in the furrow by using the dimensions of the region in which T_2 is approximately zero as representative of the furrow region. Figure 11 shows the relevant free body diagrams for the calculation of the driving force F_f. We first calculate θ^* from the equilibrium equation using the results of our numerical calculations:

$$\sin \theta^* = \frac{p_n \rho^*}{2T_1^*} \tag{28}$$

We then use the following relation obtained from the equilibrium of an element of the furrow region

$$F_f = (T_1^* \cos\theta^* + p_n \hat{\eta}) 2\hat{\rho} \tag{29}$$

The superscript (*) in Equations (28,29) indicates the value at the intersection of the furrow region with the rest of the membrane, and $\hat{\rho}$ and $\hat{\eta}$ represent, respectively, the radius of the deformed membrane at the equator and the width of the furrow zone. The values of $\hat{\rho}$ and $\hat{\eta}$ were obtained from the dimensions of the region for each stage at which $T_2 = 0$. A value of C = 2.0 dyne/cm in the constitutive relations and r_o = 46.25 µm were used to calculate the physical force in the furrow. The force values are linearly dependent on C.

Hiramoto [49] estimated the forces applied by the cortex of the dividing egg to a drop of ferrofluid in the center of a dividing sea urchin egg. He considered these forces to be measures of the forces in the furrow. He found that the forces exerted by the cortex increased during early stages of cleavage and decreased during later stages. Yoneda and Dan [47] estimated the constriction forces from calculated values of the average surface ten-

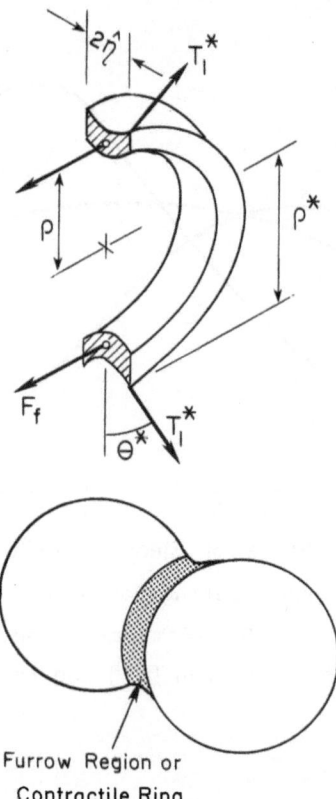

Fig. 11. Free body diagrams of the furrow region showing the direction of the constricting force, F_f, developed in the furrow cortex of the sea urchin egg during cleavage.

sion and the displacements at the equator. The values of the constriction force calculated by Hiramoto [49] and by Yoneda and Dan [47] are compared in Figure 12 with the values for F_f obtained from our calculations. The calculated results do exhibit the same qualitative changes as shown by the results of Hiramoto [49] and Yoneda and Dan [47]. These calculated values of the forces are also comparable with average forces measured directly $(25(10)^{-4}$ dynes) for cleavage in echinoderm eggs (Rappaport, [50]).

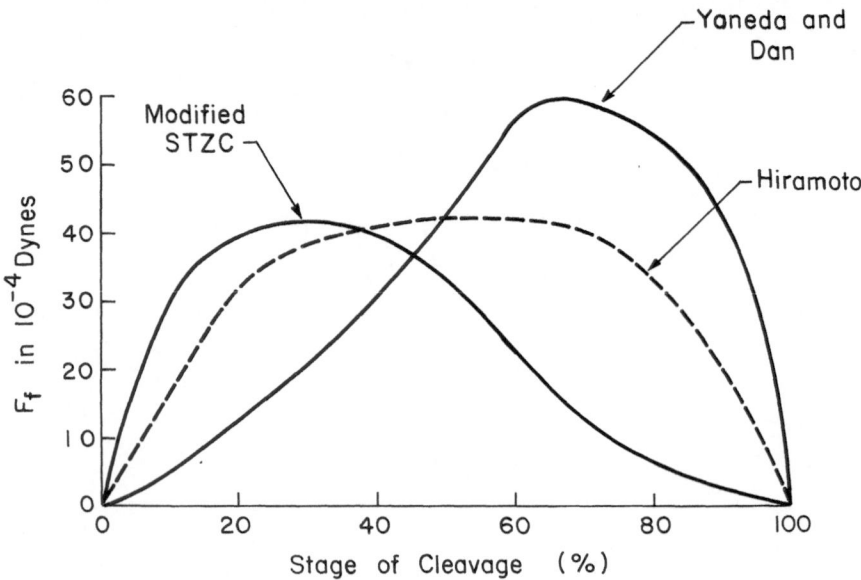

Fig. 12. Variation of the constricting force, F_f, with stage of cleavage for $\Gamma = 0.25$, $C = 2$ Dyne/cm, modified STZC material. Results calculated by Yoneda and Dan [47] and by Hiramoto [49].

The calculation yielding the results in Fig. 12 was carried out using our numerical method up to the 15% stage and the curve of the force was then extrapolated to zero. We extrapolated this force to zero guided by the intuitive observation that at the end of division, the force in the membrane should drop to zero. However, as was noted by Akkas [31], this extrapolation is not justified by the model. Again, we note that the model is not appropriate to the later stages of division. The selection of values at the membrane location shown in Figure 11 was to account for not only the dimensions of the contractile ring, but also to decrease the very high forces calculated when the size of the contractile ring was not considered. In the same spirit, a second consideration for the use of the modified STZC material was to avoid the very large values of T_1 at the later stages of division; see, for example, Figure 5.

As was pointed out by Akkas [32], any passive model with fixed elastic constants will suffer from the defect of unrealistic pressures and furrow forces in the later stages of division. Akkas, in his investigation, postulates that the numerical predictions can be brought into agreement with the experimental results by allowing the elastic constants to vary as a function of time and furrow radius. Initially, the cell membrane becomes stiffer during the early stages of cleavage and after reaching a maximum value, the stiffness value begins to decrease as cleavage continues. Akkas [32] states that the stiffness change "... is the only way that the numerical predictions of the ... model can be brought to a qualitative (and even quantitative) agreement with the available experimental results." The same statement is applicable to the elastic models developed in [34] and discussed here and to the viscoelastic model discussed in [33]. A means of including a variable stiffness and a changing initial or reference state (as was considered by Akkas [32] by allowing for an initial inflation) is to consider constitutive relations containing contractile elements as developed in Eqs. (26),(27), (A-13) and (A-15). In this way, it will be possible to model part of the initial state of furrow development and the subsequent cleavage with more realistic estimates of the cell pressure and furrow force. This is work for the future.

CONCLUSIONS

The cell division of a sea urchin egg cell was formulated as a deformation problem in nonlinear membrane elasticity in which the sea urchin egg was modeled as a spherical nonlinear elastic membrane enclosing an incompressible fluid. The model was "passive" in that it did not take into account the initiation of the furrow and/or the formation of a contractile ring and the active driving mechanism. We discussed the applicability of different constitutive relations to the description of the cell membrane and discussed the results from our investigations [34,35] and those of Akkas [31-33].

In view of the good agreement of the calculations with the experimental results for geometrical variables, we have provided additional support for the contractile ring theory of cell division. Further, from the results of the analysis we find that a description of the elastic properties of the cell membrane outside the region of the active contractile mechanism in terms of the STZC material (which has been used successfully for red blood cell membranes [17,18,21] and for other experiments with sea urchin cell membranes [19]) may be appropriate. In addition, we have suggested how a driving mechanism for active contraction can be incorporated into the proposed model. The inclusion of an active contractile mechanism into the constitutive relations is the next step in better understanding the biomechanics of cytokinesis.

REFERENCES

1. Y. Hiramoto, A Quantitative Description of Protoplasmic Movement During Cleavage in the Sea Urchin Egg, Expt. Biol., 35:407-424 (1958).

2. R. Rappaport, On the Rate of Movement of the Cleavage Stimulus in Sand Dollar Eggs, J. Expt. Zool., 183:115-120 (1973).

3. T.E. Schroeder, The Contractile Ring II, Determining its Brief Existence, Volumetric Changes and Vital Role in Cleaving Arbacia Eggs, J. Cell Biol., 53:419-434 (1972).

4. R. Rappaport, Establishment and Organization of the Cleavage Mechanism, in: "Molecules and Cell Movement", edited by S. Inoue and R.E. Stephens, Raven Press, N.Y., 287-304 (1975).

5. J.M. Mitchison, Cell Membranes and Cell Division, Symp. Soc. Expt. Biol., 6:105-127 (1952).

6. M.M. Swann, The Nucleus in Mitosis Fertilization and Cell Division, Symp. Soc. Expt. Biol., 6:89-104 (1952).

7. L. Wolpert, The Mechanics and Mechanism of Cleavage, Int. Rev. Cytol., 10, 163-216 (1960).

8. Y. Hiramoto, The Mechanics and Mechanism of Cleavage in the Sea-Urchin Egg, Symp. Soc. Expt. Biol., 22, 311-327 (1968).

9. Y. Hiramoto, Rheological Properties of Sea-Urchin Eggs, Biorheo. 6, 201-234 (1970).

10. Y. Hiramoto, A Photographic Analysis of Protoplasmic Movement During Cleavage in the Sea Urchin Eggs, Develop. Growth and Differentiation, 13, 3, 191-200 (1971).

11. T.E. Schroeder, Cytokinesis: Filaments in the Cleavage Furrow, Exp. Cell Res., 53, 272-276 (1968).

12. L.G. Tinley and D. Marsland, A Fine Structural Analysis of Cleavage Induction and Furrowing in the Eggs of Arbacias Punctulata, J. Cell Biol., 42, 170-184 (1969).

13. R. Rappaport, Cytokinesis in Animal Cells, Int. Rev. Cytol., 31, 169-213 (1971).

14. R. Rappaport, Cleavage Furrow Establishment -- A Preliminary to Cylindrical Shape Change, Amer. Zool., 13, 941-948 (1973).

15. H.W. Beams and R.G. Kessel, Cytokinesis: A Comparative Study of Cytoplasmic Division in Animal Cells, Amer. Scientist, 64, 270-290 (1976).

16. B.A. Finlayson and L.E. Scriven, Convective Instability by Active Stress, Proc. Roy. Soc., A, 310, 183-219 (1969).

17. R.P. Zarda, Large Deformations of an Elastic Shell in a Viscous Fluid, Ph.D. Thesis, Columbia University (1974).

18. R. Skalak, Modelling the Mechanical Behavior of Red Blood Cells, Biorheo., 10, 229-238 (1973).

19. T.J. Lardner and P. Pujara, Compression of Spherical Cells, Mechanics Today, 5, Pergamon Press, Ed. by S. Nemat-Nasser, 161-176, (1980).

20. E. Richardson, Analysis of Suction Experiments on Red Blood Cells, Biorheo., 12, 39-55 (1975).

21. R. Skalak, A. Tozeren, R.P. Zarda, and S. Chien, Strain Energy Function of Red Blood Cell Membranes, Biophys. J., 13, 245-2643 (1973).

22. P. Pujara and T.J. Lardner, Deformations of Elastic Membranes; Effect of Different Constitutive Relations, ZAMP, 29, 315-327 (1978).

23. E.A. Evans and R. Skalak, "Mechanics and Thermodynamics of Biomembranes," CRC Press, Boca Raton, FL (1980).

24. H. Demiray, Large Deformation Analysis of Some Basic Problems in Biophysics, Bull. Math. Biology, 38, 701-712 (1976).

25. E.A. Evans, R. Waugh, and L. Melnik, Elastic Area Compressibility Modulus of Red Cell Membranes, Biophys. J., 16, 585-596 (1976).

26. J.W. Prothero and R.T. Rockafellar, A Model of Cell Cleavage, Biophys. J., 1, 659-673 (1967).

27. H.P. Greenspan, On the Dynamics of Cell Cleavage, *J. Theor. Biol.*, 65, 79-99 (1977).
28. H.P. Greenspan, On Fluid-Mechanical Simulations of Cell Division and Movement, *J. Theor. Biol.*, 70, 125-134 (1978).
29. K. Doerner, Some Stress-Strain Relationships of Cellular Cleavage and Noncleavage, *J. Theor. Biol.*, 61, 205-209 (1976).
30. G. Catallno and J.C. Eilbeck, A Mathematical Model for Embryonic Cell Division Based on a Surface 'Cleavage Field', *J. Theor. Biol.*, 75, 123-137 (1978).
31. N. Akkas, On A Model for Cell Division, letter to editor, *J. Biomechanics*, 13, 459-460 (1980).
32. N. Akkas, On the Biomechanics of Cytokinesis in Animal Cells, *J. Biomechanics*, 13, 977-988 (1980).
33. N. Akkas, A Viscoelastic Model for Cytokinesis in Animal Cells, *J. Biomechanics*, 14, 621-631 (1981).
34. P. Pujara and T.J. Lardner, A Model for Cell Division, *J. Biomechanics*, 12, 293-299 (1979).
35. T.J. Lardner and P. Pujara, Reply to Letter on a Model for Cell Division, *J. Biomechanics*, 13, 460-461 (1980).
36. A.E. Green and J.E. Adkins, "Large Elastic Deformations," Second Edition, Oxford University Press (1970).
37. J.E. Adkins and R.S. Rivlin, Large Elastic Deformations of Thin Shells, *Phil. Trans. Roy. Soc.*, London, A., 244, 505-531 (1952).
38. L.J. Hart-Smith and J.D.C. Crisp, Large Elastic Deformations of Thin Rubber Membranes, *Int. J. Engng. Sc.*, 5, 1-24 (1976).
39. W.H. Yang and W.W. Feng, On Axisymmetric Deformations of Nonlinear Membranes, *J. Appl. Mech.*, 37, 1002-1011 (1970).
40. W.W. Feng and W.H. Yang, On the Contact Problem of an Inflated Spherical Nonlinear Membrane, *J. Appl. Mech.*, 209-214 (March 1973).
41. N. Akkas, On the Dynamic Snap-Out Instability of Inflated Non-Linear Spherical Membranes, *Int. J. Nonlinear Mechs.*, 13, 177-183 (1978).
42. H.O. Foster, Very Large Deformations of Axially Symmetrical Membranes Made of Neo-Hookean Materials, *Int. J. Eng. Sci.*, 5, 95-117 (1967).
43. A. Libai and J.G. Simmonds, Nonlinear Elastic Shell Theory, *Advances in Applied Mechanics*, 23, Section VI, C, 355 (1983).
44. A.D. Kydoniefs, Finite Axisymmetric Deformations of Elastic Membranes, *Int. J. Eng. Sci.*, 10, 939-946 (1972).
45. R.T. Shield, On the Stability of Finitely Deformed Elastic Membranes, Part 2: Stability of Inflated Cylindrical and Spherical Membranes, *ZAMP*, 23, 16-34 (1972).
46. Y. Hiramoto, Mechanical Properties of Sea Urchin Eggs, I, Surface Force and Elastic Modulus of the Cell Membrane, *Expt. Cell Research*, 32, 59-75 (1963).
47. M. Yoneda and K. Dan, Tension at the Surface of the Dividing Sea-Urchin Egg, *J. Expt. Biol.*, 57, 575-587 (1972).
48. R.P. Zarda, S. Chien, and R. Skalak, Elastic Deformations of Red Blood Cells, *J. Biomechanics*, 10, 211-221 (1977).
49. Y. Hiramoto, Force Exerted by the Cleavage Furrow of Sea Urchin Eggs, *Develop. Growth and Differentiation*, 17, 1, 27-38 (1975).
50. R. Rappaport, Cell Division: Direct Measurement of Maximum Tension Exerted by Furrow of Echinoderm Eggs, *Science*, 156, 1241-1243 (1967).
51. C.H. Wu, The Wrinkled Axisymmetric Air Bags Made of Inextensible Membrane, *J. Appl. Mech.*, 41, 4, 963-968 (1974).
52. C.H. Wu, Nonlinear Wrinkling of Nonlinear Membranes of Revolution, *J. Appl. Mech.*, 45, 535-538 (1978).
53. P.J. Blatz, Applications of Large Deformation Theory to the Thermomechanical Behavior of Rubberlike plymer-Porous, Unfilled, and Filled, Eirich, F.R., Ed., *Rheology-Theory and Applications*, 5, 1-56 (1969).

54. Bao-Qing Yu, Finite Deformations, Thermal Stresses and Inflation
     Instability of Axisymmetric Membrane Structures, Ph.D. Disserta-
     tion, University of Massachusetts, Amherst (February 1986).

APPENDIX

Constitutive Relations

The purpose of this appendix is to obtain a form for the constitutive relations for a membrane containing contractile elements. We first present a series of arguments following Blatz [53] and Yu [54] for the effect of temperature change on the deformation of nonlinear membranes and then we argue by analogy to account for the presence of contractile elements in a membrane undergoing isothermal deformation.

In a reversible non-isothermal processes we consider the membrane to be subjected to first a change in temperature with no stress, and second to a deformation under stress at constant temperature.

In the first step, the membrane is stress free and no work is done. The change of dimensions of an element of the membrane is

$$\ell_1^* = \ell_{10}^* \quad , \quad \ell_2^* = \ell_{20}^* \exp\left(\int_{To}^{T}\alpha dT\right) \tag{A-1}$$

where α is the coefficient of thermal expansion; however, the membrane only extends in the two direction . The superscript * represents the stress free state at temperature T and the subscript o represents the stress free state at temperature T_o. The change in volume is given by

$$V^* = V_o^* \exp\left(\int_{To}^{T}\alpha dT\right) \tag{A-2}$$

In the second step, the membrane is deformed slowly and isothermally into the final state under mechanical loads. The increment in the internal energy U at end of the processes is equal to the heat energy TdS added in both steps plus the work done on the membrane in step two:

$$dU = TdS + V^* W_i dI_i \tag{A-3}$$

where the subscript i in W_i represents the partial derivative of the strain energy function with respect to I_i. Introducing the Helmholtz free energy $A = U - TS$, we have

$$dA = -SdT + V^*W_i dI_i \tag{A-4}$$

It follows that

$$-S = \frac{\partial A}{\partial T} \quad , \quad V^*W_i = \frac{\partial A}{\partial I_i} \tag{A-5}$$

with

$$\frac{\partial}{\partial T}\left(\frac{\partial A}{\partial I_i}\right)_T = \frac{\partial}{\partial I_i}\left(\frac{\partial A}{\partial T}\right)_{I_i}$$

from which it follows that

$$-\frac{\partial S}{\partial I_i} = \frac{\partial V^*W_i}{\partial T} = \alpha V^*W_i + V^*W_{iT} \tag{A-6}$$

Therefore, the change in internal energy can be written in the form

$$dU = T\frac{\partial S}{\partial T}dT + V^*\left\{(1-\alpha T)W_i - TW_{iT}\right\}dI_i \tag{A-7}$$

For an ideal membrane [53], the internal energy is a function only of temperature, from which it follows that

$$(1-\alpha T)W_i = TW_{iT} \tag{A-8}$$

These equations are a system of two linear second order partial differential equations in the variables T, I_1, I_2. The solution is obtained upon integration in the form

$$W(I_1, I_2, T) = \frac{T}{T_o}\exp\left(-\int_{T_o}^{T}\alpha dT\right) W(I_1, I_2, T_o) \tag{A-9}$$

which gives the strain energy density function at temperature T in terms of the strain energy density at temperature T_o.

The strain energy density function can be used to calculate the stress resultants for step two in the form:

$$T_1 = 2h_o \left(\frac{\tilde{\lambda}_1}{\tilde{\lambda}_2} - \frac{1}{\tilde{\lambda}_1^3 \tilde{\lambda}_2^3} \right) \left(W_1 + \tilde{\lambda}_2^2 W_2 \right)$$

$$(A-10)$$

$$T_2 = 2h_o \left(\frac{\tilde{\lambda}_2}{\tilde{\lambda}_1} - \frac{1}{\tilde{\lambda}_1^3 \tilde{\lambda}_2^3} \right) \left(W_1 + \tilde{\lambda}_1^2 W_2 \right)$$

where ~ corresponds to the isothermal step in the process.

We define the total stretch ratio as measured from the initial state:

$$\lambda_1 = \frac{\ell_1}{\ell_{10}^*} = \frac{\ell_1}{\ell_1^*} \frac{\ell_1^*}{\ell_{10}^*} = \tilde{\lambda}_1 \quad ; \quad \lambda_2 = \tilde{\lambda}_2 \, \exp\left(\int_{T_o}^{T} \alpha \, dT \right) \qquad (A-11)$$

If we assume that the membrane material is a Mooney-Rivlin material, we have

$$W(I_1, I_2, T_o) = C_1(I_1 - 3) + C_2(I_2 - 3) \qquad (A-12)$$

where the I_1 and I_2 are expressed in terms of $\tilde{\lambda}_i$.

Finally, the constitutive relations can be written in the form

$$T_1 = 2h_o C_1 C(T) \left(\frac{\lambda_1 a}{\lambda_2} - \frac{a^3}{\lambda_1^3 \lambda_2^3} \right) \left(1 + \Gamma \frac{\lambda_2^2}{a^2} \right)$$

$$(A-13)$$

$$T_2 = 2h_o C_1 C(T) \left(\frac{\lambda_2}{a \lambda_1} - \frac{a^3}{\lambda_1^3 \lambda_2^3} \right) \left(1 + \Gamma \lambda_1^2 \right)$$

where

$$C(T) = \frac{T}{T_o} \exp\left(-\int_{T_o}^{T} \alpha \, dT \right)$$

$$a = \exp\left(\int_{T_o}^{T} \alpha \, dT \right)$$

$$\Gamma = C_2 / C_1$$

Equations (A-13) are the constitutive relations for a membrane in the presence of a temperature change which causes thermal elongation in the two direction only.

We now relate these equations to a material with contractile properties in the two direction.

We redefine

$$a = \exp\left(\int_{T_0}^{T} \alpha dT\right) \rightarrow \exp(-\int \hat{\alpha} dc)$$

$$(A-14)$$

$$C(T) \rightarrow C(c)$$

where c is the local value of the contraction in the two direction at a given point on the membrane and $C(c)$ and $\hat{\alpha}$ are known (empirical) functions of c or functions derived from arguments on the distribution of contractile elements in the membrane surface.

The interesting conclusion from this derivation is that we arrive at a set of constitutive relations for which the elastic constant C can be considered as a function of the contraction, a result postulated by Akkas [32] for the modeling of cell division.

For the case of the STZC material, we can argue by analogy and postulate that the constitutive relations will take the form

$$T_1 = \frac{C}{2} \frac{\lambda_1 a}{\lambda_2} \left[\Gamma(\lambda_1^2 - 1) + \frac{\lambda_2^2}{a^2} \left(\lambda_1^2 \frac{\lambda_2^2}{a^2} - 1\right) \right]$$

$$(A-15)$$

$$T_2 = \frac{C}{2} \frac{\lambda_2}{a\lambda_1} \left[\Gamma\left(\frac{\lambda_2^2}{a^2} - 1\right) + \lambda_1^2 \left(\lambda_1^2 \frac{\lambda_2^2}{a^2} - 1\right) \right]$$

where a and C are functions of c.

With these constitutive relations, the character of the boundary value problem for cell division is changed. Instead of specifying the value of equatorial constriction $\hat{\lambda}_2$, we now specify a distribution of c and $\hat{\alpha}$ on the surface. This contraction distribution drives the original sphere into a contracted shape with a specified radius at the equator.

The distribution of c and $\hat{\alpha}$ over the original sphere depends on the diffusion or movement of substances in or next to the cell membrane. At this time, the knowledge of these substances is not clear so that calculations using the modified constitutive relations above is not warranted.

FLUID MECHANICAL SIMULATIONS OF CELL FURROWING

DUE TO ANISOTROPIC SURFACE FORCES

Daniel Zinemanas and Avinoam Nir

Department of Chemical Engineering
Technion-Israel Institute of Technology
Haifa 32000, Israel

INTRODUCTION

Cell division is a complex phenomenon ultimately leading to the formation of two new daughter cells. Accomplishment of this important cell function involves two different but related processes: mitosis and cytokinesis. The former is the process of nuclear division in which the continuity of the chromosomal set is maintained. Once the necessary cytoplasmic constituents and organelles are separated and redistributed, daughter cells are formed by physical division of the cytoplasmic matrix. This process of cell cleavage is known as cytokinesis.

Even though the biochemical and physicomechanical mechanisms governing this phenomenon are not yet completely understood, it is now generally agreed that the process is accomplished by a series of steps involving the mitotic apparatus (MA) and the cell cortex.

The MA consists of a pair of asters and a spindle and its role in cytokinesis seems to be purely stimulatory. This nature of the MA activity follows from several experiments showing that absence of a MA leads to no formation of a furrow and therefore such cells do not cleave (Beams and Evans, 1940). On the other hand, when the MA is disrupted or eliminated between anaphase and the onset of furrowing the cells generally cleave normally. If these interventions are performed at earlier steps, a furrow is formed but cleavage is only partial (Hiramoto, 1956, 1971). The exact nature of the MA stimulatory effect is still not known but its source is apparently located in the pair of asters. This conclusion is suggested by an experiment conducted by Rappaport (1961) who showed that a furrow appears also when the interacting asters belong to different MA's without even being connected by a spindle. It is therefore likely that the stimulus propagates from the asters in all directions, influencing the cell surface in a quantitative rather than a qualitative manner (Conrad and Rappaport, 1981).

Apart from the fact that the MA is bisected by the cleavage plane there is also an evident geometrical relationship between the location of the asters and the formation of the cleavage furrow. When the asters are driven toward the poles closer than their normal position, a reduction or even an inhibition of the furrowing activity may result. Similarly, if the cell is manipulated so that there is a shortening of the aster-equator distance and an increase of the aster-polar distance, furrowing takes place even in cells

281

that due to chemical inhibition of the mitotic apparatus would normally not
cleave. (Rappaport 1961,1984). In addition, it has been shown (Rappaport and
Ebstein 1965) that changing the MA position with the cell at early stages of
the mitotis by rotating the spindle axis, changes correspondingly the
position of the furrow formation and preserves the perpendicular direction
of the cleavage plane. These indicate an important linkage between the
location of the asters and the stimulatory nature of the MA leading to a
non-homogeneous surface furrowing activity.

A number of hypotheses were proposed to explain the effect of the
stimulus on the cell surface. The most accepted are the polar and the
equatorial stimulations. There is, however, no clear evidence to favour one
of them and both played roles in previous models of the process. The polar
stimulation hypothesis assumes that the effect of the stimulus is concentrated at the poles and the subpolar regions, and causes a relaxation of the
surface tension there. (Wolpert 1960;Borisy and White 1978; White 1985).
On the other hand the hypothesis of equatorial stimulation assumes the main
effect to be concentrated at the equatorial region leaving the rest of the
cell surface unaffected (Conrad and Rappaport 1981).

The important role of the cell surface and in particular the role of
the surface forces, in cytokinesis is in fact an old hypothesis (Rappaport
1971). Hiramoto (1964,1965) and Rappaport (1978) showed that chemical,
physical or mechanical disturbances induced in the internal cytoplasm
during stages of cytokinesis do not avoid a normal cleavage. These include
continuous stirring of the cytoplasm and replacement of the latter with
physiological solution or oil droplets. The above findings indicate that
once the stimulus, which originates at the asters, reaches the cell surface,
cleavage proceeds autonomously independent of any cytoplasmic event. There
is also evidence that there are no specific regions on the surface which are
more predisposed to develop a furrow, as is demonstrated by the rotation of
the MA and consequently the cleavage plane (Rappaport and Ebstein 1965).

Measurements of the surface forces at different stages of cleavage
(Hiramoto 1968) show an overall increase in tension prior to furrow formation and a sudden decrease just before it. As the process proceeds the
tensions in the equator and polar regions grow continuously, keeping the
tensions in the equator higher than those at the poles. Equatorial forces in
the circumferential direction are also higher than those in the meridional
direction and this difference grows steadily. At a certain point, when the
furrow is visually well defined, all surface forces come to a maximum and
then decrease gradually. Similar findings were obtained by Schroeder (1981)
who observed an increase in overall tension before the onset of furrowing
which is attributed to a surface contraction independent of mitosis.

The source of these surface forces lies in the cortex, a firm gelled
layer immediately beneath the plasma membrane, which contains a network of
microfilaments. This layer is believed to account for most of the surface
tractions. The plasma membrane is generally folded in abundance (Hiramoto
1981) and therefore is not assumed to contribute to surface tensions. Its
excess can provide easily the necessary addition required for a rapid
change in surface area. It is also believed that the plasma membrane provides points of attachment to the microfilaments keeping them confined to
the cortex. In these respects the combination membrane-cortex behaves
somewhat similar to an interface separating fluids with its special
rheological properties.

Ultrastructural studies have shown the presence of actin and myosin
(Schroeder 1973; Fujiwama and Pollard 1976) in the cortical layer. These
findings indicate a muscle-like mechanism responsible for the actively
changing surface forces. There is also experimental evidence that at an

early stage of cytokinesis, the muscle-filaments are randomly and homogeneously distributed all over the surface (Opas and Soltynska, 1978). As
cleavage proceeds these filaments become clearly oriented and their concentration seems to grow at the furrow leading edge. The orientation in the
equatorial region is parallel to the cleavage plane (Schroeder 1975),forming
what is known as the contractile ring, CR, while in other regions the
orientation is mostly parallel to meridian lines, (Opas and Soltynska 1978;
Bluemink 1970; Forer and Behnke 1972).

The correspondence between the macroscopic measurements of surface
tensions during cleavage and the nature of the muscle-like filaments is
clear. Once formed, the CR is responsible for the anisotropy observed in
surface forces near the furrow, e.g., the differences between circumferential and meridional components, and is the principal agent to attain cleavage.
However, the dynamics governing the ultrastructural reorientations and the
formation of the CR, as well as its dependence on the macroscopic deformation of the cell cortex and the mitotic stimulus, are still unexplained.

It was suggested that the CR formation is due to agglomeration of
filaments (Bluemink 1970) and reorientation by mutual interaction (Schroeder
1975). Support of these ideas can be found in experiments showing that
inactive particles or chemicals attached to the surface move passively to
the equator during furrow formation and accumulate there (Rappaport 1976;
Koppel et al. 1982). The fact that surface forces diminish at the last
stages of cleavage and that the cross section of the CR remains constant
(Schroeder 1975) suggests that, apart from this probable agglomeration of
filaments, a biochemical mechanism participates in controlling the surface
activity continuously. The similar behavior of the forces at different sites
in the cell cortex indicates that such biochemical activity takes place
all over the cell surface.

A few models were proposed to examine various theories on the mechanism
and the relationship between active and passive processes during cytokinesis,
and to explain experimental observations. These models can be classified in
two categories: The hydrodynamic approach (Greenspan 1977a,1977b; White and
Borisy 1983) and models based on solid mechanics (Pujara and Lardner 1979;
Akkas 1980a,1981). Generally, in view of the many uncertainties in the
present knowledge of the cytokinetic process, these models provide explanations mainly to specific aspects of the phenomenon.

Reproducing an early experiment (Spek 1918), Greenspan (1978) showed
that a neutrally buoyant oil droplet, experiencing a surface tension gradient with tension lower at the poles, can undergo considerable deformations
closely resembling those of the living cell during cytokinesis. These
observations emphasize the importance of the surface forces variations in
furrow formation and cleavage, although the conditions were not hydrodynamic-
ally equivalent. The Reynolds number in Greenspan's experiments were much
higher than those typical for cell division. Greenspan (1977a) suggested
that the cell surface forces can be considered as effective surface tensions
similar to those present in fluid surfaces and the cytoplasm as a viscous
fluid.

The effective surface tension is assumed to be dependent on the presence of surface tension elements (contractile filaments) and grows with an
increase in their concentration. Therefore, if an initial homogeneous concentration of elements is perturbed, by decreasing it at the poles, a
tension gradient is formed and causes a surface flow from the poles to the
equator. This hydrodynamical hypothesis is supported by the experimental
observations of such surface flows (Hiramoto 1968; Rappaport 1976; and
Koppel et al. 1982). Furthermore, if the diffusion of the contractile
elements is relatively small the flow concentrates them in the vicinity of

the equator, increasing the tension there and the overall tension gradient on the surface.

Greenspan (1977a) suggested that this unstable process continues indefinitely and can explain the agglomeration of contractile elements which forms the CR at the equator. He also proposed that during their motion toward the equator the filaments tend to rotate due to surface contraction thus orientating themselves parallel to the equatorial plane. The fluid dynamic equations of this model with isotropic surface tension were solved using asymptotic methods, continuously maintaining a low tension at the poles. Although the results show the expected initial shape and surface concentration changes, the mathematical analysis is limited to small concentration gradients and surface deformations and therefore applies only to the initial events of furrowing.

The limitations of the use of asymptotic methods were removed by Sapir and Nir (1985) who employed integral equation representation to the equations of motion and obtained a numerical solutions to the moving boundary problem. The allowance for large shape deformations does not, however, lead to droplet cleavage, independent of the initial tension element distribution. Similar conclusions, although differently calculated, were obtained by White and Borisy (1983). The essence of this behaviour resides in the isotropy of the assumed surface tension forces which implied that the normal tractions are proportional to the scalar tension and the mean curvature. As the furrow develops and advances, the mean curvature at its leading edge varies continuously and ultimately changes sign. Soon after, the normal force at the furrow is reversed and cleavage stops. Finally, the drop returns to a spherical shape with a small local contraction at the equator.

Based on Greenspan's model, White and Borisy (1983) further supposed that the surface tension is also dependent on the orientation of the contractile elements on the surface. These elements are assumed to move freely in the plane of the cortex and have a given ratio of lateral to vertical mobilities. The initial surface tension profile was assumed to be determined by a stimulus which originates at the asters and causes a relaxation of the surface tension. Quantitatively, the effect of the asters stimuli on the surface is assumed additive and follows an arbitrary inverse power law depending on the distance from the centrioles. The simplified mathematical formulations was not explicitly presented, nor was suggested a sound physical justification for the specific ratio of mobilities assumed for the filaments. Quantitatively, however, in view of their calculations and the limitation of previous models, the work of White and Borisy (1983) emphasizes some of the most important features in the cleavage process, i.e. the role of anisotropy in surface forces and the dependence of the evolution of this anisotropy on the surface dynamics. For a mobility ratio of 4:1 White and Borisy (1983) obtained shapes which closely follows the shapes observed by Hiramoto (1968) for symmetrical cleavages of the sea-urchin egg, and showed that even unilateral division or cleavage of deformed cells could be explained by means of the same model.

The second approach used in models of cytokinesis is based on solid mechanics, and describes the passive deformation of a spherical shell, enveloping an incompressible fluid, due to a force concentrated at the equator. This approach supports the hypothesis that the main effect of the stimulus would be concentrated at the equator leaving the rest of the cell surface passively reacting to the equatorial forces. Pujara and Lardner (1979) solved the static problem of the deformation of an initially spherical membrane of non-linear elasticity for different imposed furrows radii. They found good agreement with the cell shapes observed by Hiramoto (1968). However, the internal pressure did not follow qualitatively the path observed experimentally (Akkas 1980b).

Using basically the same model Akkas (1980a,1981) tried to solve these discrepancies by allowing for volume changes and giving the membrane other elastic and viscoelastic properties. Although shape deformations were again successfully simulated the qualitative discrepancies concerning the internal pressure and the equatorial constricting force remained unsettled. The remedy proposed in order to obtain a satisfactory qualitative behaviour was allowing a dynamic change in cell stiffness during cleavage. This could suggest that the basic assumption that the entire surface is passively reacting to a singular force at the equator is not an appropriate description of the surface forces and surface reactivity involved in the deformation.

In summary, the various models described above provide important aspects of the cytokinetic process yet no single model presents a comprehensive description of the phenomenon. The hydrodynamic models (Greenspan 1977a, 1977b) demonstrate the importance of polar tension relaxation for the onset of cell deformation and the initiation of the furrowing process. The induced surface flow provide a mechanism for the formation of the contractile ring. The elastic membrane deformation models (Pujara and Lardner 1979; Akkas 1980a,1981) show the significance of a circumferential equatorial force and the passive response of the cortex in obtaining correct cell deformation during cleavage. The simulation proposed by White and Borisy (1983) emphasizes the essential role of anisotropic tensions at the surface, previously observed by Hiramoto (1968), as one of the key elements in the process of cell division.

In the following sections we shall present a model integrating the above important elements and linking the macroscopic observation of the force and shape evolution during cytokinesis to the ultrastructure of the cortical layer. The model will also simulate the dynamics of these changes from the onset of deformation due to an astral stimulus until a complete cleavage.

BIOPHYSICAL MODEL

A comprehensive model of cytokinesis should treat the process as driven by highly organized surface forces which start from an initial uniform state and respond to a stimulus originating at the asters. The model must, therefore, be able to describe the essential events taking place during cleavage, i.e., the stimulating process, the dynamic evolution and differentiation of surface forces and the relationship between these forces and the biochemical processes in the cortex.

We present here a hydrodynamic model which takes into account these aspects of cell division and is based on the following assumptions:

a. The cell cortex consists of a thin layer containing a network of microfilaments confined in a continuous matrix.

b. The filaments are initially randomly oriented and homogeneously distributed on the surface and surface tractions are uniform.

c. Surface forces depend on the concentration and orientation of activated contractile filaments as well as on the passive deformations of the cortical matrix.

d. A filament exerts a unidirectional force parallel to its longitudinal axis and the surface concentration of active filaments is determined by the local biochemical composition of the cortex which follows the surface motion and a given kinetic scheme.

e. Following an earlier surface activation, a stimulus, diffuses from the asters between early stages of mitosis and late anaphase and alter the initial uniform biochemical composition of the cortex. The stimulus induces a decrease in the concentration of active filaments thus lowering the cortical tensions. Quantitatively, the effect at each point of the surface depends on the total amount of stimulus received there.

f. Filaments move on the surface following the surface motion and reorient themselves due to rotations caused by velocity gradients and mutual interactions due to a strong lateral affinity (as postulated by Schroeder, 1975).

g. The motion of the surface and reorientation of surface filaments results in a non-homogeneous distribution of anisotropic interfacial tension.

In the following sections we shall show that the above model and the resulting behaviour are sufficient to simulate satisfactorily processes occurring at the surface such as: accumulation of microfilaments in the cell equatorial region, the formation of a contractile ring with a sharp lateral orientation, the relaxation of cortical tension at the cell poles and, finally, the total dynamic deformation and cleavage of the cell. The model can also provide, using appropriate kinetics for the stimulus-cortex chemical interaction, a dynamic relaxation of interfacial forces and internal pressures similar to those evident experimentally.

Hydrodynamic Simulation

This work assumes as previous models of biological cells (Greenspan 1977a,1977b; White and Borisy 1983; Barthes-Biesel 1980) that the cell is described as a small viscous droplet. Consider then a droplet B with surface ∂B of viscosity μ enclosed in a two dimensional membrane and surrounded by an infinite region B^* filled with a fluid of viscosity μ^*. It is assumed that under isothermal conditions the fluids are Newtonian and have constant physical properties. Thus, neglecting inertial and gravity effects, the equations of motions determining the velocity $\underline{v}$ the stress $\underline{\underline{\sigma}}$ and the pressure p fields in B and B^* take the quasi-steady form

$$\nabla \cdot \underline{\underline{\sigma}} = 0 \tag{1}$$

$$\nabla \cdot \underline{v} = 0 \tag{2}$$

where

$$\underline{\underline{\sigma}} = -p\underline{\underline{I}} + \mu(\nabla\underline{v} + \nabla\underline{v}^\dagger) \tag{3}$$

and μ^* replaces μ in B^*.

The complementary boundary conditions on ∂B are the continuity of velocities

$$\Delta\underline{v} = 0 \tag{4}$$

and continuity of surface tractions,

$$\Delta\underline{\underline{\sigma}}\cdot\underline{n} = -\nabla_s \cdot \underline{\underline{\gamma}} + (\underline{\underline{\gamma}}:\nabla\underline{n})\underline{n} \tag{5}$$

where allowance for a general surface tension, $\underline{\underline{\gamma}}$ is made. Δ denotes a difference across ∂B (outer minus inner), $\underline{n}$ is a unit vector on ∂B pointing into B^* and ∇_s is a surface operator.

In equation (5) the first term on the RHS accounts for tangential

surface traction due to variation in interfacial tension. These forces provide the major contribution for surface flow. The second term expresses the
normal components of these forces which constitute the main cause for
surface deformation.

When the surface shape is given in the form $\phi(\underline{x},t)=0$, $\underline{n}$ is defined by
$\underline{n}=\nabla\phi/|\nabla\phi|$ and the dynamic droplet deformation is described by the kinematic
condition

$$- \frac{1}{|\nabla\phi|} \frac{\partial\phi}{\partial t} = \underline{v}\cdot\underline{n} \tag{6}$$

In pure fluid interfaces local thermodynamic equilibrium requires that
surface tension be an isotropic interfacial property. In this work the
cortical surface forces can vary locally and can exhibit anisotropy depending
on the local dynamic distribution of contractile microfilaments. Thus, the
total anisotropic surface tension, $\underline{\gamma}$, is an ensemble of the contributions of
the contractile elements, denoted by $\underline{\gamma}_p$, and the passive tensions arising
from the cortex deformations $\underline{\gamma}_m$. $\underline{\gamma}_p$ depends on the local concentration, c,
and the orientation distribution function, $N(\underline{d})$, of the active filaments,
while $\underline{\gamma}_m$ stems from the rheological properties of the cortical matrix. The
total tension is of the form

$$\underline{\gamma} = \underline{\gamma}_p(c,N(\underline{d})) + \underline{\gamma}_m \tag{7}$$

where $\underline{d}$ is a director along a filament axis.

Since the filaments are confined to ∂B the mass conservation equation
has the form

$$\frac{\partial c}{\partial t} + \nabla_s \cdot (\underline{v}_s c) = D_T \nabla_s^2 c + \hat{R} \tag{8}$$

where D_T is a surface translational diffusion coefficient, $\hat{R}$ is the rate
of active filaments production and v_s is the velocity field on ∂_B. Other
chemical species involved in the surface biochemical process follow similar
balances. The convective term in (8) accounts for changes in concentration
due to surface flow as well as surface contraction or expansion. The relative
importance of concentration variation due to diffusion is anticipated to be
small since it is known that diffusion coefficients of macromolecules on
cellular surface is small (Edidin, 1977).

According to assumption (f) the filaments follow the local translation
and rotation of fluid at the interface. The orientation may also be influenced by translational and rotational diffusion and the filament's mutual
interaction. Assuming a kinematics of a rigid slender body the equation for
the conservation of the orientations is given by

$$\frac{\partial N}{\partial t} + \underline{v}_s \cdot \nabla_s N = D_R \nabla_d^2 N + \nabla_{\underline{d}}(\dot{\underline{d}}N) + \frac{\hat{R}}{c}(\hat{N}-N) + \frac{D_T}{c}(\nabla_s N \cdot \nabla_s c) \tag{9}$$

Here, D_R is the rotational diffusion coefficient, ∇_d is an operator with
respect to the filament direction and $\hat{N}$ is the orientation distribution
function of the filaments produced by reaction at the net rate $\hat{R}$. The two-
dimensional equation (9) is an extended version of Burgess equation (Burgess
1938) accounting for additional convective, reactive and translational diffusion effects. The first term on the RHS represents the contribution to the
change of N due to the rotational motion of the filaments, while the second
term accounts for the changes due to rotational diffusion which, as before,
are expected to be of secondary importance. The nonlinear coupling between

tangential gradients of concentration and orientation distribution arising from translational diffusion is manifested in the last term. Appropriate initial and boundary conditions to equations (6)(8) and (9) must be provided.

Since the filaments are small compared to the cell dimension, the local velocity field on ∂B, $\underline{v}_s$, can be considered linear, i.e., $\underline{v} = \underline{v}_0 + \nabla \underline{v} \cdot \underline{x}$ and thus the angular velocity $\underline{d}$ of a filament is given by

$$\underline{\dot{d}} = \underline{d} \times \nabla_s \underline{v}_s \cdot \underline{d} \times \underline{d} + \underline{w}(\underline{d}) \qquad (10)$$

where $\underline{w}$ is an additional rotation arising from interaction between adjacent filaments which causes orientational aggregation along the principal direction of the tension tensor.

Equations (1),(2),(6),(8) and (9) constitute the set of balances which must be solved simultaneously together with appropriate conditions and a model for the biochemical activity of the filament within the cortex. Mathematically it is a highly complex non-linear set which requires a numerical approach to obtain a solution.

Surface Forces, Stimulatory Process and Surface Kinetics

Filament and muscle contractions may not be completely equal. However, the similarity in their constituents and the factors influencing their activity could suggest that their mechanical behavior can be somehow comparable. It is, therefore, interesting to look briefly at the mechanics of muscle contraction and specifically, at the isometric sarcomere length-tension behaviour. The tension exhibits a maximum for a specific length known as the reference length and diminishes for lower or higher sarcomere dimension. In addition to these active tensions the muscle can react to external loads exhibiting increasing passive tensions with further length increase. It is also known that, when electrically or chemically stimulated, the muscle develops a tension which later relaxes.

In view of these facts we can expect that the force exerted by each filament will not be constant in time but will change according to the local instantaneous biochemical picture.

It is anticipated that the force of the active filament will grow as the filament starts to contract and will later relax in a manner similar to that described for muscles. This assumed behaviour is in agreement with Hiramoto's (1968) observations and Akkas' (1981) hypothesis for membrane changing stiffness. Passive filaments and those which are fully extended will then contribute to the passive forces of the cortical surface. Their influence together with that of additional passive forces arising from other components of the cortical matrix can be of importance, particularly in the polar zones and the meridional direction of the equator where significant surface stretching occurs.

The exact nature of the stimulatory process and the surface biochemical activity, ultimately responsible for the surface forces and their differentiation, is still unknown. In fact, these are likely to be quite complex since it is known that the biochemistry of the muscle contraction involves a rather high number of steps and biochemical species (Goody and Holmes 1983).

On the other hand, relatively simple enzymatic cycles were shown to be able to produce hydrochemical instabilities on a droplet surface. These can lead to the formation of surface tension gradients which can model successfully force variations similar to those expected by the polar relaxation theory at the onset of cytokinesis (Sorensen 1980; Gallez 1984). Hence, for the purpose of simplification, we introduce here an arbitrary kinetic scheme which is capable of providing the essential aspects of the stimulatory and biochemical processes and allows also for the biochemical modulation of the surface forces. This modulation feature is important to give a description of the dynamic force behaviour in view of the experimental observation that forces have a similar qualitative evolution over the entire surface and they decrease during the last stages of cleavage. The modulation of surface properties was suggested by Akkas (1981), however, no dynamic connection to surface processes was proposed.

We consider the following scheme:

$$F + E \xrightarrow{k_1} FE \xrightarrow{k_2} A + E$$

$$S + E \xrightarrow{k_3} SE \xrightarrow{k_4} B + E$$

where E denotes the inactive filaments, FE the active ones, S a stimulatory agent and F another specie which induces filament activation and whose formation is independent of mitosis. According to this scheme there are two competitive processes. The first induces the production of active filaments while the other inhibits it with the inhibition higher where stimulus quantity is larger. The sequential decomposition reactions contribute to the inactive byproducts A and B and provide the decrease in surface tension observed during the last stages of the division.

The above model does not intend to represent real biochemical events occurring in the asters and the cortical layer. It is, however, an instrument brought to emphasize the importance of a simultaneous analysis of the biomechanical and biochemical activities at the surface.

The stimulus, S, is assumed to diffuse to the surface, originating at two-point sources located at the asters. The duration of this diffusion is finite and the total amount arriving at the cortex is the integral of the flux of S during this time. Since the reaction of S at the surface is relatively fast it is assumed that its concentration at the surface is always zero. The flux is then calculated from a pseudosteady state concentration distribution due to the two point sources with the above surface condition. This distribution is obtained by the solution of

$$\nabla^2 S = 0 \tag{11}$$

subject to

$$S = 0 \quad \text{on} \quad \partial B$$

and a constant flux emerging from each point source located at the astral centers. The solution to (11) gives on a sphere of radius unity,

$$-\frac{\partial S}{\partial n} = \frac{1-a\cos\theta}{(1+a^2-2a\cos\theta)^{3/2}} + \frac{1+a\cos\theta}{(1+a^2+2a\cos\theta)^{3/2}} + \sum_n nA_n P_n(\cos\theta) \tag{12}$$

where θ is a meridional angle, $2a$ is the distance between the centers of the mitotic spindle and P_n are Legendre Polynomials. A plot of the stimulus flux at the initial spherical surface is shown in Figure 1.

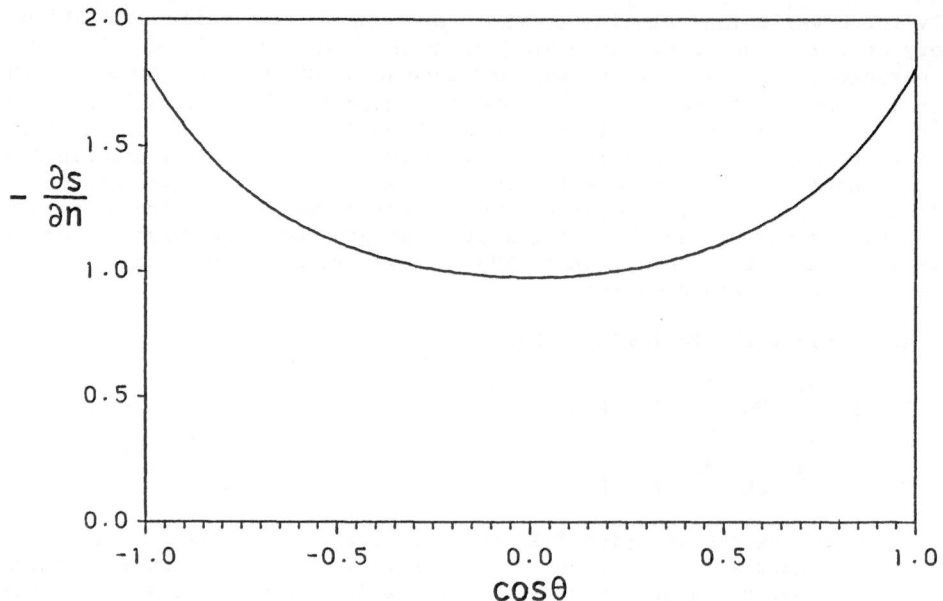

Fig. 1. Stimulus flux distribution on initial spherical
surface distance between asters, 2a=2/3.

METHOD OF SOLUTION

Boundary Integral Equation

Following the work of Ladyzhenskaya (1969) the Stokes equations of
motion can be formulated and solved via a boundary integral equations (BIE)
representation. This method was used successfully by Rallison and Acrivos
(1978) to study large deformations of an interface of a droplet embedded in
an arbitrary external creeping flow, and by Hiram and Nir (1983) to simulate
the dynamics of the coalescence of two spherical fluid particles in an
otherwise quiescent medium.

When the viscosities of the droplet and the surrounding fluid are
equal the BIE equation representation obtains the form

$$\underline{v}(\underline{x}) = \frac{1}{8\pi\mu} \int_{\partial B_y} \underline{\underline{J}}(\underline{x}-\underline{y})\,(\nabla_s \underline{\underline{\gamma}} - (\underline{\underline{\gamma}}:\nabla\underline{n})\underline{n})\,dS_y \tag{13}$$

Here, the tension γ, distances and velocities are non-dimensionalized by
the initial isotropic tension, γ_0, the initial droplet radius and by $\gamma_0/\mu a$
respectively. The single layer potential kernel for the Stokes equations,
$\underline{\underline{J}}$, has the form

$$\underline{\underline{J}} = \frac{\delta_{ij}}{|\underline{x}-\underline{y}|} + \frac{|x_i-y_i||x_j-y_j|}{|\underline{x}-\underline{y}|^3} \tag{14}$$

where $\underline{x}$ and $\underline{y}$ denote position vectors. The assumption of equal inner and outer viscosities does not exactly coincide with the biological situation. It is, however, known (Hiram and Nir, 1983) that different viscosities do not change significantly the qualitative results regarding surface deformation, and the choice of equal viscosities is made here for the sake of mathematical simplification only.

Equation (13) has an innocent form but in fact it is a rather complex nonlinear equation for the surface velocities. Yet it has the advantage of involving only surface variables which in our case are the most interesting and significant for the purpose of the simulation. The source of the non-linearity lie in the dependence of the tension and the shape on the evolution of active filaments concentration and orientation distribution, which in turn depends on the surface velocity and its gradients as is manifested in equations (6),(8),(9) and (10). Evidently, large surface deformations can only be handled using a numerical procedure in the solution of the above equations.

We have employed a scheme which uses the quasi-static nature of Stokes flows where velocity components can be calculated for a given surface shape and interfacial tension distribution. In addition, since the system is axisymmetric the BIE equation can be further simplified to involve integration along a single variable, i.e., the axis connecting the two asters. A similar approach was employed by Sapir and Nir (1985) who used BIE representation to study the deformation of droplet interfaces subject to sharp gradients of isotropic surface tensions.

Numerical Scheme

For the axisymmetric droplet, equation (13) can be integrated along the circumferential direction to obtain the simplified integral of a single variable

$$
\begin{bmatrix} v_r \\ v_x \end{bmatrix} = \int_{-\ell}^{\ell} \begin{bmatrix} A_{rr}(x-y) & A_{rx}(x-y) \\ A_{xr}(x-y) & A_{xx}(x-y) \end{bmatrix} \begin{bmatrix} f_r(y) \\ f_x(y) \end{bmatrix} dy
$$

$$
\begin{bmatrix} f_r \\ f_x \end{bmatrix} = \begin{bmatrix} (\underline{\underline{\gamma}}:\nabla\underline{n})n_r - |\nabla_s \cdot \underline{\underline{\gamma}}| t_r \\ (\underline{\underline{\gamma}}:\nabla\underline{n})n_x - |\nabla_s \cdot \underline{\underline{\gamma}}| t_x \end{bmatrix}
$$

(15)

$$
n_r = \frac{1}{\sqrt{1+R'^2(y)}} \qquad\qquad t_r = \frac{R'(y)}{\sqrt{1+R'^2(y)}}
$$

$$
n_x = \frac{R'(y)}{\sqrt{1+R'^2(y)}} \qquad\qquad t_x = \frac{1}{\sqrt{1+R'^2(y)}}
$$

x and y are coordinates measured along the axis and $R(x)$ denotes the cylindrical droplet shape. The cylindrical components A_{rr}, A_{xr}, A_{rx}, A_{xx} of the tensor $\underline{\underline{J}}$ are given by Rallison and Acrivos (1978).

To obtain a numerical solution M+1 points were distributed along the contour of the axisymmetric surface. The spacing between neighbour points is quite arbitrary and can be chosen to anticipate crowding of points or otherwise during surface flow and deformation. In our scheme we have spaced the points at equal initial meridional intervals along the surface. M=70 was used as a practical compromise between accuracy and computer resources. All derivatives at the surface, e.g., velocities, forces and surface shape, were calculated using three and five points second order accuracy finite difference schemes for unequal intervals. The kernels J_r and J_x become singular as $\varepsilon = x - y$ approaches zero. The singularity is logarithmic except at the poles where it is of $O(\varepsilon^{-1/2})$, hence, it is integrable everywhere. For each point x_n the leading orders of the singular kernels, upto $O(\varepsilon^2)$, were substracted from the integrand in the interval between x_{n-1} and x_{n+1}. The regular integration which resulted was calculated using a trapezoidal rule. An analytic integral of the asymptotic singular terms between x_{n-1} and x_{n+1} was then added.

Once the instantaneous surface velocities were calculated the evolution of the shape was followed by moving each point according to its velocity i.e.,

$$\underline{x}(t+\Delta t) = \underline{x}(t) + \underline{v}(t)\Delta t \qquad (16)$$

with the time increment Δt given by

$$\Delta t = C \frac{MIN|\Delta x|}{MAX|\Delta v|} \qquad (17)$$

where C is an $O(1)$ constant and ΔX and ΔV are the distance and velocity differences between two neighbour points. There is no obvious criterion for the stability of this quasi-steady evolution of the surface. We have arbitrarily chosen the above time step which yielded satisfactory numerical accuracy.

The condition of incompressibility was maintained by correcting the normal velocity, at each time step, by a small constant value to keep the drop volume constant. This correction was typically about 10^{-3} of the average surface normal velocity. If the correction was not used the total volume change at any single run did not exceed a few percent of the initial value.

The calculated surface velocities are used in the integration of equations (8) and (9) for the mass and orientation distribution balances. The solution of these equations and their use in the evaluation of surface forces (7) involve a number of simplifying assumptions without necessarily losing important information. As indicated in the previous section we have assumed that changes due to translational and rotational diffusions are negligible compared to those by convection and reaction. Thus, equations (8) and (9) can be solved using a straightforward Newtonian integration on a Lagrangian basis, e.g.,

$$f(t+\Delta t) = f(t) + \frac{D_s f}{Dt}\Delta t \qquad (18)$$

where D_s/Dt denotes the surface material derivative. A check of the accuracy of this procedure was made, for the case of the concentration, by calculating the total mass in the absence of reaction at each time increment. The cumulative variation was never higher than 10% with the bulk of this change occurring during the very last stages of cleavage where high surface deformation prevails.

With the absence of diffusion in equation (9) and assuming that the orientation distribution of newly activated filament conforms to that already present, i.e. $\hat{N}(\underline{d})=N(\underline{d})$, the equation admits the general solution

$$N(\alpha,t,s) = \frac{1}{\cos^2\theta \cdot e^{-\int_o^t (\frac{\partial v_s}{\partial s} + k)dt} + \sin^2\theta \ e^{\int_o^t (\frac{\partial v_s}{\partial s} + k)dt}} \qquad (19)$$

Here, as before, s indicates the meridional direction while α denotes the angle between the orientation of the filament and the tangential velocity v_s, and $\underline{d}=(\cos\alpha, \sin\alpha)$. $k=k(\gamma_{\theta\theta}/\gamma_{\phi\phi})$ and reflects the effect of particle interaction.

The integral in (19) was carried out numerically at each Lagrangian point on the surface. The tangential derivative of the surface velocity, v_s, emerges from the first term on the RHS of equation (10). To evaluate the interaction between active filaments we have assumed that two adjacent filaments oriented at angles α and α' exert on each other a moment proportional to the difference in their corresponding orientation $(T(\alpha-\alpha'))$. Hence, the total addition to the filament rotational motion is given by

$$w(\alpha) = \int_o^{\pi/2} k'cN(\alpha')T(\alpha-\alpha')d\alpha \qquad (20)$$

where k' accounts for the cortex viscous resistance. To avoid cumbersome numerical integrations we have further used the forms

$$T(\alpha-\alpha') \sim \sin2(\alpha-\alpha') \qquad\qquad N(\alpha') \sim \delta(\alpha'-\alpha_o) \qquad (21)$$

The first expression assumes a dipole moment while the second accounts for the influence of adjacent filaments at the angle α_o only. We have chosen α_o to be the orientation with maximum N at each meridional point. This implies that aggregation enhance filaments orientation in the principal directions of the tension tensor. The combination k'c was taken as independent of concentration since interaction is assumed with close neighbours only. Also, although the total torque on a single filament may rise with increase in concentration, overcrowding of the finite size particles may diminish the rotational mobility. The direction of rotation, however, has dependence on the degree of reorientation through the ratio of the components of the tension tensor at the surface. Here, the coefficient k' was arbitrarily chosen proportional to $[\frac{4}{\pi} \text{arctg} \frac{\gamma_{\theta\theta}}{\gamma_{\phi\phi}} -1]$.

Finally, in the evaluation of the surface tension tensor through equation (7) we neglect the passive cortical resistance, γ_m, and use only the primary tension due to active filaments contraction. A consideration of the complete tension on the surface should involve a rheological model of the surface layer. Such a model involves constitutive equations describing different rheological behaviours as reviewed in a comprehensive treatise on the mechanics of biomembranes by Evans and Skalak (1979). These eliminations of special cortical rheological properties intensify the effect of anisotropic surface forces which is the primary theme of this work. The additional effect of introducing viscoelastic surface properties will be the subject of a future communication.

If all filaments exert an equal contraction force, f, the integral tension at the surface in the principal directions has the form

$$
\begin{pmatrix} \gamma_{\theta\theta} \\ \gamma_{\phi\phi} \end{pmatrix} = fc \int_{0}^{\pi/2} N(\alpha) \begin{pmatrix} \cos\alpha \\ \sin\alpha \end{pmatrix} d\alpha \tag{22}
$$

The nondimensional filament muscular force dynamics is chosen to obey

$$
\frac{df}{dt} = \left(\frac{c_1}{1+c_1 t} - c_2 \right) f \tag{23}
$$

This first order kinetic agrees qualitatively with the variation of isometric sarcomere tension. c_1 and c_2 were arbitrarily chosen as 2 and 0.7, respectively, to follow the time scale of the cleavage process.

The steps of the numerical procedure are summarized as follows:

(a) Given shape and surface forces.
   Calculate surface velocities.
(b) Choose time step.
(c) Solve change of concentration and orientation distribution.
(d) Calculate new surface forces.
(e) Calculate surface deformation.
(f) Proceed to (a).

calculation were terminated when the furrow radius became 15 percent of the initial size where it is believed that additional instabilities and processes are involved in membrane fusion during the ultimate separation of the blastomers.

RESULTS AND DISCUSSION

We start the discussion of our results with a brief presentation of the main findings of previous models. Greenspan (1977a,1977b) introduced polar sources of active agent causing time dependent surface tension modulation and followed the surface flow and deformation. He considered the possibilities of isotropic tensions increase or decrease with surface concentration. The instability of the initial state was followed by the onset of furrowing at the equator. However, Greenspan's asymptotic calculations did not show the evolution of the furrow beyond about 95% of the initial spherical radial dimension. Sapir and Nir (1985) extended Greenspan's calculations to study large deformations using a numerical procedure and examined the effect of various destabilizing surface force distributions. In all cases considered onset of furrowing was obtained. However, cleavage was not completed and furrow dimension was not reduced to less than 80% of the initial radius. Sapir and Nir concluded that isotropic tension is incapable of overcoming the high negative curvature which develops at the neck and cannot lead to the ultimate division. The importance of anisotropic forces as a primary parameter of cleavage was demonstrated by White and Borisy (1983), who included such forces in their model to obtain complete cell division.

The results presented henceforth reflect the incorporation of such forces in a rigorous and comprehensive model. Asymptotic evaluations of the equations of section 2 were presented by Zinemanas and Nir (1986) considering reorientation of microfilaments due to surface velocity gradients with no interaction. Their results indicate that for a given initial surface concentration profile, causing a lower tension at the poles than at the equator, the initial surface flow in the anisotropic case is slower than that in the isotropic tension case. The total expected deformation, however, is higher. These conclusions obtain further amplification in the numerical solution

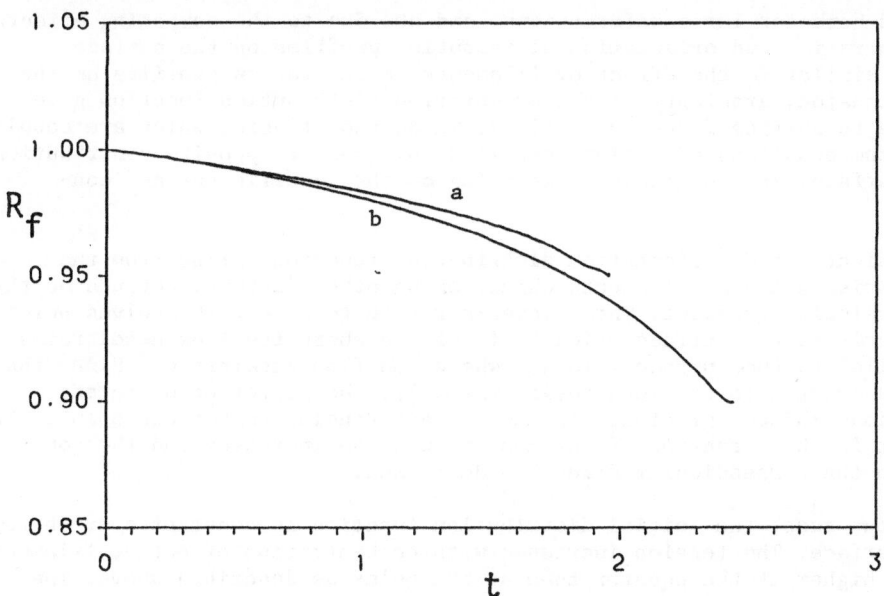

Fig. 2. The change of furrow radius, R_f, with time:
(a) Isotropic tension case; (b) Anisotropic tension.

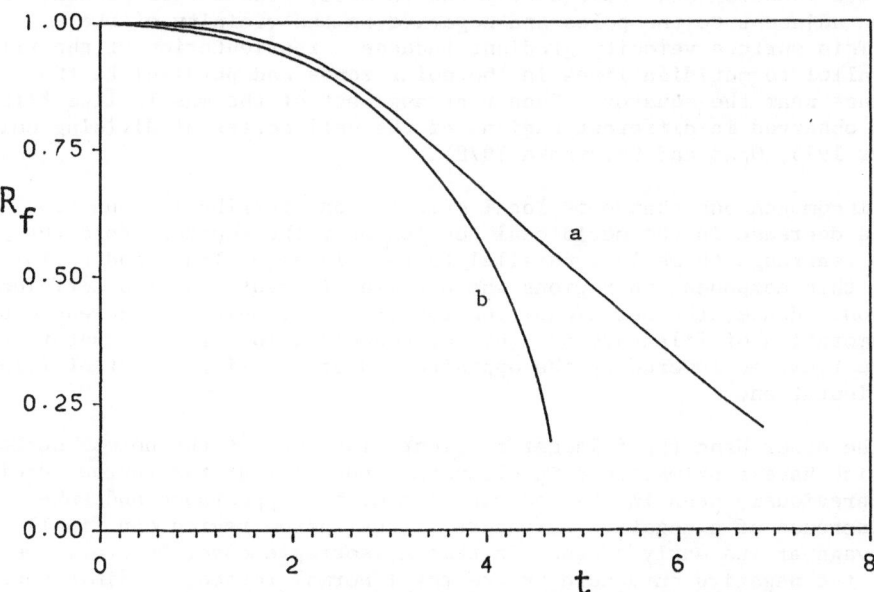

Fig. 3. The change of furrow radius, R_f, with time for
anisotropic surface tension: (a) With filaments
interaction; (b) With filaments interaction and
surface kinetics.

presented here for large deformations, and are due to the competing effect of concentration and orientation distribution profiles on the surface tension. Similar to the effect of filaments concentration profiles on the surface tension, gradients of the orientation distribution function give also rise to surface forces variations. These two effects, which are coupled through the equations of motion, can lead to equal or opposite contributions to the surface tension gradient depending on the specific surface conditions.

Gradients of N, orientation distribution function, arise from the local reorientation of filaments which, among other factors, depends on the surface velocity gradient. The filaments rotate to orient themselves perpendicularly to the surface velocity in places where the flow decelerates and parallel to the surface velocity where the flow accelerates. Since the filaments exert a force along their axis only, the preferred different orientations induce gradients in the surface tension tensor components. The component in the direction of maximum orientation increases and the component in the perpendicular direction decreases.

In our case, the initial distribution function is constant everywhere on the surface. The tension increases with concentration of active filaments which is higher at the equator than at the poles as described above. The initial concentration is determined by the stimulatory scheme described in section 4. The higher concentration and tension at the equatorial region induce two effects: a surface flow from the poles toward the equator and a simultaneous shape deformation. At this initial stage the effect is still similar to the isotropic case since no reorientation of the initial uniform distribution occurred. The surface velocity v_s, has stationary regions at the poles and at the equator. Thus, it must exhibit a maximum at some intermediate location and consequently the velocity gradient is positive in the region adjacent to the poles and negative in the vicinity of the equator. This surface velocity gradient induces a reorientation of the filaments parallel to meridian lines in the polar zones and parallel to the furrow plane near the equator. Such rearrangement of the muscle like filaments was observed in different regions of the cell cortex of dividing cells (Schroeder 1975, Opas and Soltynska 1978).

The inhomogeneous change in local orientation distribution function leads to a decrease in the meridional tension near the equator where the filaments rearrange themselves parallel to the cleavage plane, and an increase of this component in regions where these filaments adopt a meridional orientation. Hence, the meridional tension gradient, which is increased by the agglomeration of filaments to a higher concentration at the equator due to surface flow, is lowered by the opposite and stabilizing effect of filament reorientation.

On the other hand the filament reorientation augment the normal surface forces which have a primal role in cleavage, specially at the furrow leading edge. As previously seen in the isotropic case, the appearance and subsequent increase of a negative curvature at the furrow region can finally halt cleavage at its early stages. In this anisotropic case, however, the weight of the negative curvature in the total normal forces, is diminished by an increase of circumferential tensions and a decrease of the meridional ones due to filament reorientation. Thus, maintaining the total normal forces positive as required to accomplish full cleavage. The surface flow and the filament agglomeration at the equator in the anisotropic case are then lower than in the isotropic case for the same initial concentration profile but the final deformation is higher.

In Figure 2 the evolution of the furrow radius with time is shown for the isotropic and anisotropic cases. These results were obtained for a given

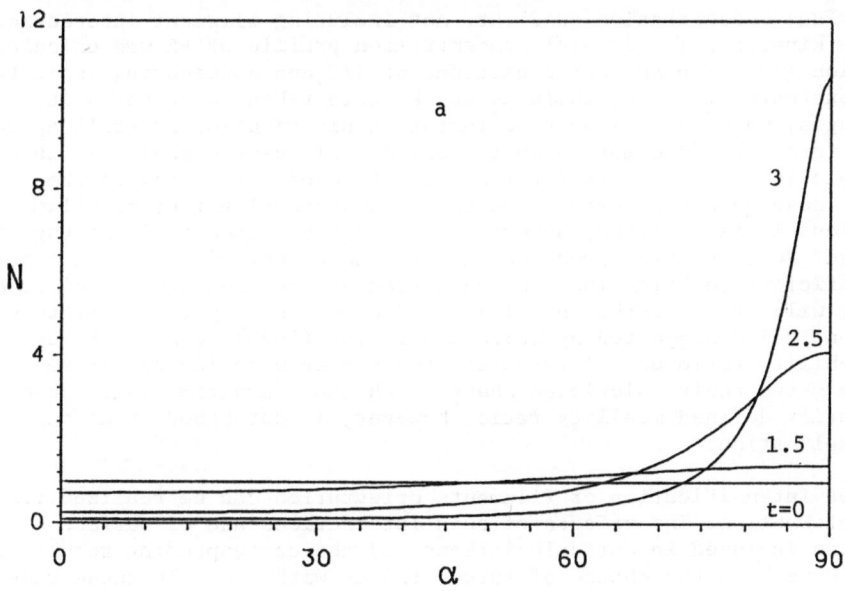

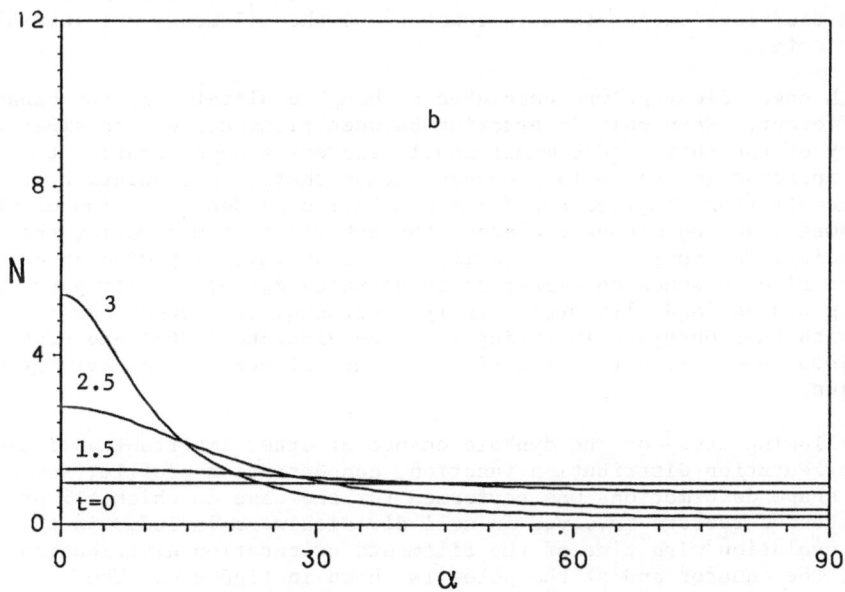

Fig. 4.   The evolution of the filaments orientation
          distribution function, $N(\alpha)$: (a) At the furrow;
          (b) At the pole.

initial surface concentration gradient, not including filament interactions and surface kinetics. The initial concentration profile of FE was calculated from equation (12) with an astral distance of 2/3 and considering infinite reaction constants k_1 and k_3 while k_2 and k_4 were taken to be zero. At initial stages, when the filaments orientation distribution is still approximately uniform, the deformation in the anisotropic case closely resemble the isotropic tension case. At later stages the cleavage furrow in the anisotropic case advances further due to the additional effect of filaments reorientation. It is observed, however, that in both cases full cleavage is not attained. Reorientation produced only by the effect of surface motion are not sufficient to bring the reorganization at the cleavage furrow to an extent that will overcome the inhibiting effect of the negative curvature. The same conclusion suggested by White and Borisy (1983), who had to use a filament mobility ratio of 4:1 to obtain further orientation of filaments in order to match their calculated shapes with those experimentally observed. The arbitrarily defined mobility ratio, however, is not based on a sound physical explanation.

Further intensification of filaments orientation can be realized via filament interaction. The effects of such interactions and of surface kinetics were included in our calculations and the corresponding results are shown in figure 3 as the change of furrow radius with time. In these runs and the following data presented, the initial stimulus concentration, S, was calculated from equation (12) with $a=1/3$. For the initial concentration E,F and FE the values 1, 2, 0.5 were used respectively, and the kinetic constants were chosen to be $k_1=1$, $k_2=0.05$, $k_3= \infty$ and $k_4=0$. The initial orientation distribution function is constant as the filaments are initially randomly oriented.

In both cases cleavage was completed although qualitatively the behaviour is different. When only interaction between filaments was considered, the velocity of the furrow edge monotonously increases demonstrating the continuous increase in the surface concentration there. Incorporation of surface concentration kinetics and force modulation by decomposition of the active filaments during cleavage changes the behaviour of moderating the furrow velocity. The furrow edge velocity, after an initial period of continuous slow rise, reaches an almost constant value during a major part of the cleaving process and ultimately slowly decreases. This behaviour is in agreement with that observed in living cells by Hiramoto (1968) and emphasizes the importance of surface kinetics and the filament force activity in cell division.

The following study of the dynamic change of other important variables, e.g., the orientation distribution function, concentration profile, surface forces and shape deformations was performed for the case in which all previously discussed factors involved in cell division were included in the model. The evolution with time of the filaments orientation distribution function at the equator and at the poles is shown in figure 4. The reorientation of the filaments in the respective regions is clearly seen with the majority of filaments oriented parallel to the furrow plane in the cell equatorial region and in a meridional direction at the poles. Clearly, this evolution depends on the particular function chosen to evaluate the interactions between filaments and on the other assumptions made. Nevertheless, although these results do not match exactly the orientation distribution function in the cell process, they are in good qualitative agreement with the experimental findings (Opas and Soltynska, 1978).

The concentration profile (Figure 5), as anticipated by the previously cited models and the above discussion, shows a sharp agglomeration of filaments at the equator. This, together with the reorientation of filaments parallel to the cleavage plane, provides an explanation for the contractile

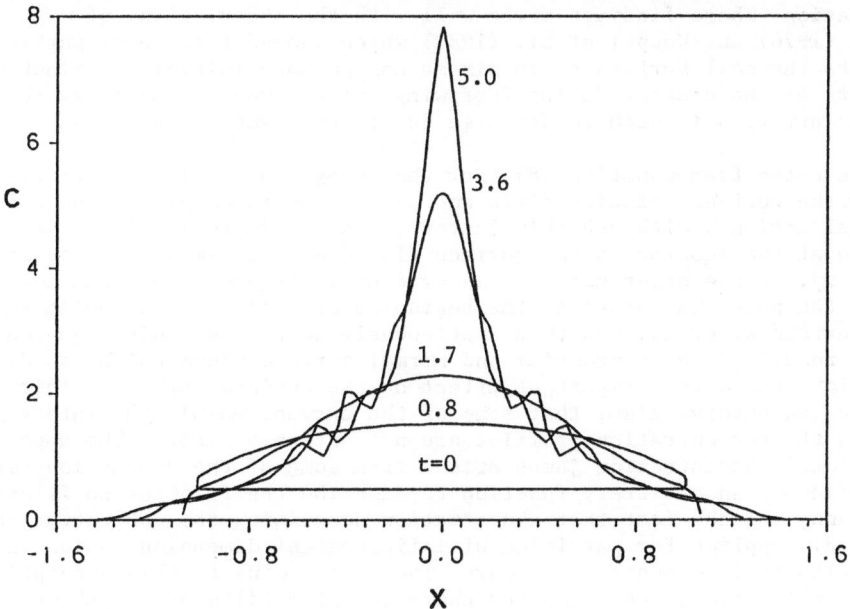

Fig. 5.   The evolution of active filaments concentration
on the deforming surface.

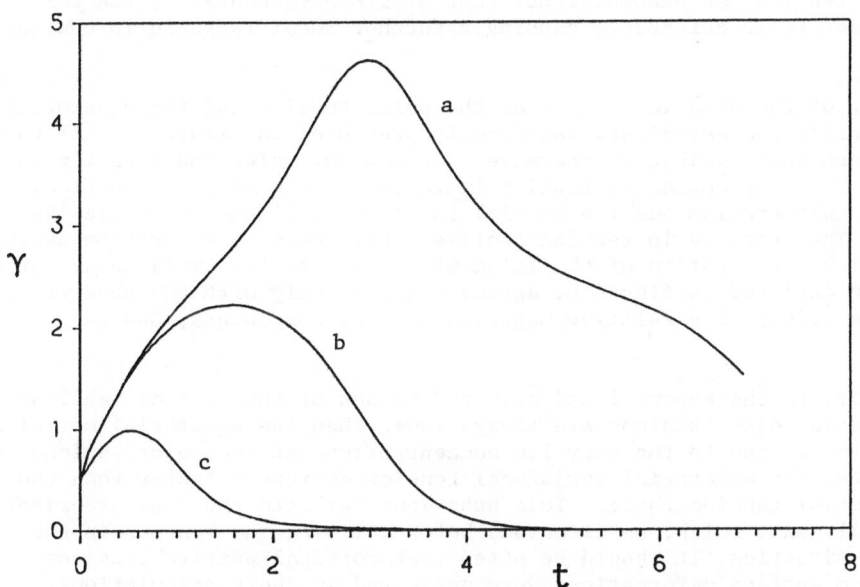

Fig. 6.   Tension dynamics: (a) Equatorial circumferential
component, $\gamma_{\phi\phi}$; (b) Equatorial meridional
component, $\gamma_{\theta\theta}$; (c) Polar tension.

ring formation. These findings agree well with the observations of Rappaport (1976) and Koppel et al.,(1982) which showed that inert particles. attached to the cell surface at an almost homogeneous initial distribution agglomerate at the equator during furrowing and cleavage. Similarly, the polar concentration is seen to decrease due to the same surface flow.

It is noted from equation (8) that the change in local concentration is caused by the surface velocity field and also by surface deformation arising from normal motions. Although this latter effect tends to increase the concentration at the equator as the surface flow does, its importance seems to be secondary. On the other hand, it is evident in figure 5 that the concentration at the poles increases at the beginning of cleavage, following the assumed surface kinetics, and then continuously decreases. This behaviour is again the result of the tangential and normal surface flows which, as discussed below, has a very important effect on the surface forces in this region. We can observe also, that some of the curves, mainly the latest ones, describing the concentration profiles are not entirely smooth. The appearance of these local concentration jumps arises from some of the assumptions made in the choice of an arbitrary function to describe the filament to filament interactions, and the fact that the equation describing the concentration changes, (8), applies for particles of infinitesimal dimensions which interact only with their closest neighbours. Thus, according to this assumption two points which are closer than the characteristic filament length can exhibit very different concentrations. Since filaments are of finite dimension such clustering can occur only in regions where orientation is high, e.g., the vicinity of the contractile ring, but generally should be viewed as a numerical artifact. The arbitrary choice of the function describing the interaction between filaments can also lead to errors in the local evolution of the surface tension gradients. At certain intermediate points, the local meridional tension can become higher that of its neighbours. These points then, attract their neighbours causing a further local increase in concentration.

A plot of the dynamic changes of the polar tension and the equatorial circumferential and meridional tensions is presented in figure 6. All three tensile components exhibit an increase   a maximum value and then a gradual decrease. The main causes of tension increase are the surface kinetics, filaments agglomeration and the initial increase in filaments muscle-like activity. The decrease in tension profiles stems from concentration depletion and delayed relaxation of the filament force activity. This behaviour of the tension depicted in figure 6, agrees qualitatively with the observation of Hiramoto (1968). The relative magnitude of the components, however, differs.

Contrary to the experimental measured values of the surface tensions, the calculated polar tensions are always lower than the equatorial meridional ones which is due to the very low concentrations at the polar regions. We also see that the equatorial meridional tension decreases faster than the circumferential tension there. This behaviour reflects the rapid reorientation of filaments which, as discussed before, lowers the tension in the meridional direction. It should be noted that cortical passive tensions arising from surface deformations were neglected in these calculations. Incorporation of these effects would cause an increment in surface tension both in the polar region and in the equatorial meridional direction. However, this additional effect is expected to be more significant at the poles since the stretching of the surface there is higher. The study of these effects will appear in a later communication. Again, we see the important role of force modulation and surface kinetics in obtaining the expected quantitative behaviours. These suggest that the arbitrary dynamic variation of surface rheological properties, proposed by Akkas (1981), could be linked in a similar manner to microscopic structural dynamic changes.

300

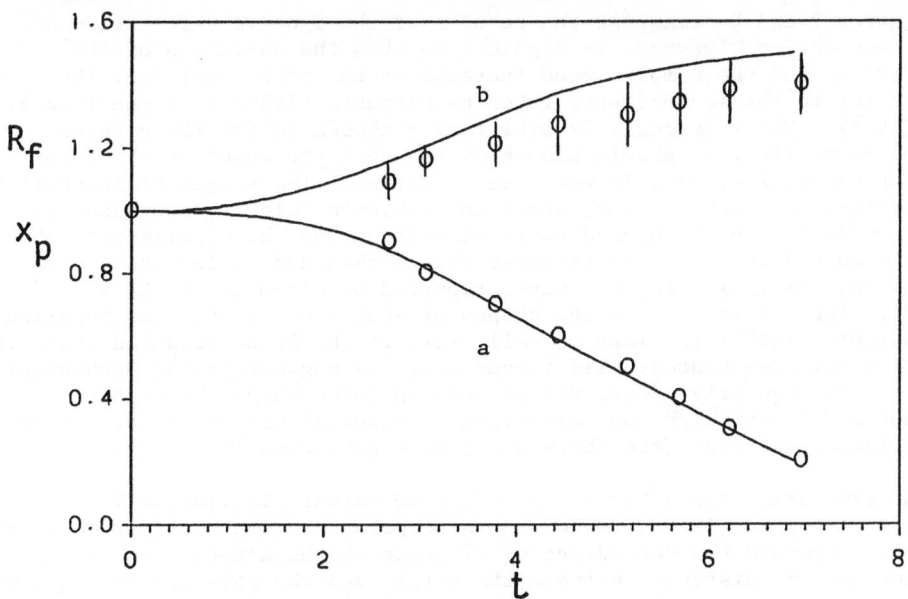

Fig. 7.   The dynamic change of geometrical dimensions during
          cleavage; (a) Furrow radius, R_f ; (b) Polar axial
          distance, x_p. O denotes experimental results of
          Hiramoto (1958).

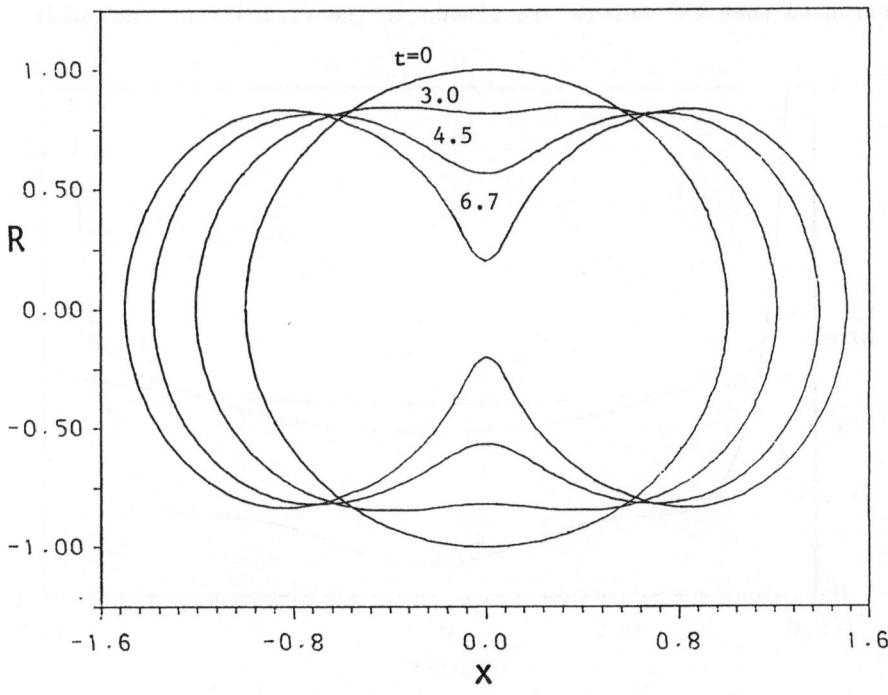

Fig. 8.   Consecutive surface shapes at different stages of
          cleavage.

Figures 7 and 8 summarize the results of the dynamic deformation of the surface during cleavage. In figure 7 we show the shrinkage of the furrow radius and the simultaneous increase of the polar distance. The points refer to the measurements taken by Hiramoto (1958) as summarized by Akkas (1981). The time scale is arbitrarily chosen to fit the unknown physical properties. It should also be noted that the onset of the cyto- kinetic process, i.e. t=0, is not exactly defined. The stages of incubation, acceleration, constant cleaving speed and ultimate velocity decrease are clearly reflected in the hydrodynamic simulation and the experimental obser- vation. Figure 8 depicts instantaneous shapes obtained during the cleavage process. The similarity to the shapes reported by Hiramoto (1968) is apparent. The resemblance to the shapes at stages of cytokinesis reported by Karasiewicz (1981) is clear as well. Yet, it should be recalled that our simulation involves hydrodynamic forces only and neglects solid mechanical influence. Consequently, local differences in shape should be expected where curvatures are high, and migration of material points on the surface can considerably differ from those observed experimentally.

The simulation described above refers to normal cleaving cells. Rappaport (1984) in his search for tests of polar or equatorial stimulation hypotheses examined the dependence of cleavage on the mitotic apparatus dimension and the distances between the asters and the pole and the equator. His findings were inconsistent with a pure polar stimulatory hypothesis. His results, however, are not in contradiction with the simulation proposed in this communication. We have tested the dependence of cleavage on the astral location, i.e., on the surface distribution of the stimulus flux. In figure 9 we show the flux to the initial spherical surface of the stimulat- ing agent. Evidently, when the mitotic apparatus is confined to the proxi- mity of the equatorial plane (a=0.167) the surface distribution is nearly uniform and therefore small surface tension gradients can be expected. On the other hand when the asters are placed in the vicinity of the poles

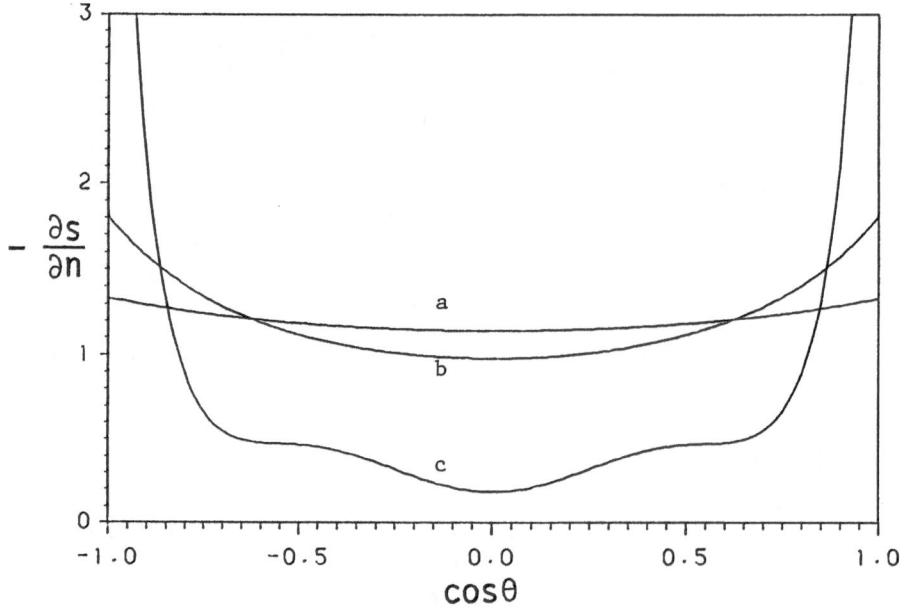

Fig. 9. Stimulus flux distribution on initial spherical surface as function of distance between asters: (a) a=0.167 ; (b) a=0.333 ; (c) a=0.9.

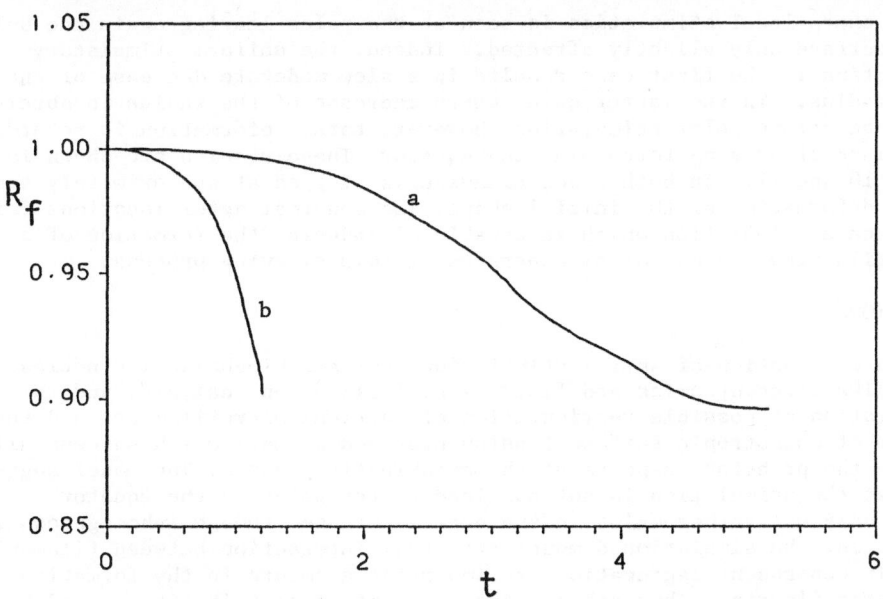

Fig. 10. The change of furrow radius, R_f, with time for
various astral locations: (a) a=0.167 ; (b) a=0.9.

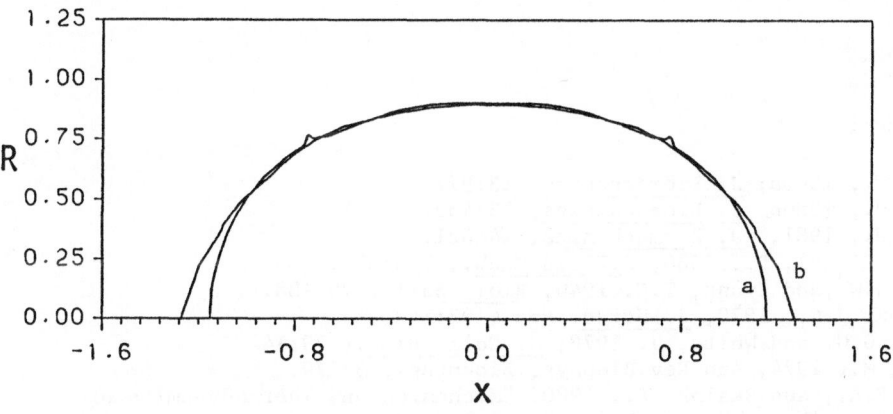

Fig. 11. Maximum surface deformation for various astral
locations: (a) a=0.167 ; (b) a=0.9.

(a=0.9) sharp local stimulation is seen at the poles leaving again the bulk of the surface only slightly affected. Indeed, the uniform stimulatory distribution in the first case results in a slow moderate decrease of the furrow radius. In the latter case, sharp decrease of the radius is obtained due to the strong polar stimulation. However, total deformation is retarded by the lack of driving force near the equator. These results are shown in figures 10 and 11. In both cases cleavage is stopped at approximately ten percent deformation of the initial shape. The abnormal aster locations fail to produce a stimulation which is capable of inducing the formation of a contractile ring, an essential apparatus of this cleaving process.

CONCLUSIONS

The combination of astral stimulation, surface biochemical kinetics, muscle-like force dynamics and fluid mechanical motion analysis, and the incorporation of possible reorientation of cortical microfilaments and the creation of anisotropic surface tension provided a model which successfully simulate the principal aspects of the cytokinetic process. The model suggests that the stimulation is not confined to the poles on the equator regions only but rather exist in the entire surface, though inhomogeneously distributed. The simulation demonstrates that interaction between filaments and their consequent aggregation are important elements in the formation of the contractile ring. They enhance the orientation distribution caused by the surface motion and thereby facilitate the cleaving process. The neglect of cortical passive reaction due to its rheological properties intensifies and distorts various results of the simulation. Their incorporation in future modelling is important and is currently under investigation.

Acknowledgement - This work was supported in part by the Fund for the Promotion of Research at the Technion. D.Z. would also like to acknowledge a fellowship from the Wolf Foundation.

REFERENCES

Akkas, N., 1980a, J. Biomechanics, 13:977.
Akkas, N., 1980b, J. Biomechanics, 13:459.
Akkas, N., 1981, J. Biomechanics, 14:621.
Barthes-Biesel, D., 1980, J.Fluid Mech., 100:831.
Beams, H.W., and Evans, T.C.,1940, Biol. Bull., 79:188.
Bluemink, J.C., 1970, J. Utras. Res., 32:142.
Borisy, G.G. and White, J. 1978, J. Cell Biol., 79:14a.
Edidin, M., 1974, Ann Rev.Biophys. Bioengng., 3:179.
Evans, E.A., and Skalak, R., 1980, "Mechanics and Thermodynamics of
    Biomembranes", CRC Press Inc., BOCA Raton, Florida.
Forer, A. and Behnke, O., 1972, Chromosoma, 39:175.
Fujiwara, K. and Pollard, T.E., 1976, J. Cell Biol., 71:848.
Gallez, D., 1984, J. Theor. Biol., 111:341.
Goody, R.S., and Holmes, K.C., 1983, Biochimica et Biophysica Acta, 726:13.
Greenspan, H.P., 1977a, J.Theor. Biol., 65:79.
Greenspan, H.P., 1977b, Stud. Appl. Math., 57:45.
Greenspan, H.P. 1978, J. Theor. Biol., 70:135.
Hiram, Y. and Nir, A., 1983, J. Coll. Interface Sci., 95:462.
Hiramoto, Y., 1956, Exp. Cell. Res., 11:630.
Hiramoto, Y., 1965, J. Cell Biol., 25:161.
Hiramoto, Y., 1968, 22nd Symp. Soc. Exp. Biol., 311.
Hiramoto, Y., 1971, Exp. Cell Res., 68:291.

Hiramoto, Y., 1981, in: "Mitosis/Cytokinesis", Eds. A.M. Zimmerman and
	A. Forer, Academic Press.
Karasiewickz, J., 1981, in: "Mitosis/Cytokinesis", Eds. A.M. Zimmerman
	and A. Forer, Academic Press.
Koppel, D.E. Oliver, J.M., and J.M. Berlin, R.D., 1982, J.Cell Biol., 93:950.
Opas, J., and Soltynska, M.S., 1978, Exp. Cell Res., 113:208.
Pujara, P., and Lardner, T.J., 1979, J. Biomechanics, 12:293.
Rallison, E., and Acrivos, A., 1978, J. Fluid Mechanics, 89:191.
Rappaport, R., 1961, J. Exp. Zool., 148:81.
Rappaport, R., 1969, J. Exp. Zool., 171:59.
Rappaport, R., 1971, Int. Rev. Cytol., 31:169.
Rappaport, R., 1976, Devel. Growth. Diff., 18:189.
Rappaport, R., 1978, J. Exp. Zool., 206:1.
Rappaport, R., 1981, in: "Mitosis/Cytokinesis", Eds. A.M. Zimmerman and
	A. Forer, Academic Press.
Rappaport, R., 1984, J. Exp. Zool., 231:81.
Rappaport, R. and Ebstein, R.P., 1965, J. Exp. Zool., 158:373.
Sapir, T. and Nir, A., 1985, Physicochemical Hydrodynamics, 6:803.
Schroeder, T.E., 1973, Proc. Natn. Acad. Sci.,USA, 70:1688.
Schroeder, T.E., 1975, in: "Molecules and Cell Movement", Eds. S. Inoue
	and R.E. Stephens, Raven Press, New York.
Schroeder, T.E., 1981, Exp. Cell. Res., 134:231.
Sorensen, T.S., 1980, J.C.S. Faraday II, 76:1170.
White, J., 1985, BioEssays, 2:267.
Wolpert, L., 1960, Int. Rev. Cytol., 10:163.
Zinemanas, D., and Nir, A., 1986, in: "Variation Methods for Free Surface
	Interfaces", Eds. Concus, P., and Finn, R., Springer-Verlag, New York.

BIOMATHEMATICS OF CELL DIVISION:

INVERSION OF THE HIRAMOTO MAP (*)

Giorgio Catalano

Dipartimento di Matematica, Facoltà di Ingegneria
Università della Calabria
87036 Arcavacata (Cosenza)   (Italy)

I. INTRODUCTION

The central target of contemporary biology is still the problem of
cell differentiation. In many animal species - spanning a wide range from
echinoderms to mammalians - the multi-cell organism originates, ultimately, from a single-cell: the egg. The process whereby an egg starts dividing to give rise to an embryo, i.e. 'cleavage', can be triggered, in different species, by a variety of different treatments: fertilization by paternal sperm, artificial activation by exposure to acids or to temperature
shock or even, in some case, simply by the pricking of a needle. Thus, it
is the 'egg-system', and not only the DNA contained in the egg, that has
the 'know-how' necessary for the making of an embryo. The egg, to say it
with Max Hamburgh[1], "far from being an 'undifferentiated' cell, is perhaps
the most highly specialized cell of any organism".

There is a well established experimental evidence that, during cleavage of the egg, when DNA appears to be simply duplicated at a very fast
rate and none of the cells resulting from the progressing divisions is still
differentiated, some cells are already committed to give rise to particular
differentiated cells in the embryo. In the case of the sea-urchin (see reviews by Horstadius[2]), if four particular cells (the so called "micromeres"
are removed after the 4th cleavage (i.e. at the 16-cell stage), then the
resulting embryo is deprived of the skeleton. Moreover, in a famous experiment on a frog's egg, it was proved by J.B. Gurdon[3] that the DNA of somatic cells, introduced into an egg the nucleus of which was previously removed,
is able to produce a complete organism.

Therefore, cleavage it is not just a passive partition of a cell into a number of cells but, rather, an active   dynamical process bearing a
decisive relationship with cell differentiation. Such a process can be considered under many aspects, ranging from transcription and duplication of
DNA, genes activation, spatial orientation of mitotic spindles, membrane
contractions and expansions, cytoskeleton, microfilaments, microtubules,
cleavage furrow, and so on. In the present paper, the attention is focused
mainly, but not exclusively, on the geometrical aspects of cleavage.

The main concern one has when devising a model of some process is, of
course, to distinguish among those details of the real process that can be

---

(*) This paper is dedicated to my wife, Nenne, alive in my mind.

regarded as 'unessential' in the model and those that are truly essential. Now, is geometry an essential feature of cleavage or is it just a secondary accident? Various experiments [4],[5] demonstrated the normal development of early embryos subjected to compression, stretching or other physical treatments modifying the natural, genuine geometrical set-up of the cells, so that one is tempted to consider the geometrical aspects as unessential. When the egg makes the first cleavage its spherical shape is soon lost and it assumes gradually the typical shape of an hourglass: its geometry has changed. The question is: what do we mean, precisely, by a 'change in its geometry'?

If we start to make measurements of different areas or of different rays of curvature, then we can say, yes, something has changed. A sphere, after all, is a geometrical object very different from an hourglass shaped surface. But this is true only if we look at the <u>metrical</u> properties (briefly, properties, such as distances and angles, that are invariant under space rotations or translations). The study of the rheological properties of an egg membrane (Hiramoto[6] and Rappaport[7]) allows us to visualize the deformation of the membrane - regarded as a mathematical surface - as a homeomorphism between the surface and its deformed image at any given stage of the cleavage. What is left invariant are the so called <u>topological</u> properties, so that, for instance, a sphere and an hourglass are, topologically, perfectly equivalent objects. Therefore, we shall neglect here the metrical differencies between the real shape of the blastomeres during cytokinesis and spherical surfaces or portions thereof obtained during an ideal process. In sect. 5. where the kinematics of cytokinesis is considered, we shall see that both processes (the ideal process and the real one) can be described by differential equations obeying the same set of functional conditions.

On the other hand, there exist certain metrical properties of cleavage that cannot definitely be ignored. In the sea-urchin, for instance (see reviews by Horstadius [2], Giudice [8], Rossi; et.al. [9]) the first two divisions cut the egg along its "animal-vegetal" axis and are mutually orthogonal. The plane of the third cleavage is orthogonal to the first two planes, namely it is 'equatorial', and the resulting eight cells (blastomeres) are almost equal in size. The fourth cleavage is 'meridional' in the top quadruple (i.e. the planes of division are parallel to the animal-vegetal axis of the egg) and gives rise to a crown of eight blastomeres of equal size (called "mesomeres"). In the downward quadruple, however, the cleavage plane not only is 'horizontal' (i.e. parallel to the equatorial plane of the third division) but its position is strongly shifted downward, giving rise to an intermediate ring of four large blastomeres ("macromeres": the largest in size) and four small ones ("micromeres": the smallest in size) at the bottom. Now, "orthogonality" and "size" are metrical (not topological) properties. Therefore, we are in a situation whereby sometimes the metrical properties are relevant, while, in other cases, the topological properties prevail.

Let us consider the following questions:

Why have the cells at the end of the fourth cleavage a different size?
Why are the micromeres so small?
Why does the mitotic index drop abruptly at hatching time?
Why are the descendants of micromeres at hatching time so few?

As far as we know, no reasonable answer has ever been given to the above questions. The first two questions have a definite geometrical flavour, while in th last two time enters the picture: no surprise, time is still geometry. In this paper we consider the problem of the difference in size of the blastomeres, of the drop of the mitotic index at hatching time and of the exact number of cells descending from the micromeres in the primary mesenchime.

The main experimental findings on which our work is based are:

i) the constancy of the egg volume during cleavage, as determined whithin an extremely small experimental error by Yukio Hiramoto[10];

ii) the remarkable elastic properties exhibited by the membrane of the sea urchin egg, as ascertained by Hiramoto[11,12,13];

iii) the surprising mitotic stop of the first four descendants of micromeres, up to the hatching stage, as found by Elio Parisi et.al.[14].

The central mathematical device which we use is a homeomorphism that maps an hemispherical surface onto a sphere with an arbitrarily small hole. This mapping, which we shall refer to as the "H-map", is a mathematical idealization of a diagram shown in one of the previously cited paper[10] by Y. Hiramoto. It has been already introduced by J. Chris Eilbeck and the author[15] and used in a mathematical model where a surface scalar field was assumed as a non-uniform boundary condition for Laplace equation, in order to obtain computer simulations of generical two-dimensional patterns of cleavage. In sect.2 of the present paper, a somewhat generalized version of the H-map is explained in detail.

The first important attempt at a theorethical model of cleavage, in the case of the sea-urchin, was the hypotesis advanced by Runnstrom[16] of a pair of chemical gradients running in opposite directions in the endoplasm along the animal-vegetal axis of the egg. In the present work, we explore the implications of the assumption of a single gradient on the membrane of the egg. To be more specific, it is assumed that, physically, there exists only one pole, here referred to as the "uni-pole", which may be identified whith the animal pole. Accordingly, in sect.3, a simple linear scalar field is defined on a spherical surface representing a (constant) unipolar gradient of a surface density (a concentration), axially symmetric, of an ideal chemical substance on the egg membrane.

In order to follow the evolution of such a concentration after the first cleavage, the surface Jacobian of the map is calculated and the contour lines of the concentration on the image-sphere are plotted. This process, which is a sequence of composed H-maps, can be continued, as is done in sect. 3 for the second and third cleavages. One point must be stressed: while the H-map is always axially symmetric, its compositions in different directions have no symmetry at all, so that the resulting contour lines at the end of the second cleavage become functions of three variables, instead of one variable only, as it was at the outset; therefore, an explicit expression of the concentration as a function of the coordinates becomes increasingly difficult. The present work has been limited to a qualitative evaluation of the contour lines. Should this approach survive, there will be ample room for improvements.

The formation of micromeres at the end of the fourth cleavage is indeed a striking enigma. Since the sphere is the solid with the minimal surface enclosing a given volume, any variation in surface shape performed at constant volume necessarily yields an increase in surface area. So, one may have the feeling that when an elastic sphere is partitioned into two spheres by a constricting circle, maintaining a constant volume of the two parts, there should be a final increase in surface area of both parts.

The point is that, in general, this is false. Such an elementary geometrical topic offers a clue to the possible meaning of the size of the micromeres.

Now, if the additional assumption of a minimal threshold for the concentration is made, then, as it is discussed in sect.4 , each one of the micromeres may divide unequally and the resulting 'micromeres of the micromeres' may go under the threshold and stop dividing, while the 'macromeres of the micromeres' shall continue to divide equally until they eventually reach the minimal threshold and stop too. It is generally accepted that at some point around hatching a control mechanism of cell division different from the one operating during cleavage is switched on: the cell count shown in sect.4 gives a precise number that can be compared with the values found experimentally.

The possible kinematics of cytokinesis, as previously mentionned, is briefly examined in sect.5, where a set of functional conditions are imposed on the time-dipendent generatrix of a solid of revolution. As an example, a differential equation is obtained that can be explicitly solved, which supplies a possible kinematics by spherical segments. Other, more realistic differential equations can be obviously derived from the same set of conditions, but we do not dig into this subject. Finally, no attempt has been made at a physical or molecular interpretation of the hypotetical substance that could constitute the concentration.

2. THE HIRAMOTO MAPPING

As a first step, let us immagine to cover the upper half ($z \geqslant 0$) of a solid sphere of radius r_0 (see Fig.1) with an elastic hemispherical surface, say a rubber calotte. Then, by stretching the calotte downward, while contracting its base circle until the point $(0, 0, -r_0)$ is reached, without any torsion or rotation, we cover the whole sphere. Now, in terms of a geometrical pointwise application of the unstretched upper hemisphere onto the whole stretched sphere, it is clear from Fig.1 that the mapping sending the point $P(\rho, \phi, z)$ in cylindrical coordinates, onto the point $P'(\rho', \phi', z')$, if the stretching is uniform, has the following equations:

$$\phi = \phi' \quad ; \quad z' = 2z - r_0 \quad ; \quad \rho' = (r_0^2 - z'^2)^{\frac{1}{2}}$$

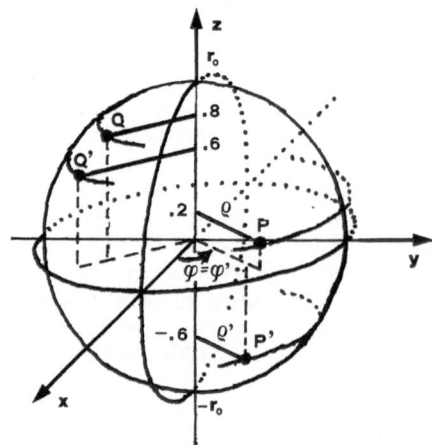

Fig.1. Basis for the definition of the Hiramoto mapping

As a second step, let the image sphere be shrunk until it reaches a radius:

$$r_1 = \alpha \, r_0 \tag{1}$$

where $0 < \alpha < 1$. Then, the first equation above does not change, while the last two ones become, respectively:

$$z' = \alpha \, (2z - r_0) \quad \text{and} \quad \rho' = (r_1^2 - z'^2)^{\frac{1}{2}}$$

Therefore, in cartesian coordinates, our application becomes as follows:

$$x' = 2 \, \alpha \, (z / (z + r_0))^{\frac{1}{2}} x$$

$$y' = 2 \, \alpha \, (z / (z + r_0))^{\frac{1}{2}} y$$

$$z' = \alpha \, (2z - r_0)$$

It may be observed that the above application is <u>not</u> invertible. In fact, as it is clear also from Fig.1, all points of the circle $x^2 + y^2 = r_0^2$ map onto the single point $(0, 0, -r_0)$. Since we need to invert the mapping, we consider its restriction to the domain:

$$- \delta \leqslant x \leqslant \delta \quad ; \qquad - \delta \leqslant y \leqslant \delta \quad ; \qquad \varepsilon \leqslant z \leqslant r_0$$

where $\varepsilon > 0$ is arbitrarily chosen and $\delta = (r_0^2 - \varepsilon^2)^{\frac{1}{2}}$. Thus, the application maps a spherical bowl of radius r_0 and base circle of radius $r_0 - \varepsilon$ (very near to an hemisphere when ε is very small), onto a sphere with a small hole, the radius of which turns out to be $2 \, \alpha \, (2 \, \varepsilon \, (r_0 - \varepsilon))^{\frac{1}{2}}$.

Clearly, the application just defined has an axial symmetry with respect to the z axis. As already mentionned in sect.1, we shall need to consider similar applications also with other axes of symmetry and to compose them. Accordingly, we use subscripts, writing $P_0 \, (x_0, \, y_0, \, z_0)$ instead of $P \, (x, \, y, \, z)$; $P_1 \, (x_1, \, y_1, \, z_1)$ instead of $P' \, (x', \, y', \, z')$ and so on. Now, if we define:

$$f_i \, (\xi) = (\xi / (\xi + r_i))^{\frac{1}{2}} \tag{2}$$

$(i = 0, \, 1, \, 2)$ and perform cyclic permutations on the variables x, y, z, we obtain from the previous application three mappings which we write down explicitly for future references:

X	Y	Z
$x_1 = \alpha \, (2x_0 - r_0)$	$x_2 = 2 \, \alpha \, f_1 \, (y_1) \, x_1$	$x_3 = 2 \, \alpha \, f_2 \, (z_2) \, x_2$
$y_1 = 2 \, \alpha \, f_0 \, (x_0) \, y_0$	$y_2 = \alpha \, (2y_1 - r_1)$	$y_3 = 2 \, \alpha \, f_2 \, (z_2) \, y_2$
$z_1 = 2 \, \alpha \, f_0 \, (x_0) \, z_0$	$z_2 = 2 \, \alpha \, f_1 \, (y_1) \, z_1$	$z_3 = \alpha \, (2z_2 - r_2)$

We now define the <u>H-map along the x-axis</u> as the mapping X (and similarly for the other axes). Then, the following composed mappings are of interest:

$$H_1 = X \quad ; \qquad H_2 = XY \quad ; \qquad H_3 = XYZ$$

where the composed mapping XY applied to point P_0 means:

$$(XY) \, (P_0) = Y(X(P_0)) = Y(P_1) = P_2 \, , \text{ and similarly for XYZ.}$$

Finally, since the initial configuration to be simulated is the division of a sphere of radius r_0 into two spheres of radius r_1, at <u>constant volume</u>, from (1) and the condition $r_0^3 = r_1^3 + r_1^3$ we obtain for α the following numerical value:

$$\alpha = 2^{-\frac{1}{3}} \tag{3}$$

3. THE UNIPOLAR GRADIENT: FIRST THREE CLEAVAGES

Let $x_0^2 + y_0^2 + z_0^2 = r_0^2$ be the sphere representing the egg membrane, where the subscripts have the meaning explained in sect.2. The hemisphere Σ_0 with $x_0 > 0$ is shown in Fig.2, and the point $(0, 0, r_0)$ represents the unipole. We define the scalar field σ_0 on the spherical surface as follows:

$$\sigma_0 = \sigma_0 (x_0, y_0, z_0) = \sigma_0 (z_0) = \tfrac{1}{2} (z_0 + r_0) \tag{4}$$

representing the concentration of an ideal substance correlated, and possibly controlling, cytokinesis. Without loss of generality, we can take a unit radius ($r_0 = 1$), although, when necessary to avoid ambiguities, we shall continue to denote it by r_0. Thus, the concentration σ_0 decreases linearly from its maximum value of 1 at the unipole to zero at the antipode $(0, 0, -r_0)$. Some contour lines are shown in Fig.2. The unipolar gradient is, in this example, the constant vector:

$$\nabla \sigma_0 = (0, 0, \tfrac{1}{2})$$

Let us take the plane $y_0 z_0$ as the plane along which the first cleavage takes place. The H-map H_1 transforms Σ_0 into a sphere $\Sigma_1 = H_1 (\Sigma_0)$ of radius $r_1 = \alpha r_0$. Let dS_0 and dS_1 be the elements of area of Σ_0 and Σ_1, respectively. Then it is not difficult, although somewhat laborious, to see that the following identity holds:

$$dS_1 = 2 \alpha^2 dS_0 \tag{5}$$

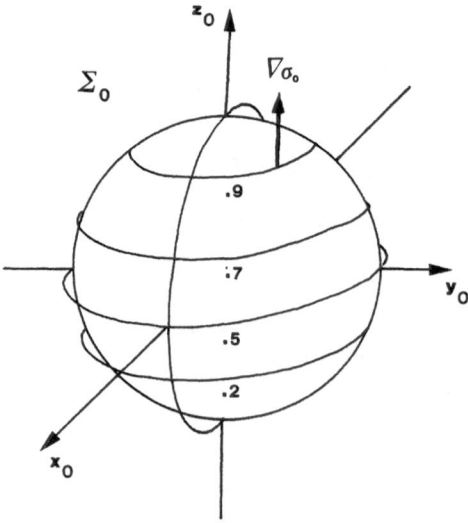

Fig.2. Unipolar gradient on the egg surface

so that the conservation law for the concentration σ_o becomes:

$$\int_{\Sigma_1} \sigma_1 \, (P_1) \, dS_1 = \int_{\Sigma_0} \sigma_o \, (P_o) \, dS_o = 2 \, \alpha^2 \int_{\Sigma_0} \sigma_1 \, (P_1) \, dS_0 \qquad (6)$$

which implies:

$$\sigma_1(P_1) = \sigma_o \, (P_o)/(2\alpha^2)$$

and, since (3) yields $1/(2\alpha^2) = \alpha$, we obtain:

$$\sigma_1(P_1) = \alpha\sigma_o \, (P_o) \qquad (7)$$

at the first cleavage. In general:

$$\sigma_n(P_n) = \alpha^n \sigma_o \, (P_o) \qquad (n = 1, 2, 3) \qquad (7')$$

at the n-th equal cleavage.

It may be noted, en passant, that (3) and (7') imply that the concentration halves at any three consecutive equal divisions.

Now, let k be an assigned value for the concentration σ_n. In order to find the corresponding contour line on Σ_n we must solve for P_n the equation:

$$\sigma_n \, (P_n) = k.$$

From (7') we have:

$$\alpha^n \sigma_o \, (P_o) = k,$$

and, from (4) (with $r_o = 1$), we obtain:

$$z_o = 2k/(\alpha^n) - 1 \qquad (8)$$

So, the problem is solved pointwise for any x_o such that $\varepsilon < x_o \leqslant r_o$: z_o is determined by (8), $y_o = (x_o^2 + y_o^2 + z_o^2)^{\frac{1}{2}}$, and $P_n = H_n \, (P_o)$.

Following the above procedure, a simple computer program can be implemented, and some of the resulting contour lines of σ_1 on Σ_1 are shown in Fig.3a, while in Fig.3b the orthogonal projection of the same lines on plane $x_1 \, y_1$ is plotted. The sphere Σ_1 has been enlarged by a factor α^{-1}, and a scale factor α^{-n} shall also be used for Σ_n, so that we can always draw spheres of the same radius. The projection shows clearly that the original symmetry of σ_o with respect to the plane $z_o \, x_o$ is preserved by σ_1 with respect to the plane $z_1 \, x_1$. The expression of $\sigma_n \, (P_n)$ is known, by (4) and (7'), as a function of z_o, while it would be useful to obtain it explicitly as a function of the coordinates x_n, y_n, z_n. In principle, this is no problem, since H_n is invertible and equation (7') can be re-written as follows:

$$\sigma_n \, (P_n) = \alpha^n \sigma_o \, (H_n^{-1} \, (P_n)) \qquad (n = 1, 2, 3) \qquad (9)$$

The problem is that the calculations involved become more and more difficult as n increases. However, when n = 1, H_1 can be very easily inverted and one obtains (with $r_o = 1$):

$$\sigma_1 \, (P_1) = \frac{1}{4} \, ((x_1 + 3\alpha)/(x_1 + \alpha))^{\frac{1}{2}} + \alpha/2 \qquad (10)$$

The sphere Σ_2, enlarged by the factor α^{-2}, is the image of the H_2 mapping and is shown in Fig.4a. The contour lines of σ_2, at the end of the second cleavage, have lost any kind of symmetry, as can be clearly seen from the orthogonal projection on the plane $x_2\, y_2$ (Fig.4b).

The behaviour of the contour lines of σ_3, at the end of the third cleavage, is shown in Fig.5. The sphere Σ_3, enlarged by the factor α^{-3}, is the image of the mapping H_3 and represents one of the four upper blastomeres (Fig.5a). The sphere Σ'_3 represents one of the four lower blastomeres (Fig.5b). The contour lines of σ_3 have no symmetry whatsoever within Σ_3 and within Σ'. From a geometrical viewpoint, however, for each contour line in Σ_3 there exists a perfectly symmetrical contour line in Σ'_3. In both cases, these contour lines start at the (local) unipole, turn around it and come back forming a twisted cusp. While the geometrical pattern in Σ_3 and Σ'_3 is the same, the value of the contour lines is, of course, widely different. The maximum value of σ_3 is 0.5, exactly a half of what it was on the 'egg', and is attained at the upper blastomere, where σ_3 has a positive minimum between 0.3 and 0.2. The minimum value of σ_3 in Σ_3' is zero. It is perhaps of interest to observe that the tangent at the cusp forms an angle of about 15-20 degrees with the x axis in Σ_3 (the same thing, however, happens in Σ'_3). The top view is shown in Fig.6.

4. THE MICROMERE'S MICROMERE

Our exploration of the behaviour of the hypothetical unipolar gradient at the end of the first three divisions has revealed a strong similarity of the geometrical patterns of the contour lines in all cases, even at the end of the third cleavage.

Therefore, within the limits of our assumptions, one is led to imagine that the extreme difference in size between the macromeres and the micromeres might be charged to a threshold for the value of the concentration, rather than to its spatial distribution. If this is the case, such a threshold could very well be at the bottom of the lower blastomeres at the end

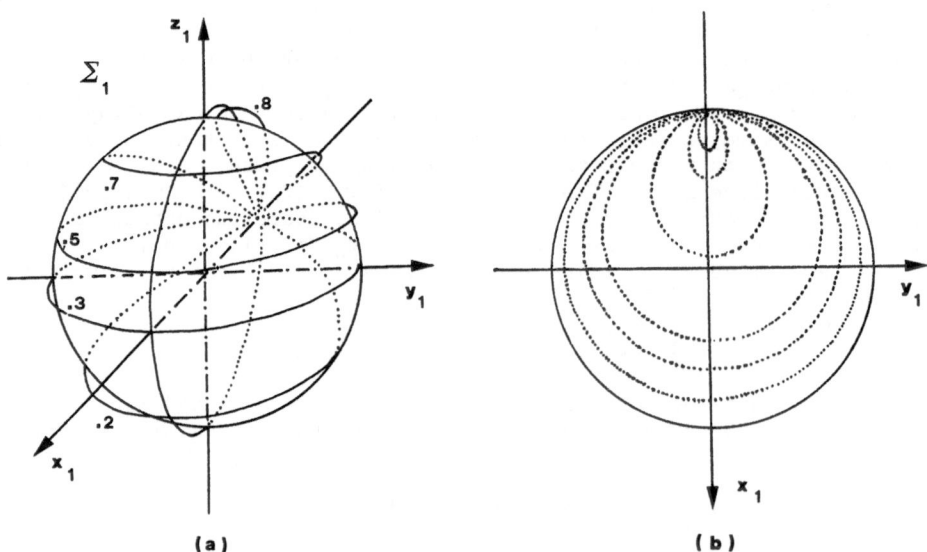

(a) (b)

Fig.3. (a) Contour lines of σ_1; (b) top view.

of the third cleavage (Fig.5b). Moreover, we can imagine that, when the concentration is below this threshold, cytokinesis takes place according to a principle of <u>minimum surface increase</u> of the membrane.

Consider the following elementary problem: to find a single based spherical segment $\overline{S}$ of a sphere S of radius r and a sphere S' of radius r' having the same volume and the same surface area, $\overline{V} = V'$ and $\overline{A} = A'$, excluding the area of the base circle of $\overline{S}$ (see Fig.7). Then the following equation holds:

$$\mu^3 - \frac{3}{2} \mu + \frac{1}{2} = 0 \tag{11}$$

where $\mu = r'/r$. The only relevant root is the following:

$$\mu_o = r' / r \cong 0.366 \tag{12}$$

and the corresponding height of $\overline{S}$ is:

$$h_o = 2 \mu_o^2 r \cong 0.268 \, r \tag{12'}$$

Accordingly, the ratio ν_o between the radius r" of the sphere S" resulting from the other spherical segment of S in the unequal division and the radius r of S is:

$$\nu_o = r'' / r = (1 - \mu_o^3)^{1/3} \cong 0.983 \tag{13}$$

Now, the ratio $\rho = \overline{A} / A'$ as a function of the height h of $\overline{S}$ is:

$$\rho = \rho(h) = h / 2(\frac{1}{4} (3-h)h^2)^{2/3} \tag{14}$$

The relevant part of the graph of this function is plotted in Fig.8a. One can see that $h > h_o$ implies $\overline{A} < A'$: the surface of the spherical bowl $\overline{S}$, in this case, must increase in order to obtain a sphere S' of equal volume. When $h < h_o$, one has $\overline{A} > A'$ and the bowl $\overline{S}$ has an <u>excess</u> of surface. Since this would imply a demolition or a permanent contraction of the membrane, we exclude the condition $h < h_o$.

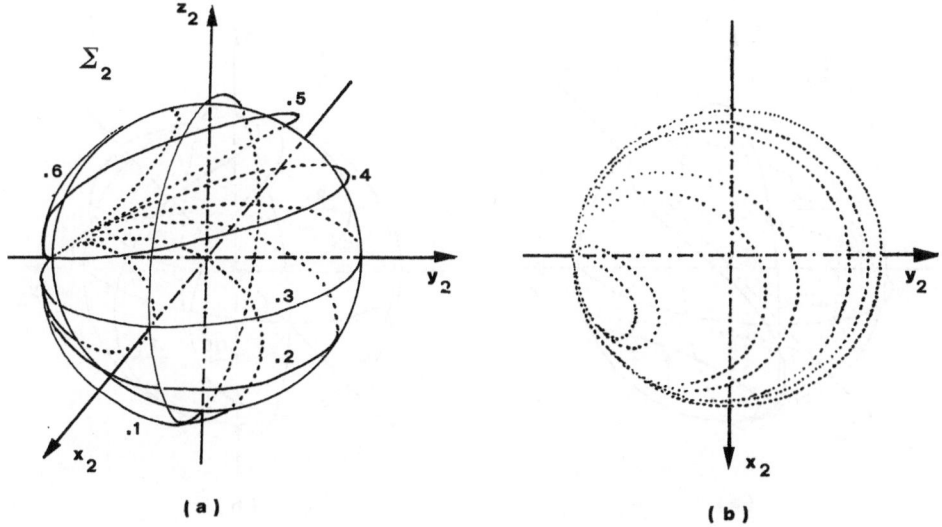

Fig.4. (a) Contour lines of σ_2; (b) top view.

Let S" be the other sphere of area A" such that V = V' + V" (Fig.7). Then the normalized area increment $\Delta A = (A' + A" - A)/\pi$ as a function of h has the following expression:

$$\Delta A = \Delta A(h) = 4(B^{2/3} + C^{2/3} - 1) \qquad (15)$$

where:

$$B = \frac{1}{4}(3 - h)h^2 \quad \text{and} \quad C = \frac{1}{4}(1+h)(2-h)^2 .$$

Since the admissible part of the graph of (15) (See Fig.8b), is that for which $h \geqslant h_o$, we see that the condition $h = h_o$ corresponds to the situation where:

i) there is no difference between the area of the small sphere S' and that of the spherical bowl $\bar{S}$ that originates it.

ii) the total area increase attains its minimum value.

Now, experimental observation appears to indicate that the ratio between the radius of a micromere and that of the generating blastomere at the end of the third cleavage is comparable with the value μ_o of (12). Should this be a coincidence, it would certainly be surprising.

Let us assume that there exists a second threshold for σ or, alternatively, a threshold for a second type of concentration associated to σ, below which the cells stop dividing. Actually, the micromeres continue to divide after the fourth cleavage, and therefore they must be over this second threshold. However, at their bottom, they are below the first threshold; thus, we may suppose that they will divide unequally and that the smallest cells (say with radius $\bar{s}$), i.e. the micromeres of the micromeres, go below the second threshold and stop dividing.

Now, if the biggest cells resulting from the hypothetical unequal division of the micromeres at the fifth cleavage, i.e. the macromeres of the micromeres, continue to divide equally, the question is how many divisions they shall undergo until the second threshold or, equivalently, the radius $\bar{s}$ will be reached.

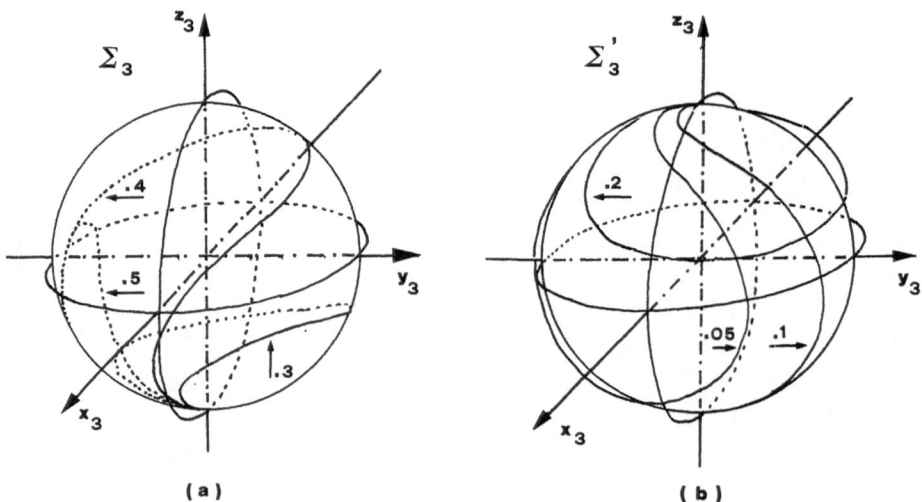

Fig.5. (a) Upper blastomere Σ_3; (b) lower blastomere Σ'_3.

In order to answer the above question, let us take one of the original micromeres resulting from the fourth cleavage, and let s_0 be its radius. If it divides unequally with the same ratio μ_0, then:

$$\bar{s} = \mu_0 \, s_0$$

and the radius s_1 of the other blastomere will be:

$$s_1 = \nu_0 \, s_0$$

After n equal cleavages the radius s_n of the resulting blastomeres is given by:

$$s_n = \alpha^n \, \nu_0 \, s_0$$

Then, solving for n the inequality:

$$\alpha^n \, \nu_0 \, s_0 \leqslant \mu_0 \, s_0 \tag{16}$$

we have:

$$n \geqslant 4.277$$

Therefore, taking into account that $\bar{s}$ may be even lower of the minimum radius necessary for σ to remain below the second threshold, the number of division must be 4 or 5. Correspondingly, one should find before hatching a minimum of 68 and maximum of 132 descendants from the micromeres. Afterwards they should all stop dividing.

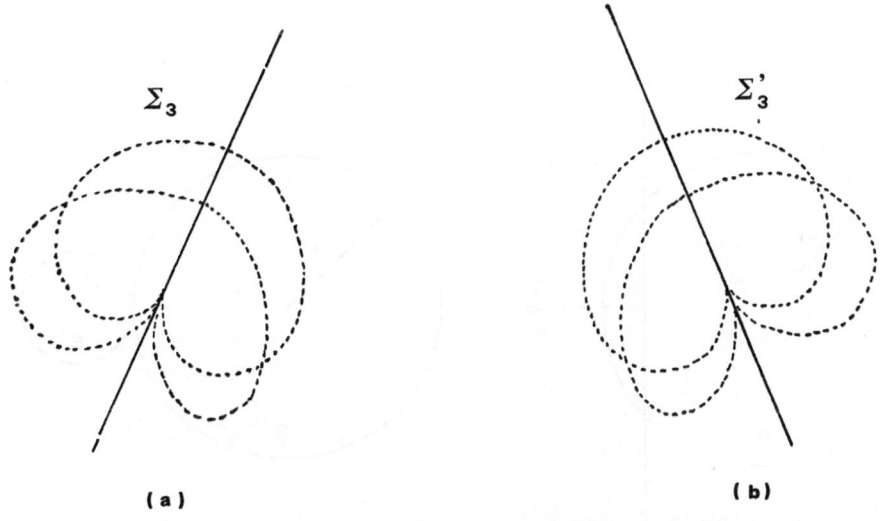

Fig.6.   (a) Top view of Σ_3; (b) top view of Σ_3'.

5. KINEMATICS OF CYTOKINESIS

The problem

Let us consider an unknown function $y = h(x, t)$ with the following initial conditions (see Fig.9):

$$h(x, 0) = h_0(x) \text{ with } 0 \leqslant x \leqslant \xi_0 = \xi(0) \qquad (17)$$

where $h_0(x) = (1 - x^2)^{\frac{1}{2}}$, with $\xi_0 = 1$, representing one quarter of the section of a sphere of radius 1 with the plane xy;

$$h(0, t) = \phi(t) \qquad (18)$$

where $\phi(t)$ is an assigned function representing the law of motion of the furrow.

Moreover, the function h must satisfy the following additional conditions:

$$h(x, t) > 0 \text{ when } 0 \leqslant x < \xi \qquad (19)$$

where $\xi = \xi(t)$ is a time-dependent point representing one of the poles (not to be confused, here, with the unipole) of the dividing cell;

$$h(\xi(t), t) = 0 \qquad (20)$$

which means that the pole moves along the x-axis;

$$(h_x(\xi(t), t))^{-1} = 0 \qquad (21)$$

which simply means that the tangent to the graph of h at point $(x = \xi(t), y=0)$ is vertical;

$$\pi \int_0^{\xi(t)} dx \, (h(x, t))^2 = V \text{ (constant)} \qquad (22)$$

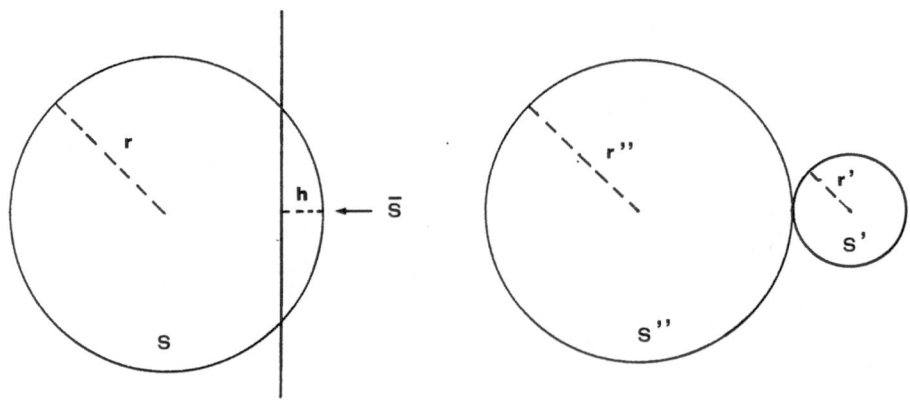

Fig.7. A simple geometrical problem.

which is the condition necessary to obtain a constant (independent of time) volume of the solid of revolution generated around the x-axis by the graph of h.

Now, it is convenient to translate the above problem into an <u>equivalent</u> one, by defining the following function:

$$F(x, t) = \pi \int_0^x d\mu \, (h(\mu, t))^2 \qquad (23)$$

so that condition (22) is equivalent to:

$$F(\xi(t), t) = V \text{ (constant)} \qquad (24)$$

Then, we shall prove that conditions (19)-(22) are equivalent to the following ones:

$$F_x(x, t) > 0 \qquad (0 \leqslant x < \xi) \qquad (25)$$

$$F_x(\xi, t) = 0 \qquad (26)$$

$$F_t(\xi, t) = 0 \qquad (27)$$

$$F(x, t) > 0 \qquad (28)$$

(where the subscripts denote partial differentiation).

Furthermore, we shall prove that the solution of the original problem is given by:

$$h(x, t) = (\pi^{-1} F_x(x, t))^{\frac{1}{2}} \qquad (29)$$

Hence, any function F satisfying (25)-(28) gives rise to a particular solution of the kinematical problem.

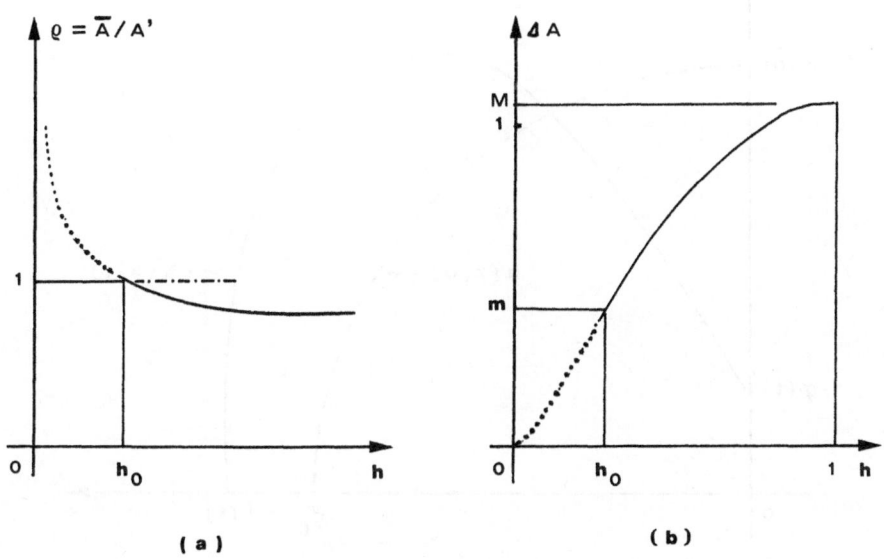

Fig.8. (a) Ratio $\overline{A}/A'$; (b) total area increment.

Proof

A. Conditions (19)-(22) imply (25)-(28). By taking the partial derivative with respect to x in (23) one obtains:

$$\frac{\partial F}{\partial x} = \pi \frac{\partial}{\partial x} \left(\int_0^x d\mu \ (h \ (\mu, \ t))^2 \right. \tag{30}$$

hence:

$$F_x \ (x, \ t) = \pi \ (h \ (x, \ t))^2 \tag{31}$$

and, since the right hand side of (31) is the square of a real number, the condition (25) is clearly satisfied. Evaluating (31) with $x = \xi$ we obtain:

$$F_x \ (\xi, \ t) = \pi \ (h \ (\xi, \ t))^2 = 0 \tag{32}$$

if condition (20) is to be satisfied, thus (26) is also satisfied.

Now, by taking the derivative with respect to t in (24) one has:

$$\frac{dV}{dt} = \dot{\xi} \ F_x \ (\xi, \ t) + F_t \ (\xi, \ t) = 0 \tag{33}$$

hence, substituting (32) for F_x (ξ, t) in (33), one obtains:

$$F_t \ (\xi, \ t) = 0$$

and condition (27) is satisfied.
Then, condition (28) follows by the definition (23), F(x,t) being the integral of a positive function.

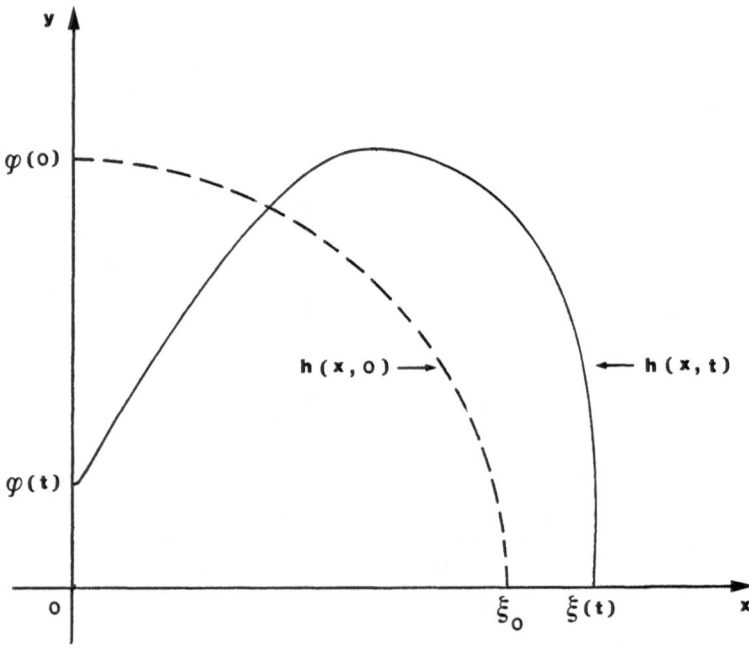

Fig.9. Kinematics of cytokinesis

B. Conditions (25)-(28) imply (19)-(22). From (31) we have:

$$h(x, t) = ((\pi^{-1} F_x (x, t))^{\frac{1}{2}} \tag{34}$$

hence (25) implies (19) and (26) implies (20). Taking the partial derivative with respect to x in (34) gives:

$$h_x (x, t) = F_{xx} (x, t) / 2 (\pi F_x (x, t))^{\frac{1}{2}}$$

hence

$$(h_x (\xi, t))^{-1} = 2\pi^{\frac{1}{2}} (F_x (\xi, t))^{\frac{1}{2}} / F_{xx} (\xi, t) = 0$$

since, by (26), $F_x(\xi, t) = 0$, so that condition (21) is satisfied.

Finally, condition (22) is equivalent to (24) by the definition (23).

The simplest solution

As far as the solutions of our problem are concerned, we only mention the simplest one. We assume the initial condition (17) and we make a particular choice for (18), taking, as an example:

$$(\phi (t))^2 = \alpha (t) = 1 - t/T$$

where T is the duration of cytokinesis. Then the solution can be obtained by taking:

$$(h (x, t))^2 = (\xi - x)(\eta + x) \tag{35}$$

where $\xi = \xi(t)$ and $- \eta = -\eta (t)$ are the time-dependent roots of the polynomial.

Then, by (29), one obtains:

$$F_x (x, t) = \pi (\xi - x)(\eta + x) \tag{36}$$

so that condition (25) is satisfied, provided $\eta > 0$, and condition (26) is satisfied as well (being $x = \xi$ one of the roots).

Making use of definition (23), we obtain:

$$F (x, t) = \pi \int_0^x d \mu (\xi - \mu)(\eta + \mu) = \pi x (\xi \eta + \tfrac{1}{2}(\xi - \eta)x - \tfrac{1}{3} x^2) \tag{37}$$

Thus, condition (28) is certainly satisfied, since F (x,t) turns out to be the integral of a positive function.

In order the verify condition (27), we can take in (37) the partial derivative of F with respect to t and evaluate it for $x = \xi$. Condition (27) then implies (since $\xi > 0$):

$$\dot{\xi}\eta + \tfrac{1}{2} \dot{\xi}\eta + \tfrac{1}{2} \dot{\xi}\xi = 0 \tag{38}$$

We observe that, by (30), one has (for x = 0):

$$(h (0, t))^2 = \xi \eta \tag{39}$$

so that $\xi \eta$ gives the radial motion of the contractile ring during cytokinesis. Hence, by the initial condition:

$$\xi\eta = \alpha \ (t) \tag{40}$$

Therefore:

$$\eta = \alpha/\xi \tag{41}$$

and

$$\dot{\eta} = (\dot{\alpha}\xi - \alpha\dot{\xi})/\xi^2 \tag{42}$$

Substituting (41) and (42) for η and $\dot{\eta}$ in (38), it is very easy to see that ξ must satisfy the following equation:

$$\xi^3 + 3\alpha\xi - 4 = 0 \tag{43}$$

which has only one real root, given by:

$$\xi = (2 + \beta)^{1/3} + (2 - \beta)^{1/3} \tag{44}$$

where $\beta = (\alpha^3 + 4)^{\frac{1}{2}}$.

Once ξ and η are found, this gives the most simple, although not too realistic, kinematics, and it can be seen that the surfaces involved are merely spherical bowls.

By choosing more accurate polynomial expansions of h, other than (35), one can certainly find other solutions with any desired degree of approximation to the real process.

6. FINAL REMARKS

The questions listed in sect.1 are those that have not been omitted. One of the missing questions is the following:

Why are the planes of the first three divisions mutually orthogonal?

This seems to be one of the most elusive problems. If one explores the possible relationship between the orthogonality of the planes of cleavage and the value of the concentration σ, an interesting suggestion pops up at the end of the first division (see Fig.10a). There is an indication that, at high values of σ_1, the great circle on plane $x_1 z_1$ in Σ_1 could be the one selected for the next cleavage. Thus the second plane would be orthogonal to the first one $(y_0 z_0)$. Unfortunately, this is an illusory conjecture, since it fails immediately at the end of the second cleavage, where the great circle on plane $x_2 y_2$ would not be the one selected for the third cleavage (Fig.10b). Similarly, the attempt at a possible relationship between the lines of flux of σ and orthogonality falls down at the end of the second division.

On the other hand, the choice of a linear gradient is, as previously stated, a mere example (the simplest one). The possibility to explore a wide range of different functions for σ certainly allows a great flexibility, specially in terms of possible applications to different biological species, other than the sea-urchin. However, the geometrical behaviour, under the H-map, of any scalar field having an initial symmetry with respect to the z-axis is, qualitatively, not too far from that of a linear field. Thus, further investigations are necessary on this problem.

In a more refined approach, the sequence of cleavage planes should be automatically determined by a 'built-in' device within the transformation, rather than being imposed from the outside. Moreover, a <u>vector</u> field, instead of a scalar one, could be introduced in order to account for a possible directional polarization of the contractile microfilaments on the egg surface. For the time being, we answer the simple question of how the H-map changes in the case of unequal division.

The first unequal division of the sea-urchin egg takes place at the fourth cleavage in each one of the four lower blastomeres resulting from the third division. Assuming that the plane of cleavage is parallel to the $x_3 y_3$ plane and that it cuts the z_3 axis at the some point $\overline{z} < 0$ (for instance, $\overline{z} = (h_o - 1) r_3$), we define the mapping H_4^+ as the composition between the mapping H_3 and the following application:

$$x_4 = 2 \alpha^+ g_3^+ (\overline{z}) f_3^+ (z_3) x_3$$

$$y_4 = 2 \alpha^+ g_3^+ (\overline{z}) f_3^+ (z_3) y_3$$

$$z_4 = \alpha^+ g_3^+ (\overline{z})(2 z_3 + r_3 - \overline{z})$$

where

$$g_3^+ (\overline{z}) = r_3 / (r_3 + \overline{z})$$

$$f_3^+ (z_3) = ((z_3 - \overline{z})/(z_3 - r_3))^{\frac{1}{2}}$$

$$\alpha^+ = r_3^{-1} (\tfrac{1}{4} (r_3 + \overline{z})(2 r_3 - \overline{z}))^{1/3}$$

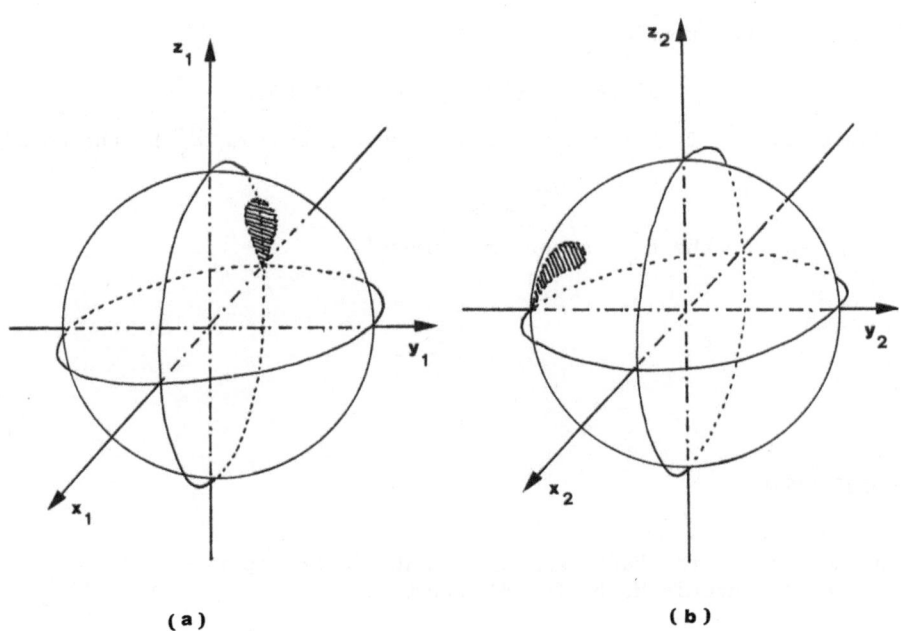

(a)   (b)

Fig.10. Example of a wrong conjecture.

and for the concentration σ_4 on Σ_4 :

$$\sigma_4 \ (P_4) \ = \ (2\alpha^{+2} \ g_3^+ \ (\overline{z}))^{-1} \ \sigma_3(P_3)$$

The domain of H_4^+ is the spherical bowl obtained from Σ_3 by the condition:

$$\overline{z} < z_3 \leqslant r_3 \quad,$$

and the image Σ_4 of H_4^+ is the 'macromere'.

Similarly, the mapping H_4^- is defined as the composition between H_3 and the following application:

$$x_4' \ = \ 2\alpha^- \ g_3^- \ (\overline{z}) \ f_3^- \ (z_3) \ x_3$$

$$y_4' \ = \ 2\alpha^- \ g_3^- \ (\overline{z}) \ f_3^- \ (z_3) \ y_3$$

$$z_4' \ = \ \alpha^- \ g_3^- \ (\overline{z})(2z_3 + r_3 - \overline{z})$$

where:

$$g_3^- \ (\overline{z}) \ = \ r_3 \ / \ (r_3 + \overline{z})$$

$$f_3^- \ (z_3) \ = \ ((z_3 - \overline{z}) \ / \ (z_3 - r_3))^{\frac{1}{2}}$$

$$\alpha^- \ = \ r_3^{-1} \ (\tfrac{1}{4} \ (r_3 + \overline{z})(2r_3 - \overline{z}))^{1/3}$$

and for the concentration:

$$\sigma_4' \ (P_4') \ = \ (2\alpha^{-^2} \ g_3^- \ (\overline{z}))^{-1} \sigma_3(P_3)$$

The domain of H_4^- is the spherical bowl obtained from Σ_3' by the condition:

$$-r_3 \leqslant z_3 < \overline{z} \ .$$

The image Σ_4'. of the H_4^- is the 'micromere'.

ACKNOWLEDGEMENTS

The author wishes to thank Professor Antonio Degasperis and Professor Antonio Machì for discussions.

324

REFERENCES

1. Hamburgh, M., 1971, _Theories of Differentiation_. Arnold Publ., London.

2. Horstadius, S., 1973, _Experimental Embriology of Echinoderms_. Clarendon Press, Oxford.

3. Gurdon, J.B., 1962, _Devl. Biol_.4: 256-73.

4. Buznikov, G.A., Zvezdina, N.D. and Markova, L.N., 1971; _Zhurn. Evol. Biochim. Physiol_. 7: 241-246.

5. Freeman, G., 1976, _Devel. Biol_. 51: 332.

6. Hiramoto, Y., 1970, _Biorheology_, 6: 201-234.

7. Rappaport, R., 1967, _Inter. Rev. Cytol_. 31: 169-213.

8. Giudice, G., 1973, _Developmental Biology of the Sea Urchin Embryo_. Acad. Press, New York.

9. Rossi, M., Augusti-Tocco, G. and Monroy, A., 1978, _Quart. Rev. of Biophysics_. 8: 43-119.

10. Hiramoto, Y., 1958, _Journ. of Exptl. Biol_. 35,2: 407-424.

11. Hiramoto, Y., 1963, _Exptl. Cell Res_. 32: 59-75.

12. Hiramoto, Y., 1976, _Develop., Growth and Differ_. 18: 379-388.

13. Hiramoto, Y., 1978, _Develop., Growth and Differ_. 20: 317-327.

14. Parisi, E., Filosa, S., De Petrocellis, B. and Monroy A., 1978, _Dev. Biol_. 65: 38.

15. Catalano, G. and Eilbeck,J.C.,1978, _Journ. of theor. Biol_. 75: 123-137.

16. Runnstrom, J., 1928, _W. Roux'Arch. Entw mech.d.Org._

THE SHAPING OF CELL SHEETS: AN APPLICATION OF MECHANICS

IN DEVELOPMENTAL BIOLOGY

Jay E. Mittenthal

Department of Anatomical Sciences
University of Illinois College of Medicine
Urbana, Illinois 61801, U.S.A.

INTRODUCTION

Living organisms deform themselves. The most dramatic deformation is morphogenesis -- the generation of form during the development of an embryo. A single cell, the fertilized egg, divides repeatedly to form a mass of cells. This mass deforms itself into more and more complex structures, and eventually develops into a new adult organism. What mechanical processes shape a multicellular organism? Evolution can use intracellular mechanisms, such as those operating in cytokinesis; it can augment these with mechanisms involving interaction between cells. We shall inquire how intracellular and intercellular mechanisms may contribute to shaping an embryo.

An embryo contains only two kinds of tissues -- three-dimensional masses of cells, or mesenchyme; and sheets of cells, or epithelia.

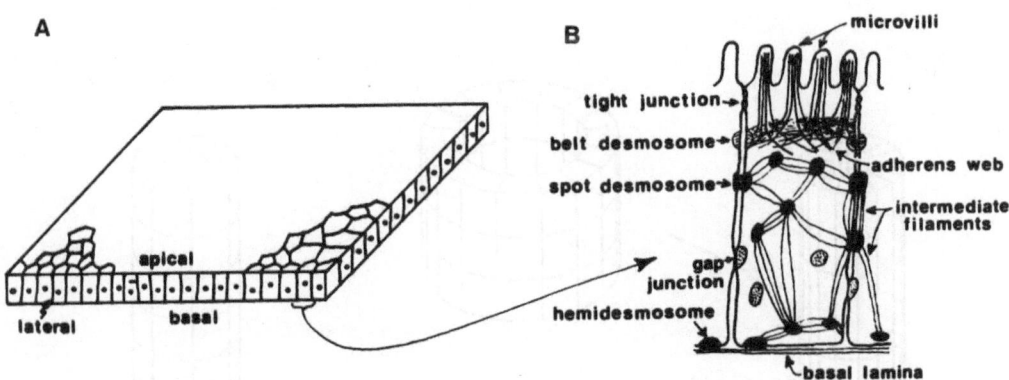

Fig. 1. A monolayer epithelium. (A) Low magnification: apical, basal, and lateral surfaces. (B) Higher magnification: intercellular junctions and intracellular filaments. Details of organization vary among epithelia.

Epithelia generate the surfaces of the body, including the skin and the lining of internal tubes. Fig. 1 shows the simplest epithelium, a mono-layer of cells. The apical surface of the epithelium faces a region with few cells -- the outside of the embryo, or the hollow lumen of a tube. The basal surface faces the interior of the embryo, a region containing many cells. The epithelium defines the interface between these regions.

An epithelium behaves as a continuous sheet because its cells are coupled together by specialized junctions on the lateral surfaces (see Staehelin and Hull, 1978; Alberts et al., 1983). The mechanical strength of the sheet comes from these junctions and from the cytoskeleton, an intracellular meshwork of filamentous molecules anchored to the junctions. Filaments in adjacent cells attach to each junction, providing a mechan-ically continuous network: The adherens web of microfilaments attaches to the belt desmosome, and intermediate filaments contact spot desmosomes. Communication between neighboring cells is mediated through the extracellu-lar matrix and through gap junctions which admit small molecules. A tight junction girdling the apex of each cell provides a barrier between the apical and basal surfaces. An epithelium is more than a passive barrier between these surfaces; it can transport molecules and so can regulate the composition of the apical and basal regions.

A FLUID ELASTIC SHELL MODEL FOR THE SHAPING OF EPITHELIA

Considerations Underlying the Model

To analyze the morphogenesis of an epithelium we need a mechanical model. The model uses equations of mechanical equilibrium which follow from Newton's laws of motion, together with constitutive equations which describe the relations between stress and strain in the epithelium. To formulate constitutive equations for an epithelium one needs to see how it deforms. Fig. 2 shows the elongation of a tubular epithelium. If the epithelium resembled an elastic material such as a thin rubber sheet, during the deformation each cell would retain the same neighboring cells

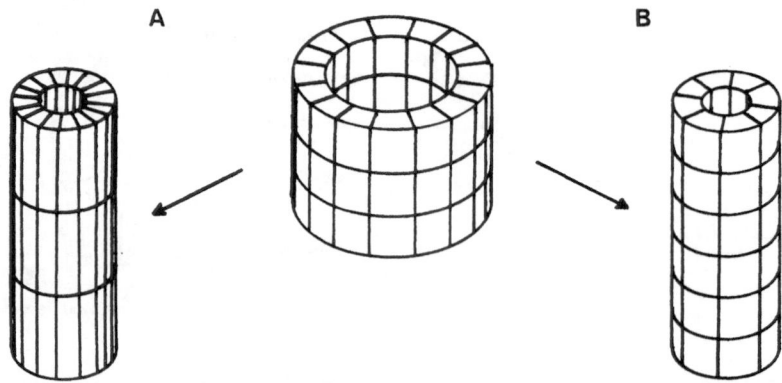

Fig. 2. Elongation of a tubular epithelium (A) without rearrangement of cells; (B) with rearrangement to relieve shear stress on cells. After Fristrom (1976).

and the cells would be deformed. This is not observed. Rather, cells rearrange so that they are only slightly deformed. During the rearrangement cells deform themselves, transiently and actively. Thus the epithelium is not wholly elastic; it shows fluid-like rearrangement of cells.

Though an epithelium is fluid-like, it differs from a soap film. A soap solution is a fluid with a very fine grain; its elementary units are the molecules of water and soap. This fine grain allows a soap film to undergo any deformation down to curvatures of molecular scale. Thermodynamics implies that the soap film will assume the shape which minimizes the energy associated with its surface tension. Thus a soap film covers a wire frame with the minimum-area surface compatible with the shape of the frame. For example, a soap film covering a circular hoop must be planar.

By contrast, an epithelium is a fluid with a relatively coarse grain; its elementary units are cells. The cytoskeleton of each cell limits its deformation, so that an epithelium can not minimize its area for an arbitrary boundary shape. If we squeeze on the perimeter of a circular patch of epithelium, the patch will thicken; the cells will change shape, becoming more columnar. Eventually the patch will buckle, forming a dome (Fig. 3). In the extreme of doming the patch forms an isolated vesicle. Thus the coarse grain of the epithelium allows it to cover a planar hoop with a non-planar surface. However, we must do work to buckle the patch; the more curved it is, the more work we do. That is, the cytoskeleton and intercellular junctions make an epithelium resist bending, as does an elastic sheet.

Thickening and doming of a patch are seen in normal development, so it is worth asking what forces could produce them. Suppose we graft a patch from a donor site into a hole of the same size at a host site. If the donor and host sites are at corresponding locations in different animals, we might expect the graft to remain undeformed as it heals in at the host site. However, suppose that the sites are not at corresponding locations, and that cells at the donor and host sites have different adhesive affinities. Because the graft and host are fluid there is an interfacial tension at the border between them, as at the interface surrounding a lens of oil in a thin sheet of water. The interfacial tensions at the graft-host boundary contribute to an adhesive disparity -- an energy associated with the mismatch of adhesive affinities. An adhesive disparity proportional to the perimeter of the interface would be reduced if the graft became more circular, and if its cells became more columnar and more densely packed. These phenomena have been observed in many experiments in which a patch was grafted to a foreign host site.

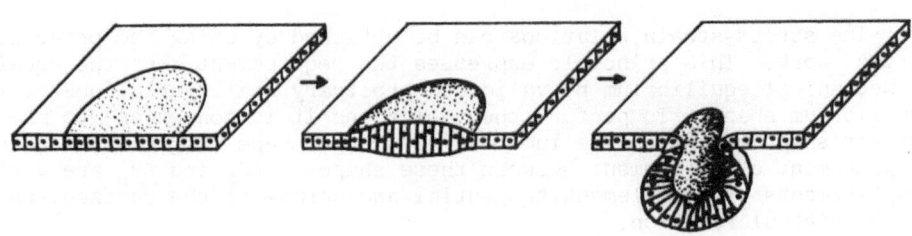

Fig. 3. Thickening and doming of a patch of epithelium.

Though there may be an interfacial tension between groups of cells, the forces between individual cells are not interfacial tensions. These forces are exerted by the cytoskeleton, and they have components perpendicular to cell surfaces. Interfacial tensions act tangent to an interface. In a fluid of molecules, surface tension emerges from the interaction of many molecules. So, in an epithelium, interfacial tensions must result from the interaction of many cells.

These considerations provide assumptions for a mechanics of epithelia. One can model an epithelium as a <u>fluid elastic shell</u> (Mittenthal and Mazo, 1983; Mittenthal and McMeeking, 1986). It is elastic in resisting bending or equal bi-axial stress, and in its transient response to shear stress. It is fluid in allowing rearrangement of cells in-plane. The fluid may be non-uniform; we assume that cells of the same type adhere together preferentially and are segregated in one region of the epithelium. Then cells cohere within each region, and there may be an interfacial tension between regions. These interfacial tensions can affect the shape of each region, and so can alter the shape of the shell.

Our model treats an epithelium as a continuum. It does not consider the details of cell motility and cell-cell interactions. It shows how differential adhesion may select a particular shape of the epithelium from among the alternative shapes which motility of its cells makes accessible.

Formalism of the Fluid Elastic Shell Model

The model treats an axially symmetric epithelium (Fig. 4) which is sufficiently thin, relative to its radius of curvature, that the equations of mechanical equilibrium for a thin shell can be used:

$$\frac{d(rM_s)}{ds} - M_\phi \frac{dr}{ds} + rQ = 0. \qquad \text{(balance of moments)}$$

$$\frac{d(rN_s)}{ds} - N_\phi \frac{dr}{ds} - rQ\kappa_s + q_t = 0. \qquad \text{(balance of tangential forces)}$$

$$\frac{d(rQ)}{ds} + N_\phi\kappa_\phi + N_s\kappa_s + q_n = 0. \qquad \text{(balance of normal forces)}$$

$$(1)$$

In these equations the stresses and moments have been integrated through the thickness of the shell to get moment resultants and stress result-ants. Inertial terms are neglected. Movements of cells and cell aggregates are dominated by viscosity; the inertial forces have been estimated to be at least 100,000 times smaller than viscous forces in cellular systems (Odell et al., 1981).

The stress-strain relations can be obtained by using the principle of virtual work. This principle expresses the requirement that the equations of mechanical equilibrium be valid for arbitrary small variations of the equilibrium shape. To perform the variations it is convenient to map the current shape of the epithelium to a reference shape (Figure 5). U is the displacement of an element between these shapes. δU_t and δU_n are virtual displacements of the element tangential and normal to the surface, and $\delta\theta$ is the virtual rotation.

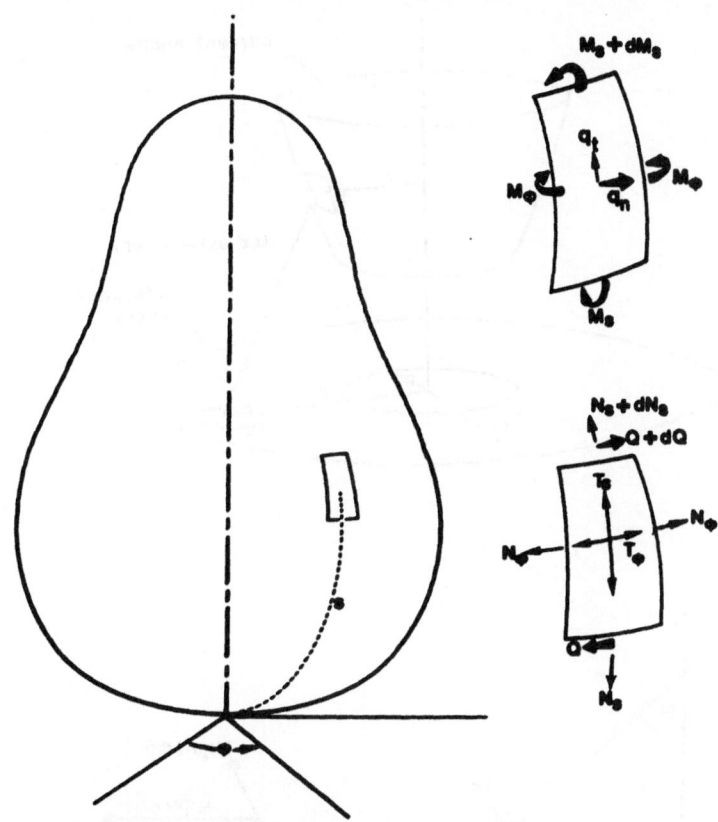

Fig. 4. Axially symmetric shell; definitions of symbols: The neutral
surface of a figure of revolution, showing coordinates (s, ϕ),
bending moments (M_s, M_ϕ) , stress resultants (N_s, N_ϕ, Q),
internal surface tensions (T_s, T_ϕ), and surface loads (q_n, q_t).

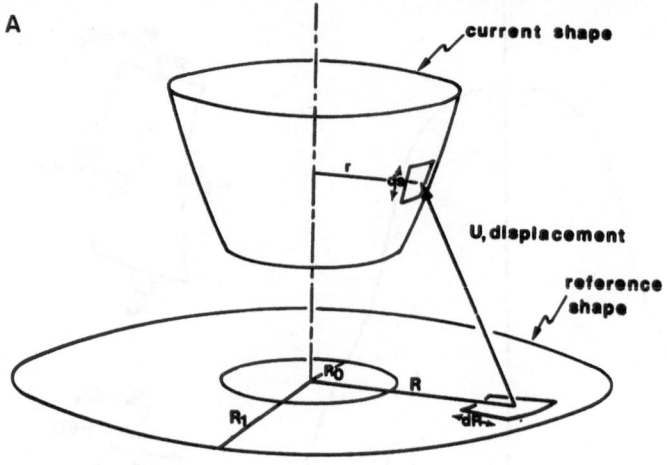

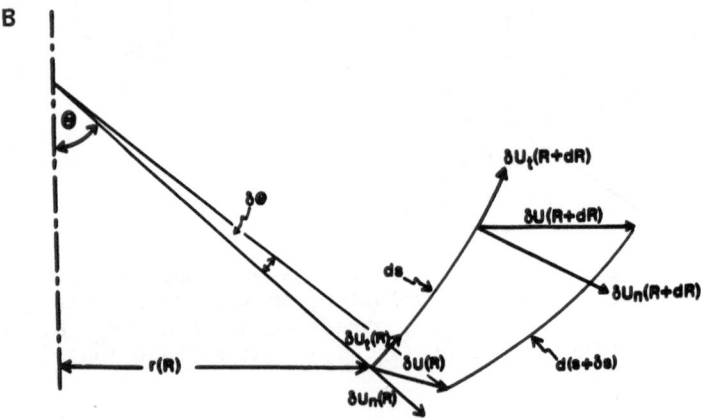

Fig. 5. Mapping to reference configuration. (A) An annulus of the
epithelium is shown in the current shape and in a planar reference
shape. (B) Variation of components of the displacement U during a
small variation in the current configuration of a shell element.

332

$$2\pi \int_{R_0}^{R_1} \left\{ \binom{\text{balance of}}{\text{momentum}} \delta\theta + \binom{\text{balance of}}{\text{tangential}}_{\text{forces}} \delta U_t + \binom{\text{balance of}}{\text{normal}}_{\text{forces}} \delta U_n \right\} dR = 0$$

(2)

This principle of virtual work can be expressed in the form

$$2\pi \int_{R_0}^{R_1} \{ \bar{M}_s \delta\bar{\kappa}_s + \bar{M}_\phi \delta\bar{\kappa}_\phi + \bar{N}_s \delta\lambda_s + \bar{N}_\phi \delta\lambda_\phi + \bar{Q}\delta\bar{\gamma}_n$$

$$[\bar{q}_t \delta U_t + \bar{q}_n \delta U_n]\} RdR - 2\pi \left[R(\bar{M}_s \delta\theta + \bar{N}_s \delta U_t + \bar{Q}\delta U_n) \right]_{R_0}^{R_1} = 0$$

(3)

Here the terms are products of a bending moment or a stress resultant with the variation of a thermodynamically conjugate strain variable. Bars over symbols designate variables in the reference configuration. The integrated terms provide the boundary conditions: At R_0 and at R_1

M_s or $\delta\theta = 0$;

N_s or $\delta U_t = 0$;

Q or $\delta U_n = 0$.

(4)

We specified the stress-strain relations by formulating an energy functional which incorporates hypotheses about the mechanical properties of the epithelium. This functional included strain energy of elastic bending and stretching, and adhesive energy. We set the coefficients of the variations $\delta\kappa_s$, $\delta\kappa_\phi$, ... in the principle of virtual work equal to the corresponding coefficients in the variation of the energy functional. This procedure gives constitutive equations for M_s, M_ϕ, N_s, and N_ϕ.

The constitutive relations between bending moments and curvatures are:

$$M_s = D(\kappa_s + \nu\kappa_\phi) + \frac{1}{2}h(\xi_a - \xi_b)$$

$$M_\phi = D(\kappa_\phi + \nu\kappa_s) + \frac{1}{2}h(\xi_a - \xi_b)$$

(5)

Because the shell is fluid-like, the bending moments depend only on the curvatures in the current configuration, not on the reference configuration. The flexural rigidity D measures the stiffness of the epithelium to bending. ν, the Poisson ratio, is $\frac{1}{2}$ for a material that maintains constant volume as it is deformed. This is assumed for the model epithelium. ξ_a and ξ_b are interfacial tensions at the apical and basal surfaces, and h is the thickness of the epithelium. If these surface tensions differ the bending moments bend an element of the shell, reducing the area of the surface with greater surface tension by making this surface concave.

The constitutive relations for the in-plane stress resultants are:

$$N_s = -K + X + \xi_a + \xi_b + T_s$$

stress resultant (meridional)

$$N_\phi = K + X + \xi_a + \xi_b + T_\phi$$

stress resultant (latitudinal)

$$K = \frac{1}{2}\left(\kappa_s^2 - \kappa_\phi^2\right)$$

$$X = -cH\lambda_h^2(\lambda_h - 1)$$

elastic isotropic in-plane stress

$$\xi_a, \xi_b$$

apical and basal interfacial tensions

$$T_s, T_\phi$$

interfacial tensions within epithelium, meridional and latitudinal

(6)

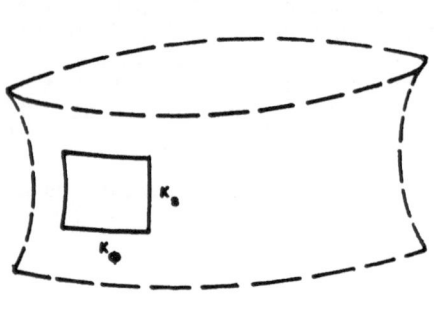 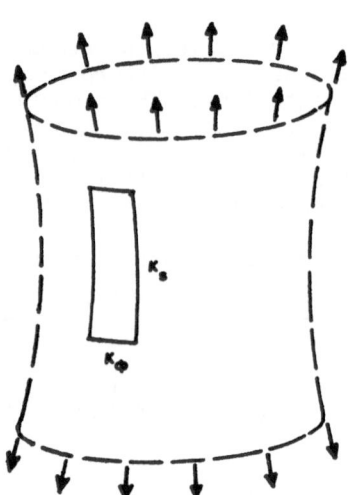

Fig. 6. Contribution from the bending strain energy to the stress resultants. A multicellular element with unequal principal curvatures can support in-plane shear stress. Increasing the meridional stress resultant elongates and narrows a tube of which the element is a part. Changes in the principal curvatures -- increasing latitudinal curvature and decreasing meridional curvature -- allow the resistance of the element to bending to oppose the increase in meridional stress.

The model gives an unusual contribution K from the bending strain energy to the stress resultants. This contribution can be understood as follows (Fig. 6). Consider an element of the shell as part of an annulus between two latitudes. Increasing N_s will tend to elongate the annulus and decrease its diameter, decreasing κ_s and increasing κ_ϕ. Because the epithelium resists bending, the increase in κ_ϕ opposes the increase in N_s, and the decrease in κ_s aids the increase in N_ϕ. This is so regardless of the sign of κ_s; hence an even power of κ_s must be in the stress-strain relation for N_s. But κ_s and κ_ϕ must enter this relation with the same functional form, because the "meridional" and "latitudinal" designations of these curvatures can be interchanged by making the element part of a shell annulus which has its symmetry axis perpendicular to the axis of the original annulus. From these arguments the constitutive relations for the in-plane stresses must contain the difference of the squares of the two curvatures. We found this term formally, from the variation of the energy functional; it seems unlikely that we would have found it by intuitive arguments. This shows the utility of deriving constitutive equations from an energy functional.

Because a multicellular element of the shell is fluid-like, the element can not support shear stresses in a static configuration. However, the term K, appearing here with opposite signs in the relations for N_s and N_ϕ, allows a curved element to support unequal residual stress resultants $N_s - T_s$ and $N_\phi - T_\phi$.

X is the contribution to the stress resultants from the stretching strain energy. Cells in a multicellular element can rearrange in-plane without generating stress. Therefore the stretching strain energy can depend only on λ_h, the ratio of the thickness of the element in the current configuration to its thickness in the reference configuration, H.

It is useful to relate this result for a fluid elastic shell to the corresponding relation for a linear isotropic elastic solid. σ_{ij} is the stress tensor, e_{kl} is the strain tensor, and E is the Young's modulus. We have assumed

$$\sigma_{zz} = \sigma_{xz} = \sigma_{yz} = 0 \qquad \text{(plane stress)}$$

$$\sigma_{xy} = 0 \qquad \text{(no in-plane shear stress)}$$

$$\sigma_{xx} = \sigma_{yy} = \sigma \qquad \text{(equal biaxial stress)}$$

$$e_{xx} = e_{yy} = e \qquad \text{(equal biaxial elastic strain)}$$

$$(7)$$

Then Hooke's law,

$$e_{ij} = \frac{1 + \nu}{E} \sigma_{ij} - \frac{\nu}{E} \sigma_{kk} \delta_{ij}$$

$$(8a)$$

becomes

$$e = \frac{1 - \nu}{E} \sigma$$

$$(8b)$$

or

$$\sigma = -\frac{E}{2\nu}\, e_{zz}.$$

(8c)

This is comparable to the above relation between X/H and $\lambda_h - 1$. The effective modulus in the nonlinear relation is $c\lambda_h^2$, replacing $E/2\nu$ in the linear relation.

The stress resultants depend on contributions from interfacial tensions. ξ_a and ξ_b are interfacial tensions at the apical and basal surfaces. T_s and T_ϕ are tensions at interfaces within the epithelium; they are derived from an adhesive disparity density function. This is an interesting function; it has the units of the apical and basal surface tensions, energy per unit of in-plane area. However, it may vary as the shape of a region of the epithelium varies; the familiar surface tension of a fluid is independent of the shape of a unit area of surface.

Using these constitutive equations with the equilibrium equations, we obtained analytical solutions for simple cases of biological interest. These solutions, with further heuristic arguments, offer an interpretation for a complex morphogenetic movement, gastrulation, early in the development of a frog embryo.

AMPHIBIAN GASTRULATION: SOME POSSIBLE ROLES FOR INTERCELLULAR ADHESION

To analyze the mechanics of gastrulation one needs information about earlier stages of development, during which processes set up and regulate the morphogenetic machinery. A survey of events following fertilization in the frog Xenopus laevis provides an adequate basis for the present model (for reviews (*) see *Gerhart, 1980; *Slack, 1983; *Keller, 1985). Before fertilization the egg is radially symmetric around an axis extending from the animal pole to the vegetal pole. Along this axis, however, the egg is quite asymmetric: The animal and vegetal hemispheres differ in composition and mechanical properties. The transition between animal and vegetal cytoplasm occurs at a relatively well-defined interface, roughly at the equator.

During the first cell cycle after fertilization the egg changes symmetry from radial to bilateral. The symmetry plane will be the plane of bilateral symmetry in the adult. This plane intersects the plasma membrane at the dorsal and ventral meridians. The embryo proceeds through several cell cycles to form a ball of cells, the blastula (Fig. 7A). The superficial layer of cells is an epithelium, with tight junctions and apical surfaces that maintain a barrier to the environment. The deeper cells are more loosely packed, and contacts among them are relatively sparse. The cells, or blastomeres, secrete fluid which collects in a cavity, the blastocoel. The blastocoel develops in the animal hemisphere, with its base at the animal-vegetal interface. During subsequent development cells in the roof of the blastocoel will form the ectoderm covering the embryo. Cells in the blastocoel floor will form endoderm, which contributes to many visceral structures. In the marginal zone at the rim of the blastocoel are cells of the prospective mesoderm, which will contribute to the skeleton and muscles.

Epiboly

At the mid-blastula stage the blastocoel roof and marginal zone begin to spread and thin (Fig. 7B, C). This process, called epiboly, carries the

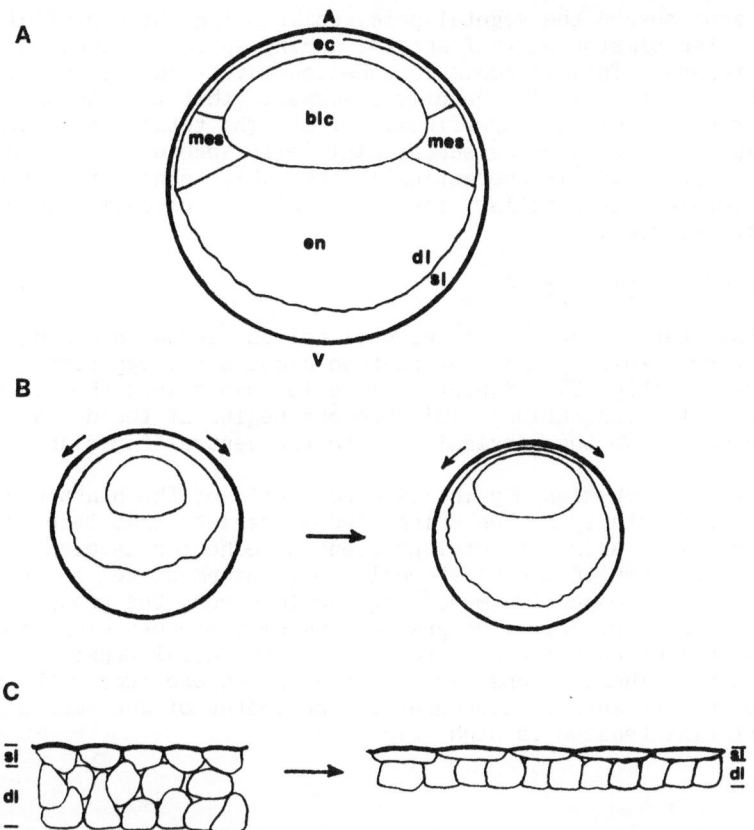

Fig. 7. <u>Xenopus laevis</u>, blastula stages of early development. (A, B)
Blastula, sections through the animal-vegetal axis. (B, C) During
epiboly the roof of the blastocoel thins and spreads. Left, mid-
blastula; right, late blastula. (C) Detail of a section through
the blastocoel roof near the animal pole, showing thinning of the
deep layer by radial intercalation of deep cells. No transit of
cells between the superficial and deep layers occurs. Though
thinning is shown with a constant number of cells, in fact cell
division occurs, so that superficial cells thin less than is
shown. (after Keller, 1980.) Symbols in this and following
figures: A, animal pole; a, archenteron; bc, bottle cells; blc,
blastocoel; blp, blastopore; d, dorsal; dl, deep layer of cells;
ec, prospective ectoderm; en, prospective endoderm; mes, prospective mesoderm; mm, mesodermal mantle; sl, superficial layer of
cells; v, ventral; V, vegetal pole.

marginal zone toward the vegetal pole, compressing the superficial vegetal cells. As the blastocoel roof spreads during epiboly, cohesion among its cells increases. This increase in cohesion may produce epiboly. The morphology of intercellular junctions suggests that the strongest contacts occur at the apex of the superficial cells. The total length of apical cell perimeters increases as superficial cells spread and divide. This increase in perimeter is energetically favorable, since it increases the work of cohesion among cells. Thus increasing cohesion may thin and spread the blastocoel roof.

Formation of the blastopore

At the onset of gastrulation, cell apices narrow in a band around the marginal zone. This apical constriction dimples the egg surface, forming the blastopore (Fig. 8). Superficial cells move toward the blastopore, in the process of invagination. This process begins at the dorsal meridian and proceeds around the marginal zone to the ventral meridian.

Changes in cohesion may initiate formation of the blastopore. Diverse evidence (*Mittenthal and McMeeking, 1986) suggests that late in the blastula stage an animal-vegetal gradient of cohesion develops: Cohesion is high and uniform in the blastocoel roof, intermediate in the marginal zone containing prospective mesoderm, and low among the prospective endoderm cells (Fig. 8A). The gradient becomes steeper with time. This steep gradient of cohesion may produce an interfacial tension in the marginal zone. There a constriction, the blastopore, would be energetically favorable because it decreases the perimeter of the embryo at which the interfacial tension is high (Fig. 8B).

Involution

Within the embryo epiboly shears the cohesion gradient, as the blastocoel roof moves past the deep endoderm (Fig. 9). However, the deep cells adjacent to the blastopore involute, turning inward and migrating back toward the blastocoel. As the deep cells migrate they carry the overlying superficial cells into the embryo, as on a conveyor belt; these superficial cells line the archenteron. The migration of mesoderm between ectoderm and endoderm restores the sequence of adhesive cell types that was present in the blastula but was disrupted by epiboly. The migrating deep cells form a layer, the mesodermal mantle, which girdles the interior of the embryo. Thus involution establishes the sequence of germ layers -- ectoderm outside, endoderm inside, and mesoderm between them -- that is energetically most favorable in terms of intercellular adhesion (Phillips and Davis, 1978).

AMPHIBIAN GASTRULATION: SOME POSSIBLE ROLES FOR INTRACELLULAR PROCESSES

The preceding discussion shows how adhesion among cells may direct the movements of gastrulation. However, similar surface movements can be seen in a single cell, an unfertilized amphibian egg, in the phenomenon of pseudogastrulation. In some species after suitable treatment of a mature egg the dark animal cortex spreads, compressing the light vegetal cortex toward the vegetal pole in a movement resembling epiboly. When the light-dark boundary reaches roughly the latitude where the blastopore would form in normal development, this boundary becomes sharp and resembles a blastopore. The dark cortex continues to spread until it has wholly engulfed the vegetal cytoplasm, and the edges of the pseudoblastopore close together. The striking similarity between these events and gastrulation suggests that intracellular mechanisms may contribute in similar ways to both phenomena.

338

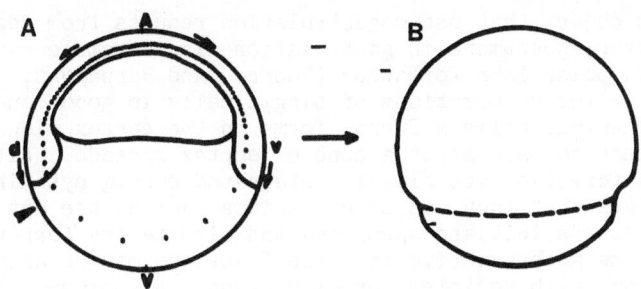

Fig. 8. Interfacial tension may initiate formation of the blastopore. (A)
Section through a blastula. As the blastocoel roof spreads
(arrows), an animal-vegetal gradient of cohesion steepens. Closer
spacing between dots represents higher cohesion. Arrowhead shows
the site where the blastopore first forms. (B) The surface of an
axially symmetric gastrula. Where the gradient of cohesion is
steepest and the interfacial tension greatest, a constriction (the
blastopore -- dashed line) develops. (Note, dorso-ventral
gradients in the timing and intensity of cohesion changes do not
alter this argument.)

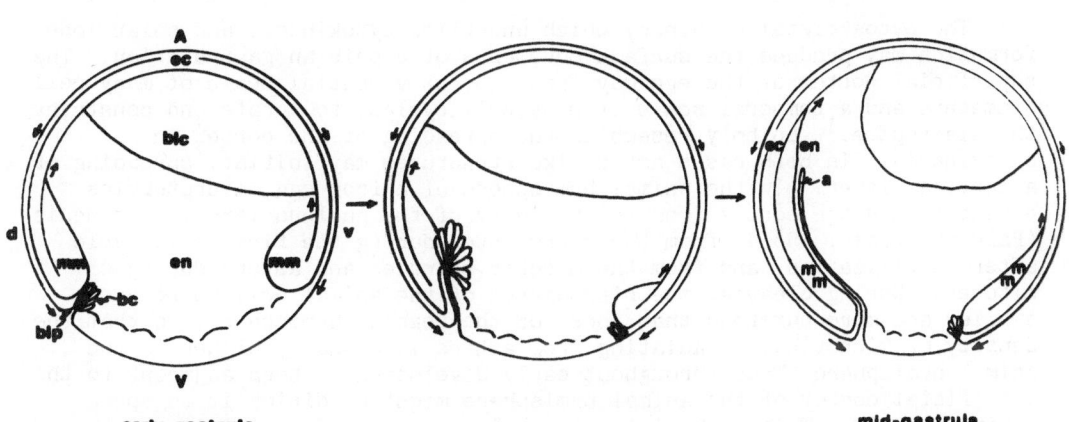

Fig. 9. Invagination and involution in <u>Xenopus</u>. The surface of the embryo
invaginates at the blastopore, where bottle cell apices contract.
The blastocoel wall continues to spread, forming an outer wall
around the endoderm. At the vegetal margin of this wall prospective mesoderm cells involute to form the mesodermal mantle, which
girdles the embryo. Small arrows show the direction of movement of
the outer wall and the mesodermal mantle. Symbols: See Fig. 7.
(after Keller & Schoenwolf, 1977; Keller, 1981.)

One might object that pseudogastrulation results from degeneration of the egg. However, phenomena in gastrulation also resemble events in cytokinesis and polar lobe formation (*Conrad and Rappaport, 1981; *Bray et al., 1986) -- two deformations of single cells in good condition. In cytokinesis of animal cells a furrow forms in the cortex midway between two asters. Adjacent to each aster a zone of cortex spreads. Although spreading and furrow formation are closely integrated during cytokinesis, some evidence suggests that they can be dissociated and so are not causally related. The asters initiate spreading and situate the furrow. The components of the asters active in these functions may be associated with the cytoskeleton, with vesicles, or with trapped cytoplasm. The furrow contains a band of microfilaments which probably helps to constrict it. Recent modelling (Greenspan, 1977) suggests that movement of cortex from sites of spreading toward the furrow drives a circulation of the deeper cytoplasm inward from the furrow and outward toward the poles. A similar movement, cortical ingression, has been inferred to occur in cleaving blastomeres of amphibian embryos (Ballard, 1955).

In early embryonic cleavages of some marine annelids and arthropods a polar lobe forms during mitosis (*Conrad and Rappaport, 1981; *Dohmen, 1983). A constriction transiently separates the polar lobe from its blastomere. A cleavage furrow and a polar lobe constriction are analogous in that they contain microfilaments and seem to depend, at some phase, on microtubules. However, a cytoplasmic clock can trigger formation of a polar lobe constriction in the non-nucleated vegetal half of a fertilized egg. It is unclear whether aster-like structures function in this process, or whether the mechanism for triggering and performing the constriction is wholly localized in the cortex.

The cytoskeletal machinery which underlies cytokinesis and polar lobe formation may produce the surface movements of amphibian gastrulation. The superficial cortex of the egg may function, at a spatial scale of many cell diameters and a temporal scale of many cell cycles, to locate and constrict the blastopore.[1] Epiboly resembles the spreading of the cortex in cytokinesis. In both cases aster-like structures may initiate spreading of a cortical layer. In the animal hemisphere of a frog egg, microtubules extend toward the cortex from the vicinity of the nucleus late in oogenesis (Palecek et al., 1985), from the sperm aster during the first cell cycle after fertilization, and from the mitotic spindles and asters during mitoses. During cleavage the blastomeres of the animal hemisphere are smaller and more numerous than those of the vegetal hemisphere, so that the density of microtubules radiating from asters is probably higher in the animal hemisphere. Thus throughout early development asters adjacent to the superficial cortex of the animal hemisphere might condition it to spread during epiboly. This mechanism may require storage of a signal from the asters to initiate a later expansion. The processes mediating the expansion of the cortex in cytokinesis and in epiboly are not known; White and Borisy (1983) and Keller (1980) discuss some hypotheses.

The blastopore seems to be a meta-furrow -- a groove produced by contraction of microfilaments in the apices of the bottle cells (Fig. 10). In cytokinesis the two asters provide a bipolar machinery which situates the cleavage furrow. In gastrulation it is unclear what processes situate the blastopore; the following section offers a hypothesis.

---

[1] If embryonic development be regarded as music, on this hypothesis gastrulation uses the thematic material of cytokinesis in a stretto by augmentation, as occurs in Bach's Art of the Fugue.

Superficial cells move toward the blastopore during invagination, as cortex moves toward the cleavage furrow. Both invagination and furrowing require an increase in surface area: The surface of a cleaving cell must enlarge to cover two daughter cells, and the surface which covered a blastula must cover both the outer surface and the lining of the archenteron in the gastrula. As a cleavage furrow deepens, new cortex from intracellular stores of membrane is added to the furrow wall to augment its area. The surface area of a gastrula increases by division and spreading of superficial cells; no deep cells enter the superficial layer in <u>Xenopus</u>.

The involution of deep cells during gastrulation may resemble the circulation of deep cytoplasm in cytokinesis. Keller (1985) has proposed that elongation of bottle cells initiates involution: Elongation of the bottle cells may exert a shear stress on the adjacent deep cells of the outer wall, turning their migration inward. This process corresponds, in cytokinesis, to the shearing of deep cytoplasm by cortex moving toward the furrow. However, in cytokinesis the contractile ring of microfilaments seems wholly responsible for constricting the neck between the daughter

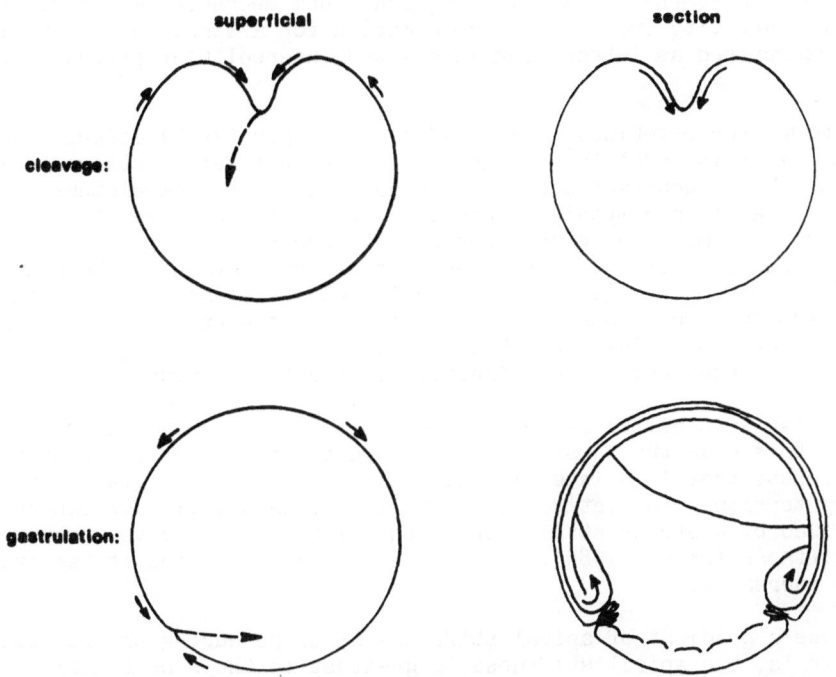

Fig. 10. Analogy between cytokinesis and amphibian gastrulation. Upper: The first cleavage in a <u>Xenopus</u> embryo. Lower: Gastrulation of a <u>Xenopus</u> embryo. Left: Superficial views. Cortex spreads toward a constriction (arrows) as the constriction elongates (arrowhead). Right: Cross-section perpendicular to the constriction. Arrows indicate ingression of subcortical cytoplasm during cleavage (upper figure) and involution of prospective mesoderm during gastrulation (lower figure).

cells. In gastrulation the bottle cells may initiate involution, but they are not necessary for its completion.

Several other parallels between cytokinesis and gastrulation are noteworthy. Before the furrow is visible in cytokinesis the cortex is under tension and its stiffness increases (*Hiramoto, 1981). By the gastrula stage the volume of a frog embryo has increased 30% as the blastocoel enlarges; an increase in tension and stiffness of the cortex probably accompanies this enlargement. The increase in cohesion which precedes gastrulation, discussed above, may also increase the surface stiffness. In a cleaving cell and an embryo, a cytoplasmic clock controls the time at which the constriction appears (Sawai, 1979; Hara et al., 1980; Kobayakawa and Kubota, 1981). In cytokinesis a cleavage furrow can form at an abnormal position if the asters are displaced. Similarly, in eggs producing a polar lobe constriction, constrictions can be produced at abnormal sites by injection of calcium or cyclic adenosine monophosphate. In Xenopus rotation or centrifugation of a zygote can alter the subsequent location of the blastopore (*Malacinski, 1983). Apparently a contractile band can form at any position on a cell or an amphibian gastrula, if earlier events set up an appropriate spatial pattern of conditions.

TOWARD MODELS FOR THE MORPHOGENESIS OF CELLS AND EMBRYOS

Constriction of a single cell and formation of the blastopore in an amphibian embryo look sufficiently similar that the same formal model might describe both phenomena. A model using continuum mechanics might relate cytoskeletal activity and cortical deformation for a single cell, and might also be interpreted as integrating intra- and intercellular processes for an embryo.

The model for cytokinesis proposed by Greenspan (1977) offers a useful framework. In this model the surface of a spherical cell contains tension elements which can contract actively. Presumably a tension element is an assemblage of actin and myosin molecules. Initially the elements are distributed uniformly. However, subcortical processes initiate a non-uniform contraction, slightly stronger at the equator than at the poles. This gradient of contractile activity pulls tension elements toward the equator, reducing their density at the poles and steepening the gradient of contractile activity. This unstable process (Fig. 11A) aggregates tension elements at the equator; their collective contraction produces a cleavage furrow.

A similar instability may produce the blastopore in an amphibian embryo. In this case the tension elements are the apices of superficial cells. Suppose that in a late blastula cell apices exert a gradually increasing tension. The active contractility of each cell apex can be characterized by a stress-strain curve; the slope of the curve measures an active stiffness (Oster, 1984). As a cell apex is activated its active stiffness increases.

Suppose a gradient of apical stiffness develops during the blastula stage. That is, the apical stiffness is greatest in the superficial cells of the blastocoel roof, intermediate in the marginal zone, and least in the large, yolky superficial cells of the vegetal hemisphere. The gradient is shallow in the blastocoel roof and the vegetal hemisphere, and steep in the marginal zone. The gradient becomes steeper with time, as the mechanical properties of cells diversify.

This non-uniformity of stiffness, with increasing tension in the surface of the embryo, will displace cell apices in-plane. To analyze this deformation one needs a mechanical model for the transient deformation of

astic medium, but gradually rearrange as in a fluid to relax shear
s (Phillips and Steinberg, 1978; Phillips, Steinberg, and Lipton,
. It seems reasonable to model the superficial layer as an elastico-
us fluid shell, since its cells show fluid-like rearrangement (Keller,
. Then the superficial layer of a late blastula should respond to a
ly-developing non-uniformity of stiffness as does an elastic shell.

Preliminary calculations (Mittenthal and Hom, unpublished) suggest
as the gradient of stiffness steepens, cells in the marginal zone will
retched anisotropically, with long axes parallel to the equator (as in
lastopore), while the less stiff cells of the vegetal hemisphere are
ched isotropically (Fig. 12). The model of Greenspan (1977) suggests
an instability localizes the blastopore. For example, rapid stretching
marginal zone cell, in response to the stiffness gradient or to
action of a neighboring cell, might stimulate it to contract more
ously, as in the model of Odell et al. (1981). Thus a broad band of
mation in the marginal zone could be focussed to a narrow line of
acting cell apices, the blastopore (Figs. 11B, 12C).

Further analysis is under way to see how epiboly and dorso- ventral
etry affect this model. Until shortly before the blastopore forms,
ly proceeds so slowly that superficial cells rearrange to maintain

age (Greenspan, 1977):

gradient of contractile
element density

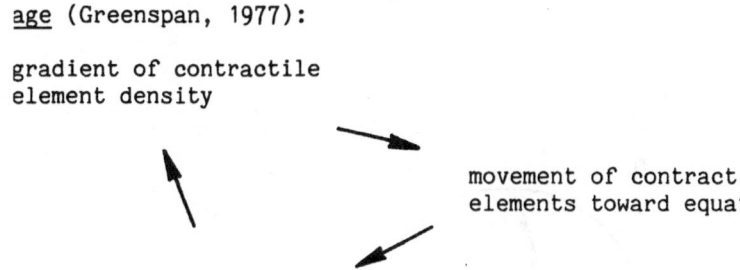

movement of contractile
elements toward equator

contractility
of furrow

ulation:

gradient (nonlinear) of
active stiffness of
superficial cell apices

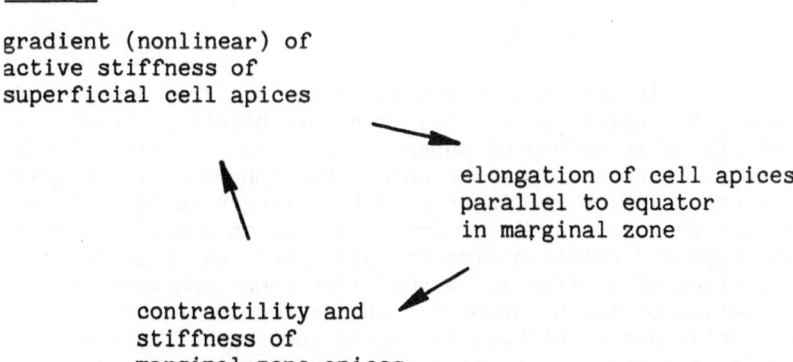

elongation of cell apices
parallel to equator
in marginal zone

contractility and
stiffness of
marginal zone apices

11. An unstable process involving a positive feedback loop may produce
    a constriction in cytokinesis (upper) and in formation of the
    blastopore during gastrulation (lower).

This essay is an inquiry into the relations between the mechanics of lls and the mechanics of embryos. In a cleaving cell and a frog embryo, ə external surface deforms actively to change the shape. The cortex of ə cell deforms; the superficial layer of cells deforms in an embryo. əse parallels suggest the possibility of a unified model for the chanics of cells and tissues. The continuum mechanics of shells offers a amework for this model. The cortex of a cell is continuous. In a sheet teractions among cells via junctions and extracellular matrix integrate tivities of the cytoskeleton over many cells, allowing the sheet to nave as a mechanical continuum.

A continuum mechanical model for a sheet of cells should treat the ⁻face as having fluid-like as well as solid-like properties. Several ⁻ces should be considered. At the apical and basal interfaces between ə cell sheet and its environment, there may be interfacial tensions. ːive contraction and passive elasticity of the cytoskeleton will ιtribute to the stiffness of the sheet. Its fluidity is associated with ιrrangement of cells. If the adhesive affinities of cells vary from ιce to place, an effective interfacial tension will arise at the lateral ːerfaces between cells of different types. This tension may be ιsotropic as well as inhomogeneous.

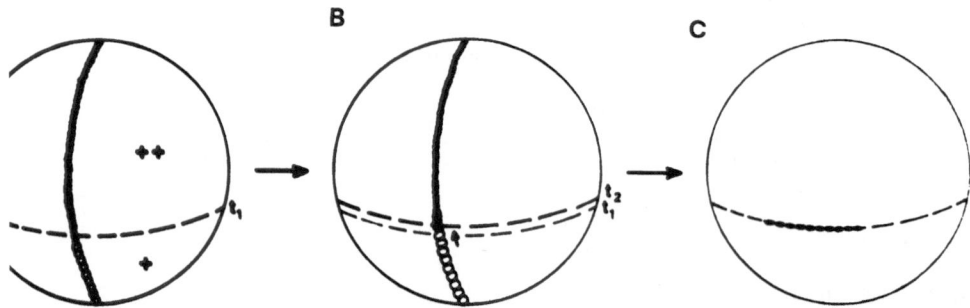

ʒ. 12. A model for the formation of the blastopore in a frog embryo. A, animal pole; V, vegetal pole. (A) In a late blastula, tension and active stiffness in apices of superficial cells increase. Small cells containing cortex from the animal hemisphere of the zygote develop greater active stiffness (++) than larger cells containing vegetal hemisphere cortex (+). Small circles on a meridian designate the shape and relative size of cell apices early in the diversification of stiffness. Dashed line shows approximate future location of the blastopore. (B) As tension rises, low-stiffness cells are stretched, enlarging the vegetal surface slightly (arrow) and elongating adjacent high-stiffness cells. (C) Stretch of activated cells stimulates vigorous contraction of cell apices (black ellipses) to form the blastopore. (As in Fig. 8 the embryo is treated here as axially symmetric.)

4

and embryos can suggest a unified phenomenological model, which may
to a deep unity in the mechanics of morphogenesis.

)WLEDGMENTS

Phil Arcuri, Jim Nardi, and Mark Sturtevant gave helpful comments on
work; I thank them. I appreciate the help of Suchada Broeren and Marj
itt in preparing the manuscript. The work was supported by U.S.P.H.S.
; HD16577.

(ENCES

'ts, B., Bray, D., Lewis, J., Raff, R., Roberts, K., Watson, J. D.,
    1983, "Molecular Biology of the Cell," Garland Publ. Co., New York.
ird, W. W., 1955, Cortical ingression during cleavage of amphibian
    eggs, studied by means of vital dyes, J. Exp. Zool., 129:77.
  D., Heath, J., and Moss, D., 1986, The membrane-associated 'cortex'
    of animal cells: Its structure and mechanical properties, J. Cell
    Sci., Suppl. 4:71.
id, G. W., and Rappaport, R., 1981, Mechanisms of cytokinesis in animal
    cells, in: "Mitosis/Cytokinesis," A. M. Zimmerman and A. Forer,
    eds., Academic Press, New York.
in, M. R., 1983, The polar lobe in eggs of molluscs and annelids:
    Structure, composition, and function, in: "Time, Space, and
    Pattern in Embryonic Development," W. R. Jeffery and R. A. Raff,
    eds., Alan R. Liss, Inc., New York.
;rom, D., 1976, The mechanism of evagination of imaginal discs of
    Drosophila melanogaster. III. Evidence for cell rearrangement,
    Develop. Biol., 54:163.
irt, J. C., 1980, Mechanisms regulating pattern formation in the
    amphibian egg and early embryo, in: "Biological Regulation and
    Development. Vol. 2: Molecular Organization and Cell Function,"
    R. F. Goldberger, ed., Plenum Press, New York.
ispan, H. P., 1977, On the dynamics of cell cleavage, J. Theor. Biol.,
    65:79.
  K., Tydeman, P., and Kirschner, M., 1980, A cytoplasmic clock with
    the same period as the division cycle in Xenopus eggs, Proc. Natl.
    Acad. Sci. U.S.A., 77:462.
ioto, Y., 1981, Mechanical properties of dividing cells, in:
    "Mitosis/Cytokinesis," A. M. Zimmerman and A. Forer, eds., Academic
    Press, New York.
ir, R. E., 1978, Time-lapse cinematographic analysis of superficial
    cell behavior during and prior to gastrulation in Xenopus laevis,
    J. Morphol., 157:223.
ir, R. E., 1980, The cellular basis of epiboly: An SEM study of deep
    cell rearrangement during gastrulation in Xenopus laevis, J.
    Embryol. Exp. Morph., 60:201.
ir, R. E., 1981, An experimental analysis of the role of bottle cells
    and the deep marginal zone in gastrulation of Xenopus laevis, J.
    Exp. Zool., 216:81.
ir, R. E., 1985, The cellular basis of amphibian gastrulation, in:
    "Developmental Biology: A Comprehensive Synthesis," ed. L.
    Browder, Plenum Press, New York.
ir, R. E., and Schoenwolf, G. C., 1977, An SEM study of cellular
    morphology, contact and arrangement as related to gastrulation in
    Xenopus laevis, W. Roux's Arch. Dev. Biol., 182:165.

lacinski, G. M., 1983, Axis specification in amphibian eggs, in: "Pattern Formation," G. M. Malacinski and S. V. Bryant, eds., Macmillan Publishing Co., New York.

ttenthal, J. E., and Mazo, R. M., 1983, A model for shape generation by strain and cell-cell adhesion in the epithelium of an arthropod leg segment, J. Theor. Biol., 100:443.

ttenthal, J. E., and McMeeking, R., 1986, A model for the shaping of epithelia, with applications to vertebrate embryos, J. Theor. Biol., in press.

ell, G. M., Oster, G. F., Burnside, B., and Alberch, P., 1981, The mechanical basis of morphogenesis. I. Epithelial folding and invagination, Develop. Biol., 85:446.

ter, G. F., 1984, On the crawling of cells, J. Embryol. Exp. Morph., 83 Supplement:329.

lecek, J., Habrova, V., Nedvidek, J., and Romanovsky, A., 1985, Dynamics of tubulin structures in Xenopus laevis oogenesis, J. Embryol. Exp. Morph., 87:75.

illips, H. M., and Davis, G. S., 1978, Liquid-tissue mechanics in amphibian gastrulation: Germ-layer assembly in Rana pipiens, Amer. Zool., 18:81.

illips, H. M., and Steinberg, M. S., 1978, Embryonic tissues as elasticoviscous liquids I. Rapid and slow shape changes in centrifuged cell aggregates, J. Cell Sci., 30:1.

illips, H. M., Steinberg, M. S., and Lipton, B. H., 1977, Embryonic tissues as elasticoviscous liquids II. Direct evidence for cell slippage in centrifuged aggregates, Develop. Biol., 59:124.

vai, T., 1979, Cyclic changes in the cortical layer of non-nucleated fragments of the newt's egg, J. Embryol. Exp. Morph., 51:183.

ick, J. M. W., 1983, "From Egg to Embryo," Cambridge University Press, New York.

iehelin, L. A., and Hull, B. E., 1978, Junctions between living cells, Sci. Amer., 238:140.

ite, J. G., and Borisy, G. G., 1983, On the mechanism of cytokinesis in animal cells, J. Theor. Biol., 101:289.

A MECHANICIAN'S VIEW OF THE

BIOMECHANICS OF CYTOKINESIS

Nuri Akkaş

Department of Engineering Physics
Ankara University
Ankara, Turkey

CYTOMECHANICS

Biology is the study of all living things that inhabit the universe. Mechanics, when employed in its classical connotation, as opposed to statistical or quantum mechanis, may be defined as that discipline which investigates the macroscopic response of systems to the action of forces or couples. The methods used are analytical, numerical and experimental. Systems may be discrete, in aggregate, or continuous. Biomechanics is the application of the laws of mechanics to a biological system.

The field of biology is almost indescribably large. Like all other sciences, the science of biology is divided into several disciplines. These, in turn, are subdivided into still more highly specialized areas of study. One of these subdivisions, cytology, is the study of cell structure and cell activities. Cytologists, in their study of cells, occasionally talk about the forces within the system, the motion of/within the system and the failure of the system. The concepts of force, motion and failure are the basic ingredients of mechanics. Therefore, mechanics of cell structure and cell activities should always be considered in cytology. As was proposed some years ago (Akkaş, 1979; Akkaş and Engin, 1981), the application of the principles of mechanics in cytology deserves to be called cytomechanics in its own right. In very general terms, a cell is a mechanical system. It is capable of passively resisting forces and actively generating forces. Obviously, the forces are of chemical origin but the result is mechanical.

Biochemistry and biophysics deal with structure and function at the molecular level. Biomechanics, in its more common applications, deals with the gross structure of the living being and with the tissue structure. This last sentence should not imply that the mechanics of cell structure and cell activities is not part of biomechanics. However, it should be stressed that the word "biomechanics" generally implies that it deals with structures at the tissue or larger level. A quick review of the journals such as Journal of Biomechanics and Journal of Biomechanical

347

Engineering would support this view. It is surprising to note that papers somehow related to cytomechanics are published more commonly in biophysics journals. Of course, there is always an overlap between two neighboring fields; However, in this case, it appears that the study of cell structure and cell activities has been mostly left to the field of biophysics, although it is part of biomechanics also. This is apparently because of the fact that, in cytomechanics, we deal with systems which are much smaller than those <u>commonly</u> treated in biomechanics. The researcher may even have to go down to the molecular level. He, then, starts wondering whether the principles of, say, continuum mechanics would be applicable at this level. This is, indeed, an important point to keep in mind. In some cases, it may be necessary to revert to theories other than that of classical continuum mechanics.

It is safe to apply the principles of continuum mechanics to the cell structure and cell activities as long as we are aware of its limitations at that level. For instance, the plasma membrane-the lipid bilayer-may be treated as a continuum if we deal with the stresses and strains in the plane of the membrane. It is obvious that it would not be reasonable to apply the continuum mechanics equations to the plasma membrane in the third (thickness) direction. The matrix of the fluid lipid bilayer has a thickness of about 7.5 nm and it consists of protein and fat molecules. The local study of the cell membrane should, therefore, fall in the domain of biophysics. On the other hand, the study of the global behavior of the membrane under the action of various internal and external forces is the topic of cytomechanics. As another example, study of tubulin, actin and myosin is the topic of biophysics whereas microfilaments and microtubules should be studied as load (stress) carrying elements by the use of the principles of cytomechanics.

The purpose of this work is to present a mechanician's view of the animal cell as a biological unit and of the biomechanics of cytokinesis as an example of various cell activities. The presentation of our understanding of the cell and the cell activities may be very simple for a cytologist who may even question some of the statements made. First of all, it should be emphasized that the statements presented in the following are mostly from the works of cytologists active in the field. If there are questionable statements, it is because there is not yet an agreement on this point among the cytologists themselves. The author was very surprised to find out during the NATO Advanced Research Workshop on Biomechanics of Cell Division that the biologists were not in a unanimous agreement on the role, and even the existence, of the contractile ring of a cleaving egg. On the other hand, for a mechanician, it is very reasonable (and easy) to accept the necessity of the contractile ring during cytokinesis because, for the mechanician, cytokinesis is the division of one sphere into two spheres. Simplification of a problem is the common approach followed by mechanicians. The resulting models-mathematical or experimental-are, at the beginning, easy to handle. Working with simple models, such as those of cytokinesis presented in this book, gives the mechanician an insight to the problem. Then, he can decide which agents play important roles during division and so he tries to incorporate them in the next generation of his model. Each simplification, when decided to be an oversimplification, is discarded from the model and, thus, the new model becomes a bit more realistic. In cytomechanics of cell division, we are at the very beginning of the modeling process. The models are still so young (and, thus, disputable) that the biologists at the Workshop were very eager to question the purpose of the modeling itself.

The presentation below of an animal cell and its ultrastructure is obviously not complete. Some seemingly important points may have been disregarded and some seemingly insignificant points may have been overemphasized. However, this, in itself, may be a tool in bridging the gap between cytologists and mechanicians. What is important for one may be negligible for the other and vice versa. The purpose, then, should be to come up with a model which is to the reasonable satisfaction of both parties concerned. In the preparation of the following sections free use has been made of the author's earlier work (Akkaş, 1980 a, b; Akkaş, 1981; Akkaş and Engin, 1980, 1981). Interested readers are referred to these works for detail and for references.

THE CELL

The cell is an integral and relatively independent body surrounded by a boundary and it is a common feature of all living organisms. It can simply be defined as a discrete mass of cytoplasm enveloped in a selective and retentive membrane and containing a nucleus. Its study is the subject of cytology. A cell lives and it carries out all the living functions such as storing and transforming energy, synthesizing molecules, dividing to reproduce, differentiating, and carrying out various mechanical motions. Cells occur in life in a wide variety, not only of sizes but of shapes and functions as well. They range in size from the smallest bacteria, only a few tenths of a micron in diameter (a pneumococcus is about $0.2\ \mu$ in diameter), to certain marine algae and to the yolks of bird eggs, with dimensions of centimeters. For all their apparent diversity, however, cells have many characteristics in common. For instance, a cell has a high-precision apparatus for energy-providing, synthesizing, and specific functional reactions, a center for the control of these reactions, and pathways of communication. Cells have the ability to continue living in the absence of any other cell. In other words, they are capable of obtaining energy from the environment and of using this energy to support essential life processes.

All cells are divided into two groups, eucaryotic and procaryotic. This is done according to whether or not the genes of the cell are contained in a well-defined nucleus. Only bacteria and the blue-green algae belong to the simpler, procaryotic group. The procaryotic cells have undeveloped nuclear areas. The well-defined nucleus of the eucaryotic cell is surrounded by nuclear membrane. All the cells from higher animals and those of many microscopic organisms belong to the eucaryotic group. Although most of what will be presented in this work is valid for both groups, it should still be noted that here we are concerned with animal cells only. A discussion of plant cells will not be included either, because plant cells are surrounded by a rigid and sturdy cell wall which animal cells do not have.

The extranuclear part of the cell is called the cytoplasm which is a translucent, colloidal material of gelatinous consistency. The cytoplasm is composed of numerous differentiated sub-cellular organelles which carry on the diverse cellular activities mentioned previously. The nucleus, the directing center of the cell, communicates with the cytoplasm through the pores contained in its membrane. The size and spacing of nuclear pores vary somewhat, but a diameter of 0.1 µ is quite common. Most eucaryotic cells have a complicated network of membranes in their cytoplasm. This network is called the endoplasmic reticulum and it extends to a variable degree from the nuclear membrane to the outer membrane of the cell. In addition to this membrane network, the cytoplasm contains various organoids and also vacuoles. Among these organoids are mitochondria, lysosomes, ribosomes and centrioles. Functions and features of these various organoids will not be discussed here. The necessarily short description given above of the cell structure as a whole is sufficient for our purposes. On the other hand, for the purposes of the present work, it is necessary to present a more detailed discussion of the outer membrane of the cell.

THE CELL MEMBRANE

The cytoplasm is separated from the external environment by a membrane which is interchangeably called the cytoplasmic membrane, or plasmalemma. This is a vital biological interface without which the cell cannot survive. Substances necessary for cellular activities are acquired through this selective and semipermeable barrier. It is no longer considered as a simple, inert interfacial boundary. It is undoubtedly involved in cell recognition, adhesiveness, inward transport of metabolites and outward transport of waste products. There is sufficient evidence suggesting that the membrane plays a critical role in the control of cell division. The cell surface contains not only the plasma membrane but also extra- and intracellular, cell-associated components that are bound to the plasma membrane. According to the contemporary fluid mosaic model of Singer and Nicolson (1972), the plasma membrane is composed of lipids, glycolipids, proteins and glycoproteins, as shown in Fig. 1. In this figure, the membrane-bound cell cytoskeletal components are also shown, and they will be discussed in detail later in the paper. Lipids and glycolipids apparently form the plasma matrix.

It is in order to point out here the difference between the concepts of the plasma membrane presented above and of the cell membrane utilized in the review article of Hiramoto (1970). The matrix of the fluid lipid bilayer shown in Fig. 1 has a thickness of about 7.5 nm, whereas the "membrane" thickness mentioned by Hiramoto is in the order of a few microns. Obviously, the latter is not the thickness of the lipid bilayer alone but that of the so-called cortical layer (or cortex) which includes both the lipid bilayer and the submembraneous region containing the cytoskeletal components. Various investigators used different values (ranging between 1 µ and 3 µ) for the thickness of the cortical layer, due to the fact that there is no definite inner boundary defining it. Moreover, a fluid-like lipid bilayer of 7.5 nm thickness can probably not show the cell surface stiffness characteristics that have been measured by various experimental techniques. Accordingly, it is of importance to have a clear view of the structural interrelationship between the lipid bilayer and the cytoskeletal components bound to the former.

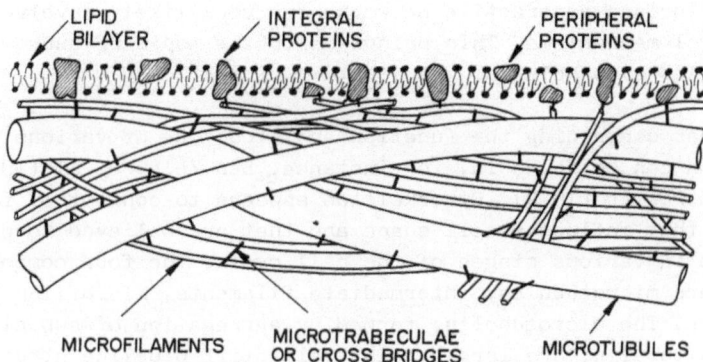

Fig. 1. Schematic drawing of plasma membrane and
       membrane-associated cytoskeletal components.
       Modified version of that given in Nicolson,
       Poste and Ji (1977). NOT DRAWN TO SCALE.

Plasma membrane proteins and glycoproteins can be considered to fall into two major classes. The integral membrane proteins have been proposed to traverse the lipid bilayer. Freeze-fracture technique revealed that these particles that penetrate deeply into the lipid bilayer are about 8.5 nm in diameter. The peripheral membrane proteins are more loosely associated with the lipid bilayer which they do not traverse. According to the fluid mosaic model, the plasma membrane has a fluid nature in that essentially all cell surface components are capable of rapid re-distribution (Nicolson et al. 1977). The membrane proteins can undergo dynamic change with respect to other membrane components within the matrix of "fluid" lipid which is itself capable of lateral motion. Since proteins are much larger than lipids, the former diffuse at a much lower rate than the latter. It is suggested that proteins behave like "icebergs" in an ocean. The evidence suggests that the membrane proteins are stiffer than the lipid matrix, so they may be acting as "stiffeners" of the membrane. For instance, the dynamics and topographic display of membrane proteins are controlled by their being clustered into a small region of the lipid domain. There is another means of controlling the motion of membrane proteins. Cell membrane associated subsurface contractile networks are very likely involved in mobility control mechanisms. This brings us to the topic of submembraneous structures called cytoskeleton.

Discussions concerning the function and structure of various components of cytoskeleton can be found in, for instance, Ben-Ze'ev et al (1979) and Wolosewick and Porter (1979). Cytoskeleton appears to consist of four major fiber systems that influence cell shape and that are believed to play an important role in various stages of the cell cycle. The four components of cytoskeleton are microtubules, intermediate filaments, microfilaments, and microtrabeculae. The microtubules, formed by aggregation of subunits of a protein called tubulin, are large, relatively stiff pipelike structures with an outside diameter of approximately 25 nm and an inner core of 15 nm diameter. They are regarded as a flexible and elastic skeleton which can undergo rapid, reversible assembly and disassembly. Microtubules are functionally linked to surface membrane glycoproteins. The intermediate filaments have a diameter of about 10-12 nm. They do not seem to occupy a submembraneous position. They apparently relate to the microtubules and serve to transmit mechanical forces within the cell. The microfilaments are thin protein polymers arranged in double helical filaments of diameter 6-8 nm and variable length. The microfilament bundles may run very close to the inside of the plasma membrane. They are less stiff than microtubules and some investigators consider them as the muscles or the cables of cytoskeleton. An actin like material is an important chemical constituent of the microfilaments. Evidence indicates that myosin is located predominantly in submembraneous positions along or within the microfilament bundles. Since actin and myosin are major components of contractile systems, it has been proposed that the microfilaments serve for this purpose in cells. The microfilament bundles, which are also called stress-fibers, are predominantly found as membrane-associated filaments and they terminate in upon ruffling areas of the cell. The other ends of the bundles penetrate deep into the cell. Immediately adjacent to the plasma membrane, the microfilaments are seen to form lattice or network of filamentous structures.

The microtrabeculae are slender strands of 3-5 nm diameter. They are interconnected so as to form an irregular, three-dimensional microtrabecular lattice. This lattice, which interconnects the other components of cytoskeleton, appears to be confluent with the submembraneous layer. The interrelationship among the various cytoskeletal components is shown schematically in Fig. 1 which is <u>not</u> drawn to scale. This is a modified version of the model presented in Nicolson et al (1977). This figure can be considered as the outermost layer of cytoskeleton attached to the plasma membrane. As seen in the figure, the microtubules and the microfilaments of the submembraneous region are connected to each other and also to the membrane proteins via various linking devices. The microtrabecular lattice of Wolosewick and Porter (1979) or "α = actinin" and "cross-bridge molecules" of Nicolson, Poste, and Ji (1977) can all serve for this purpose.

The brief description of the cell ultrastructure presented above has been, hopefully, sufficient to point out the fact that cytoskeleton and the plasma membrane constitute a very complicated and highly dynamic three dimensional structure. Although the existence of this complex space structure has been generally accepted, the author was unable to find a detailed actual micrograph or a schematic drawing of the structure in his literature survey. The closest three-dimensional schematic drawing is Fig. 4 of Wolosewick and Porter (1979) which shows the complexity of the structure very clearly. Ben-Ze'ev et al (1979) were able to obtain this cytoskeletal framework in cultured mammalian cells by removing lipids and soluble proteins by a detergent extraction. They showed that an intact cagelike network remains, consisting of the four components of cytoskeleton described above. The boundary of this skeletal framework appears to be a largely continuous surface that helps support the plasma membrane via numerous attachments. The components of this membrane-associated framework apparently help control the various functions of the plasma membrane also. The conclusion is that the complex cytoskeleton and skin like plasma membrane surrounding it should be considered simultaneously in cytomechanical studies. Note also that the microfilamentous cables present in the boundary of the cytoskeletal framework are interconnected not only by other filaments but also by being connected to the same membrane proteins. Accordingly, an increase in the mechanical stress in some region (say, the equatorial region) of this boundary layer will gradually be felt at the other regions of the cell surface also. It should be emphasized that the membrane-associated cytoskeletal components are transient in nature, dependent on cell energy systems for structural integrity. These components, and also the linkages among them, can undergo rapid polymerization and depolymerization. From a structural mechanics point of view, this can be viewed as the breaking or reconstruction of the various elements of the space structure. Finally, we should note that the microfilaments do indeed behave as cables in that they can apparently carry tensile stresses only. They are not rigid but flexible. However, they can be made relatively rigid by some chemical bridges between actin monomers. During the NATO Advanced Research Workshop on Biomechanics of Cell Division, it was stated by Rappaport and others that microfilaments can carry compressive stresses. The compressive stress that a microfilament can carry must be small. If microfilaments can indeed resist compression such as that occurring in microvilli, they must be acting as a bundle in which the single microfilaments are bridged to each other at various points along their lengths.

The description of the outer boundary of the cytoskeletal framework and its attachment to the plasma membrane discussed above is helpful in a better visualization of some of the plasma membrane characteristics. For instance, it is agreed that there is a general increase in membrane fluidity after transformation.

A study of the structure of the cytoskeleton-membrane system may give a reasonable explanation for this change in membrane behavior. If either the linkages between the cytoskeletal components and the membrane components or some of the cytoskeletal components themselves (such as the submembraneous microfilamentous cables) break (i.e., depolymerize), the stiffness of the membrane decreases. The membrane proteins etc., now not being anchored to the cytoskeleton framework, will have an increased mobility. Such an explanation for the increased fluidity of transformed cell membrane has been accepted to be more reasonable than others by, for instance, Robbins and Nicolson (1975).

There are many experiments reported in the literature in which the external force vs some characteristic deflection curves of the cell membrane were obtained. They are presented in detail in a review article by Hiramoto (1970). In the suction method, a micropipette connected to a movable reservoir of water is brought up to a cell and when the reservoir is lowered a bulge is sucked out of the cell surface. The pressure vs deformation curves obtained so are approximately linear. Mitchison and Swann (1955) calculated the Young's modulus of the cell membrane of sea urchins using this method. They found a range from 0.91×10^4 to 2.08×10^4 dynes/cm^2. In the compression method developed by Cole (1932), the cell is compressed between a pair of parallel plates. The force of compression is known and, hence, the intracellular pressure can simply be determined from a consideration of the static equilibrium. The pressure can, then, be related to the tension in the membrane. Cole's results led him to the conclusion that the cell membrane is a thin, elastic membrane. Hiramoto (1963), using this method, estimated the Young's modulus of the cell membrane as 1.2×10^3 dynes/cm^2 for the unfertilized egg. This is one order of magnitude smaller than that obtained by Mitchison and Swann (1955). More detailed information about these and similar techniques can be found in the articles by Skalak, Hiramoto and Evans in this book.

As seen, the compression and suction methods yield different elastic moduli for the cell membrane, the difference being one order of magnitude. It is thought to be of interest to discuss the possible reasons causing this apparent discrepancy. It should be realized that the conclusions concerning the mechanical properties of the cell membrane will be affected by the manipulation and interpretation of the experimental data. For instance, as emphasized in Pujara (1978), in suction experiments, whether the cell surface slips, slips partially or does not slip at all over the edge of the pipette will affect the resulting pressure vs deflection curves significantly. In case there is no slip, it is the region of the cell membrane within the pipette only that is stretched. If there is slip, the whole surface of the cell is stretched. In compression experiments, if there is no slip, it is the region of the cell membrane between the two compressing plates that is stretched. Under these circumstances, one should not expect to have good correspondence between the experimental results obtained from the two methods.

An ultrastructural study of the cell membrane may also be helpful in explaining the apparent discrepancies mentioned. It should be noted that the cell "membrane" that is studied by Mitchison and Swann (1955), Cole (1932), and Hiramoto (1970) is not the bilayer lipid membrane of Singer and Nicolson (1972). As stated previously, the thickness of the lipid bilayer is in the order of several nanometers, whereas the membrane thickness used by the former investigators in their calculations is in the order of a few micrometers. The submembraneous region, corresponding to the "cortex" of Hiramoto (1970), is considered as part of this membrane. It is this microfilamentous cortical layer that contributes to the stiffness of the cell membrane. A composite structure may require two or more moduli to characterize it completely and different loading conditions may measure different moduli. In the compression method, the cell membrane is pushed out due to the intracellular pressure increase caused by compression. In the suction method, the membrane is sucked out locally. In other words, the pressure difference is applied to most of the cell membrane in the compression test and to only part of the cell membrane by the pipette. Apparently, these two different loading conditions act on the submembraneous region in different manners which may cause some difference in the experimentally observed load-deflection curves.

The fact that the cell membrane is viscoelastic is now well established. If the cell membrane were an ideal elastic material, it would return to its original configuration immediately and completely, upon release of the deforming force. But this is not the case. When a cell is compressed between a pair of parallel plates with a constant external force, it requires some minutes to attain an equilibrium shape. Using this compression method, Yoneda (1973) determined that the relaxation time is 5.5 min for sea urchin eggs. Hiramoto (1976b) determined that, when a sea urchin egg was compressed between two parallel plates, the force required to keep the distance between the plates constant gradually decreased. When the force was removed, the relaxation consisted of a rapid phase ceasing within half a minute and a slow phase lasting for several minutes. The egg did not completely recover its original form within 5 min. after the removal of the force. Similar experiments on starfish oocytes yielded comparable results, Hiramoto (1976a). Nakamura and Hiramoto (1978) used the cell elastimeter in their suction technique to determine the mechanical properties of the cell surface of the starfish egg. When constant negative pressure was applied to a part of the cell with a micropipette closely in contact with it, the cell bulged out and the bulge rapidly increased at first and then gradually reached a steady value within 1 min.

The viscoelastic properties of red blood cell membranes were determined by Evans and Hochmuth (1976) and Chien et al. (1978) using the suction technique. The theoretical analyses given in these works are based on a viscoelastic stress-strain law suitable for the two-dimensional material of Evans (1973). In Evans and Hochmuth (1976) a red blood cell discocyte is aspirated into a small micropipette. When it is expelled from the pipette, the membrane relaxes back to its original biconcave, discoidal shape. The relaxation time turns out to be less than 1 sec. In Chien et al. (1978), the deformation of a portion of red blood cell during aspirational entry into a micropipette has been analyzed. Experimental studies showed that the deformation consisted of an initial rapid phase and a later slow phase. The time constant of the rapid phase of deformation in the micropipette ranged from 0.02 to 0.11 sec; wheras that of the slow phase was about 5 sec. Later, the aspiration pressure was removed and the time constant during recovery was determined to be about 0.15 s.

CYTOKINESIS

Cytokinesis is the division of the cytoplasm and the plasma membrane. It constitutes a relatively short stage of the cell cycle and its process is causally related to mitosis-nuclear division. The literature contains some recent review articles on the subject, i.e. Rappaport (1971), Schroeder (1975), Arnold (1976), Beams and Kessel (1976), Bluemink and de Laat (1977). A brief description of the process of cytokinesis is necessary in order to put the present work into its proper perspective.

Prior to division, animal cells attain a roughly spherical configuration. Cytokinesis appears to be the relatively simple event of the division into two of the spherical membrane enveloping the cytoplasm. According to the currently accepted mechanism, the process begins with the formation of a furrow in an equatorial plane. The division furrow has a dense ring of parallel microfilaments, oriented properly, to provide the structural basis of constriction. The microfilaments, which are actin-containing contractile fibrils 40-70 A in diameter and of an indeterminate length, are able to slide along each other. Thanks to this property, the extent of overlap between the microfilaments of the furrow ring is increased and, hence, the circumference of the ring is decreased. During his talk at the NATO ARW on Biomechanics of Cell Division, Schroeder, who was the first to confirm experimentally the existence of the contractile ring, stated that the thickness of the furrow band is about 0.1 µ and its width is about 10 µ. The edges of the band in the meridional direction occur pretty much at the inflection points of the cleaving egg. In the cross sectional view, there are seen about 5000 dots, each dot representing a microfilament. According to Schroeder, all cells have about the same size ring. It is the muscle-like contraction of this ring that causes cleavage. The micro-filaments of the ring as well as those of the other regions of the cortical layer are anchored to the plasma membrane. According to Robbins and Nicolson (1975), certain integral membrane glycoprotein components of the plasma membrane may well serve as the needed attachment points. Accordingly, as the equatorial contraction continues, the microfilamentous furrow ring pulls the attached plasma membrane down to displace the cytoplasm to either side, Arnold (1976). If it is assumed that the total cell volume remains constant during this process, then it is necessary that the surface area increase during division. It is known that tissue cells grow after cell division until they have attained the size of the parent celll, Dyson(1974). In early embryos, on the other hand, cell divisions merely partition the existing cytoplasm without any apparent change in the total volume. Divisions of this kind are called cleavages. In the present work, we are concerned with cleavage of animal cells. It is, therefore, implied that the total cell volume remains constant during cytokinesis and, hence, the total cell surface area must increase.

The fact that the total surface area increases (by about 26%) during cytokinesis constitutes a significant aspect of the problem to be considered. The surface of a dividing cell is not smooth. There are microvilli, ruffles or protrusions on the surface. As the equatorial constriction continues, such folds unfold, providing the extra amount of cell membrane needed for the daughter cells. According to Bluemink and de Laat (1977), it remains to be established whether 'new' plasma membrane is formed concomitantly with the cytokinetic event or at an earlier time. According to Arnold (1976), the formerly smooth surface of a dividing cell becomes covered with microvilli as the final mitosis which precedes cytokinesis occurs. Whatever the stage of the formation of the 'new' plasma membrane is, the fact that there are folds on the surface and they unfold during cleavage is well established.

Investigations concerned with the biomechanical models of cytokinesis are discussed in detail by Lardner and Nir in this book. Here we shall dwell upon our previous work only.

In Pujara and Lardner (1979) and in Akkaş (1980b) the animal cell is modeled as a spherical membrane of nonlinear elastic material containing a nonviscous fluid. The membrane undergoes large deformations under the action of a contractile ring force in its equatorial plane. In the former the membrane is made of the so-called STZC material which is proposed by Skalak et al (1973). Akkaş (1980b) uses the Mooney-Rivlin material to describe the behavior of the cell membrane material. Akkaş and Engin (1980) compare the Mooney-Rivlin and STZC materials using them in the same program. Later, Akkaş (1981) solved the same problem modelling the cell membrane material as a viscoelastic material. It suffices to point out that all these models essentially yield the same conclusions. In presenting their numerical findings, Pujara and Lardner (1979) emphasize mostly the geometric variables, such as displacements, which can be directly observed in experiments. Their numerical results on the geometric variables are in very satisfactory agreement with the experimental observations reported by Hiramoto (1958). On the other hand, Akkaş (1980b) showed that the Mooney-Rivlin material representation for the cell membrane also yields results which are in good agreement with the observations as far as the geometric variables are concerned. His emphasis has been mostly on the variation of those variables which are explicitly dependent on the cell membrane stiffness, such as intracellular pressure, the membrane force, and the axial ring force.

The numerical findings of Akkaş for the Mooney material showed that the intracellular pressure and the contractile ring force both increase monotonically during cleavage, if the cell membrane stiffness is assumed to remain constant during this time. These two variables are plotted in Fig. 2 as functions of the furrow radius. The non-dimensional intracellular

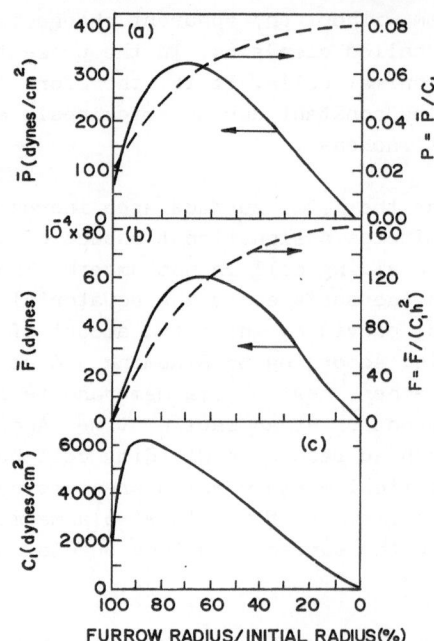

Fig. 2. Variations of intracellular pressure, contractile
ring force, and membrane stiffness during cleavage.
Mooney material.

pressure, $p = \bar{p}/C_1$, and the nondimensional total axial ring force $F = \bar{F}/(C_1 h^2)$, are shown as dashed lines in Figs. 2a and 2b, respectively. Here, $\bar{p}$ and $\bar{F}$ are the corresponding physical quantities, h is the membrane thickness, and C_1 is the membrane material constant. If C_1 is assumed not to change during cleavage, both the physical quantities $\bar{p}$ and $\bar{F}$ will increase monotonically during cleavage. This numerical result, namely that the intracellular pressure increases monotonically during cleavage, is in disagreement with the experimental result reported in Hiramoto (1968) according to which the intracellular pressure increases by about tenfold during the first half of cleavage and decreases by a similar degree during the second half. It is surprising to have such a significant qualitative discrepancy between the numerical and experimental results on the intracellular pressure, in spite of the fact that the corresponding results on the geometric variables are in very good agreement. This apparent discrepancy can be eliminated only by allowing the material "constant" C_1 of the mathematical model to change during cleavage as shown in Fig. 2c. The implication is that neither the Mooney-Rivlin representation nor any other exponential representation, in their classical forms, is appropriate for a description of the material properties of the cell membrane during cytokinesis. Under the mentioned assumption, the physical quantities $\bar{p}$ and $\bar{F}$ increase during the first half of cleavage and decrease during the second half as shown by the solid lines in Figs. 2a and 2b; these results being, now, in agreement with the experimental results. Recalling that C_1 is directly related to the stiffness of the cell membrane, Akkaş (1980b) concludes that the plasma membrane stiffness must change during cleavage. As seen in Fig. 2c, the membrane gets stiffer during the early stages of cleavage and, later, it loses its stiffness gradually or it "deteriorates". This numerical prediction concerning the plasma membrane stiffness change during cleavage is in agreement with the experimental findings as reported in Hiramoto (1970). The experiments clearly indicate that the membrane stiffness is a function of the stage of cleavage, showing a behavior qualitatively in agreement with that shown in Fig. 2c.

Although the numerical results presented in Fig. 2 are for the Mooney-Rivlin material, the conclusions are also valid for the STZC-material. This has been pointed out by Akkaş and Engin (1980). The comparison of the results for the Mooney material and the STZC material is given in Figs. 3 and 4. The responses are similar qualitatively.

MEMBRANE STIFFNESS CHANGE DURING CYTOKINESIS

In a previous section, we discussed the ultrastructure of animal cells. The conclusion, in short, was that the plasma membrane is attached to the three dimensional cytoskeletal trusswork and, moreover, that there is a network of cables (microfilaments) very closely associated with the plasma membrane. In addition to being linked to the membrane components, the cables are connected to each other and also to the microtubules. Now, we present mechanisms that may occur within this layer during cytokinesis and that may give a reasonable explanation for the cell surface stiffness change.

The model is a spherical membrane within the plane of which there

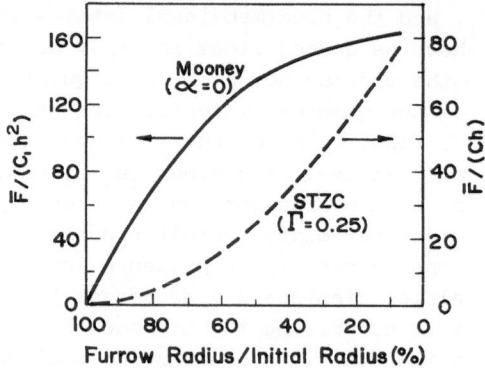

Fig. 3. Change in nondimensional equatorial ring force during
cleavage.

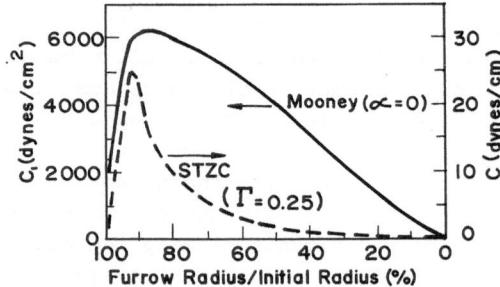

Fig. 4. Required change in membrane stiffness during cleavage
to have agreement between experimental and numerical
results.

exists a network of cables, as shown in Fig 5. Before the onset of cleavage, these cables, which can be considered to correspond to the surface tension elements of Greenspan (1977), are randomly distributed over the cell surface which, therefore, can be thought of in uniform tension. Due to some internal mechanism, which is probably the subject of the biochemistry of the cell cycle, the cables lying along the circumference of the future equatorial plane start constricting. The cleavage has now been initiated. Here, it is implicitly assumed that it is the equatorial surface where the events causing division take place. Therefore, the present work is an application in the equatorial constriction theory. The numerical findings reported in the earlier work of Akkaş (1980b) indicate that the following events take place during cleavage:

a) There is an almost monotonic increase in surface area of the cell membrane, the total surface area of the two daughter cells being about 25% larger than that of the parent cell at the end of the cleavage.

b) The stretch ratio of the cell membrane in the circumferential direction (λ_θ) is always much less than that in the meridional direction (λ_ϕ). The maximum value of λ_θ occurs near the pole. Near the equatorial surface, the membrane does not even stretch in the circumferential direction, but, rather, it is wrinkled.

c) In the meridional direction, there is always stretch (no wrinkling) and λ_ϕ near the equatorial surface is always larger than that near the polar surface.

d) λ_ϕ increases monotonically during cleavage, reaching very high values (about 300%) around the equatorial region towards the completion of cleavage.

The alignment of the microfilaments during the initial stages of cleavage explains the initial increase in the cell membrane stiffness. To be able to explain the subsequent decrease in the stiffness during the later stages of cleavage, one must consider the fact that the microfilaments and their attachments are fragile components of cytoskeleton. Or it is also possible that the chemical activities taking place within the cell during cleavage depolymerize these components. Let us first discuss the mechanical explanation, namely that the microfilaments and/or their attachments break when the stress levels in them exceed a critical level. The chemical explanation, namely that some chemical activities depolymerize the microfilaments, will be discussed later.

It was mentioned above that the cortical layer of an animal cell contains a three-dimensional network of polymers. For this network we have so far interchangeably used the terminologies such as cytoskeletal trusswork, microfilamentous layer, network of actin filaments, etc. In essence, this network is simply a spatial network of bio-polymers. Polymers may be defined to be substances formed by a more or less regular repetition of a large number of smaller molecules (monomers) which are joined by chemical bonds into long and relatively flexible chains. The chains may have side branches and they may be joined into a three-dimensional network in which case they are called crosslinked polymers. Linear polymers are those where one monomer is joined to the next in an extended chain. If such linear polymers are crosslinked so that they form three-dimensional networks, then

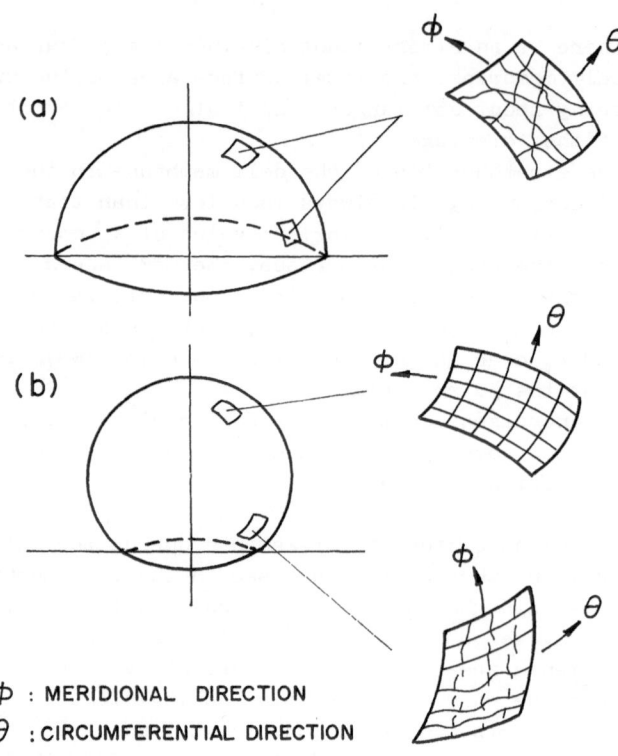

(a)

(b)

ϕ : MERIDIONAL DIRECTION
θ : CIRCUMFERENTIAL DIRECTION

Fig. 5. Realignment and subsequent breaking of membrane-associated microfilaments during cleavage.

they are called primary molecules. During crosslinking, each and every
primary molecule becomes coupled to other molecules through chemical bonds.
That portion of a primary molecule that extends from one junction to the
next is called a chain. A pull on one of these chains is transmitted to
others as with a bundle of threads that are entangled with each other. It
is important to note at this point that if the stress applied to such a
network becomes excessive the molecular chain is broken, either along its
length or at its junction. The number of crosslinks in the network is one
of the most important properties of the polymer. Stress and strain in any
network can be related to the number of crosslinks present. For the so-called
neo-Hookean material which is a special form of the Mooney-Rivlin material
the strain-energy function is

$$W = C_1(\lambda_1^2 + \lambda_2^2 + \lambda_3^2 - 3), \tag{1}$$

in which C_1 is the elastic constant and λ's are the principal extension
ratios. W represents the work of deformation or elastically stored free
energy per unit volume of the material. The following expression for W
using the Gaussian network theory is obtained:

$$W = \frac{1}{2} NkT(\lambda_1^2 + \lambda_2^2 + \lambda_3^2 - 3), \tag{2}$$

in which k is Boltzmann's constant, T is the absolute temperature, and N
is the number of crosslinked units in the network. Accordingly, the elastic
constant C_1 is directly proportional to the number of chains N,

$$C_1 = \frac{1}{2} NkT. \tag{3}$$

For the Mooney-Rivlin material, a second elastic constant C_2 is introduced.
However, C_2 does not have a clearly understood structural significance.
Although the discussion presented above was for the well-studied Mooney-
Rivlin material in particular, we are of the opinion that the elastic
constant of the STZC material can also be related to the number of chains
in the network.

The purpose of the above discussion was to emphasize the fact that the
elastic stiffness of a membrane consisting of a polymer network is directly
proportional to the number of chains or crosslinks. If this number increases
or decreases during cytokinesis, then the stiffness will also increase or
decrease. The breaking apart of the chains (microfilaments) and/or their
attachments is possible under excessive strain. It is obvious from the
numerical findings (b), (c), and (d) that this breaking apart will be much
more severe in the meridional direction than that in the circumferential
direction. Indeed, near the equatorial region the circumferential microfilaments will not break apart at all and they will be able to continue with
their constricting activities with ease. Moreover, since the meridional
stretch is the greatest near the equatorial region, the circumferential
fibers in this region can, in principle, easily slide down towards the
equator, Fig. 5b. This is quite possible because in this region, the
meridional fibers are now all broken apart and they can not restrain the
sliding of the circumferential fibers towards the equator. The resulting
clustering of the circumferential microfilaments at the equator will
further increase the magnitude of the contractile ring force which in a

self-perpetuating manner deepens the furrow. Note that not all of the circumferential cables of the cell surface are capable of sliding towards the equator.

The implication of the present mechanical model, as far as the origin of the microfilamentous contractile ring is concerned, is that the ring is the result of the local reorganization of the microfilaments pre-existing over the entire cell cortex. In a recent work, Opas and Soltynska (1978) present evidence that supports this hypothesis. They studied the changes in the ultrastructure of the filamentous sub-membraneous network in mouse blastomeres before and during cytokinesis. Before the onset of the cleavage furrow, they observed a loose network of thin filaments oriented randomly in all directions. This continuous network is closely associated with the plasma membrane and the ends of numerous filaments extend toward the center of the cell. During the early cytokinesis, the polar region still consists of randomly oriented filaments. In the remaining cell surface, the filaments tend to be oriented parallel to the long axis (pole-to-pole) of the blastomere. During the later stages of cytokinesis, the cell surface appears to consist of three distinct regions. At the polar region the microfilaments are still randomly oriented as before. At the equatorial region, they are predominantly circumferential. At the remaining transitory region, the filaments are parallel to the long axis of the cell. These experimental observations appear to support our model quite well. Reorganization of the microfilamentous network during cleavage as observed in these experiments, indeed, forms the basis of the present cytomechanical model.

Obviously, the mechanical factors discussed above are not the only ones that may affect the cell membrane stiffness change during cytokinesis. There are various chemical activities taking place in the cell during cytokinesis and they may play an important role in controlling the membrane stiffness. This chemical explanation, however, implies that membrane stiffness change is a programmed part of the cell cycle rather than its being the outcome of the geometrical and mechanical changes occurring during cleavage.

Polymerization is the reaction of linking together of several monomers. Actin is an important protein present in many nonmuscle cells. Purified actin can exist in two forms: G-actin or monomeric actin, and F-actin or polymeric actin. The G-actin can be polymerized into F-actin and it forms fibriller protein. Actin polymerization requires the presence of salts. There is a dynamic equilibrium established between G- and F- actins, and the change in the salt concentration of the medium can displace this equilibrium. Local cytoplasmic structure is usually described by using the colloidal terms sol and gel. By definition, fluid and solid states of colloidal substances are called sols and gels,respectively. Accordingly, a gel state is stiffer in tension than a sol state. The gel structure of an egg cortex breaks down during contraction of the cleavage furrow.

This decrease in membrane stiffness is apparently caused by the transformation of the gel state to the sol state. This transformation, which is called the solation of the gel, is the result of the depolymerization of F-actin to G-actin. A cell can, therefore, control

actively the stiffness of its membrane (to be more precise, its cortical
layer) by controlling the salt concentration of the medium. Thus, it is
possible that mechanical factors discussed previously may not be playing
any important role in cell membrane stiffness change during cleavage. The
controlling factors may be purely chemical. A recent and very detailed
presentation of the role of various proteins in controlling various
nonmucsle cell activities explained from a chemical point of view can be
found in Taylor and Condeelis (1979). We are of the opinion that the
mechanical explanation discussed earlier may be a valid one as well. Indeed
the chemical and mechanical models may very well be complementary. Future
work by cytologists will hopefully shed more light on this subject. In the
following, we will necessarily dwell upon the mechanics of the phenomena,
rather than its chemistry. However, it should be pointed out that the
processes that occur in the cell are mechano-chemical, and the source of
the mechanical work is the chemical energy stored within the cell.

DISCUSSION AND CONCLUSION

In this section, the experimental observations reported in the
literature which the present mechanical model appears to be able to explain
from a structural mechanics point of view will be discussed. Recall that,
according to the equatorial constriction theory, the only <u>active</u> mechanism
that initiates the cleavage is that of muscle-like contraction of the
circumferentially oriented microfilaments already existing at the furrow
base. The further increase in the contractile ring force necessary for a
successful completion of the cleavage is provided by the alignment and
sliding of the circumferential microfilaments towards the equatorial plane.
Such a mechanism has been proposed in Rappaport (1975) also. This
rearrangement of the preexisting microfilaments near the equatorial region
and the initial stiffening and subsequent deterioration of the cell membrane
components due to the realignment and then breaking down of the meridional
microfilaments are all <u>passive</u> mechanisms.

The cleavage furrow consists of a band of circumferentially aligned
microfilaments. Arnold (1976) reports that if the cell surface in advance
of an established furrow band is cut, the band width will narrow. He
concludes that this observation demonstrates that the actual contractile
tension extends beyond the surface area banded by the furrow. This result
is in total agreement with the implications of the present model. Cutting
the surface in advance of the established band means that the microfilaments
which have, by now, been aligned in the meridional direction are cut.
See Fig. 5. Recall that these meridional microfilaments, according to our
model, are subjected to tension and they try to prevent the circumferentially oriented microfilaments from sliding towards the equatorial plane as
long as they remain intact. Cutting these meridional microfilaments will
cause the circumferential microfilaments present near the circumferential
edge of the furrow band to slide towards the equatorial plane. This will,
in turn, cause a decrease in the furrow width. The present model implies
also that if the cutting of the surface in advance of the established
furrow band is affected <u>near</u> the completion of cleavage, the narrowing of
the band should be relatively much less (or even no narrowing at all),

because the meridional microfilaments near the band have already been broken apart by this time. However, at present, to the author's knowledge, there is no experiment reported concerning this point. Arnold (1976) reports also that the microfilamentous furrow band is wider and more diffuse in the early stage of cleavage and, later, it narrows and becomes more dense. The gradual sliding of the circumferential microfilaments towards the equatorial plane eased by the further breaking of the meridional microfilaments, as proposed by our model, explains the mechanics of the mentioned phenomenon.

When the furrow base is microsurgically destroyed (Arnold, 1976), the cleavage furrow disappears and the plasma membrane returns to the position of the uncleaved surface. According to the present model, the contractile force in the furrow band is the only active force that initiates and completes the cleavage. Its microsurgical destruction will surely affect cleavage. However, whether the cell returns to its initial configuration or not depends upon the stage of cleavage at wich the furrow base destruction is affected. As seen in Fig. 2b, if the destruction is affected at an early stage of cleavage (to be exact, if the furrow radius to initial radius ratio is larger than about 70%), the cell will return to its initial configuration. If, when the destruction takes place, the cleavage has already passed the critical stage proposed in Akkaş (1980b), then the cell will not return to its initial configuration because the ring force is already decreasing at this stage. The cleavage can be completed in spite of the destruction, or it can, at most, be arrested. However, here we have assumed that the microsurgical destruction did not affect anything on the cell surface but the band itself. Arnold (1976) does not report the cleavage stage at his microsurgery. On the other hand, Rappaport (1974) and Schroeder (1975) imply that there is, indeed, a critical stage of cleavage. Cytochalasin B is known to cause the very rapid dissolution of microfilaments. Exposure of cleaving eggs to this substance may result in reversal or arrest of furrowing, depending upon how far advanced the process was at the time of exposure. Although the cytochalasin B treatment is essentially different from the microsurgical destruction, Rappaport's statement does, indeed, support our view that the cleavage stage is important in interpretation of the results of such experiments.

There are many agents that have been shown to affect cleavage. Among those studied extensively are cytochalasin B and calcium ion. Before discussing the effects of these agents from the point of view of the present cytomechanical model, it is necessary to emphasize some of the implications of the model.

Recall that there is only one active mechanism in the present model and it is the constricting mechanism of the equatorial ring. The subsequent stiffening of the membrane and its deterioration later are both passive mechanisms. Let us hypothetically assume that it is possible, by some, say, external agents, to control these active and passive mechanisms so that they can both be considered as active mechanisms independent of each other. In other words, assume that equatorial constriction and membrane stiffening can both take place without affecting each other at various stages of cleavage. The following possibilities arise: At any stage of cleavage, the ring force created by the equatorial microfilamentous band can increase, remain the same, or decrease in magnitude. Similarly, the membrane stiffness can increase, remain the same, or decrease. There are nine possible combinations, but we will discuss only those that appear to be more likely. Suppose that the membrane stiffness does not decrease as shown in Fig. 2c after reaching its maximum value, but rather it remains at this value or even keeps increasing. For division to be completed, the ring force must reach large values accordingly. If the equatorial micro-filamentous band can not provide such a large force, either because there are not enough microfilaments in the band or because they break apart not being able to support the required high stresses, the cleavage can not be completed. Suppose an external agent destroys both the equatorial micro-filaments that provide the constricting force and the remaining surface microfilaments that provide the membrane stiffness before the critical stage shown in Fig. 2b has been passed. One can not make a definitive statement about the future of the cleavage. If the deterioration of the membrane occurs at a faster rate than the decrease in the contractile force, the cleavage will be completed. Otherwise, the cell will return to its initial spherical shape. In view of the explanation about the various possibilities concerning cleavage given above, it appears to be easier to interpret the results of some drug and calcium treatment of cleaving eggs.

It is reported that cytochalasin B inhibits furrowing within a few seconds. On the other hand, some other experiments indicate that the same drug has no effect on cleavage. Schroeder (1975) states that these discrepancies have not yet been satisfactorily explained. Arnold (1976) emphasizes the observed fact that cytochalasin B affects not only the contractile ring but also the plasma membrane and the submembraneous microfilament region. This probably corresponds to the second possible case discussed in the previous paragraph, namely that an external agent disrupts both the equatorial and the remaining microfilaments. In the light of the discussion presented in the previous paragraph, therefore, it is not surprising to find out that cytochalasin B treatment of cleaving cells may yield contradictory results.

It is known that calcium specifically stimulates contraction and that polymerization of microfilaments is somehow involved. It, then, can be thought of as an agent that increases the membrane stiffness and also the ring force that can be provided by the equatorial band. Rappaport (1974) states that localization of calcium under the equatorial surface could facilitate cleavage. This localization will increase the force producing capacity of the equatorial band but it will not increase the stiffness of the remaining cell surface and, hence, according to the predictions of our model, the cleavage will be facilitated. What happens if both the equatorial region and the remaining surface are affected by the calcium ions? Our model indicates that the outcome can not be predicted. Arnold (1976) showed that the treatment of the fertilized eggs with calcium under various conditions may inhibit or facilitate cleavage. Rappaport (1974) states that the role of calcium ions in cell division has not yet been identified.

In conclusion, the cleavage mechanism in animal cells was studied

from a cytomechanical point of view. The model is a spherical membrane
within the plane of which there exists a very complex network of
cytoskeletal components. It is proposed that it is this network that
passively controls the membrane stiffness change during cleavage. The only
active force that initiates and completes cleavage is that produced by the
microfilaments of the equatorial band. The model appears to explain most
of the experimental observations reported in the literature.

Before closing out this chapter, it is thought that it may be of interest
and of use to the readers to mention some of the points discussed during
the NATO Advanced Research Workshop on Biomechanics of Cell Division. Use
will be made of "The Resolution of the Participants" which was written down
by Rappaport and Skalak at the end of the workshop.

It is emphasized that

(a) the combination of biology, mathematics, physics, mechanics,
computational methods and modeling is at a stage where further progress
can be made rapidly by intimate and genuinely cooperative efforts among
these varied disciplines,

(b) that field of cell division and motility is very important
to current biological developments and applications so that very general
and far reaching theories may have wide application and importance in
biology if they are soundly and thoroughly developed.

Some specific items that need further exploration and elucidation are:

a) Theoretical modeling of the entire cell division process
needs to be carried out including the motion and movement of centrosomes
and chromosomes as well as the membrane and cleavage.

b) Ameboid locomotion, cell shape change and related active
cell motions should be modeled and studied for possible mechanisms and
related to cell division, if appropriate.

c) Models should incorporate information of motions which occur
with and without astral structures present. Models should be tested for
their ability to reproduce behavior of abnormal experimental conditions,
to predict results, and to compare to experiments. For example, models
should aim to reproduce experiments involving relations between centrosomal
and astral structures and contraction (tension generation).

d) Models should be developed to incorporate more closely the
actual detailed observed behavior of cell components and to incorporate all
types of data, spatial and temporal. For example, the known variation of
cytoplasmic properties such as variation of viscosity with position in
the cell and known microtubule architecture should be incorporated. No
models do this at present.

e) While non-dimensional models are a convenience and should
be continued in general research, there is a need to carry out detailed
comparison to particular experimental data in dimensional form to be
useful in establishing the validity of general approaches to particular
cases. This is particulary valuable in comparing different cells and
situations to define typical abnormal values that may occur, for example,

in cancer cells. This will require much more cooperative effort than has been the case to date.

  f) The effective thickness of the cortical shell and other modules of cell anatomy need to be better defined and measured to permit realistic modeling. Definitions should be established according to possible measurements; even if only operational, such data and their variation throughout a cell need to be incorporated in models.

  g) There is a need to develop channels through which relevant known experimentally derived data can be brought to the attention of modelers to insure that such experimental data are included in modeling from an early stage so as to avoid unrealistic or incomplete model development which may be otherwise not useful. This calls for exchange of information at a stage prior to formal publication.

REFERENCES

Akkaş, N.(1979) Biomechanics of cell division. Engineering Mechanics
Seminar, Ohio State University, Columbus, Ohio, USA.

Akkaş, N.(1980a) Letter to the editor. J. Biomechanics. 13, 459-460.

Akkaş, N.(1980b) On the biomechanics of cytokinesis in animal cells.
J. Biomechanics 13, 977-988.

Akkaş, N.(1981) A viscoelastic model for cytokinesis in animal cells.
J. Biomechanics, 14, 621-631.

Akkaş, N. and Engin, A.E.(1980) Nonlinear mechanics and cell division.
ASME 1980 Adv. Bioengng. 271-272.

Akkaş, N. and Engin, A.E.(1981) Ultrastructure of animal cells and its
role during cytokinesis from a structural mechanics viewpoint.
In: Mechanics of Structured Media, pp.187-207. (Edited by
Selvadurai, A.P.S.) Studies in Applied Mechanics 5A, Elsevier,
Holland.

Arnold, J.M.(1976) Cytokinesis in animal cells: new answers to old
questions. In: The Cell Surface in Animal Embryogenesis and
Development pp. 55-80. (Edited by Poste, G. and Nicolson, G.L.)
Elsevier, North Holland Biomedical Press.

Beams, H.W. and Kessel, R.G.(1976) Cytokinesis: a comparative study of
cytoplasmic division in animal cells. Am.Scient. 64, 279-290.

Ben-Ze'ev, A., Duerr, A., Solomon, F., and Penman, S.(1979) The outer
boundary of the cytoskeleton: a lamina derived from plasma
membrane proteins. Cell 17, 859-865.

Bluemink, J.G. and de Laat, S.W.(1977) Plasma membrane assembly as
related to cell division. In: The Synthesis, Assembly and
Turnover of Cell Surface Components, pp.403-461. (Edited by
Poste, G. and Nicolson, G.L.) Elsevier, North-Holland Biomedical
Press.

Chien, S., Sung, K.P., Skalak, R., Usami, S. and Tözeren, A.(1978)
Theoretical and experimental studies on viscoelastic properties
of erythrocyte membrane. Biophys. J. 24, 463-487.

Cole, K.S.(1932) Surface forces of the Arbacia egg. J.Cell. Comp.
Physiol. 1, 1-9.

Dyson, R.D.(1974) Cell Biology, A Molecular Approach. Allyn and
Bacon, Boston.

Evans, E.A.(1973) A new material concept for the red cell membrane.
Biophys. J. 13, 926-940.

Evans, E.A. and Hochmuth, R.M.(1976) Membrane viscoelasticity.
Biophys. J. 16, 1-11.

Greenspan, H.P.(1977) On the dynamics of cell cleavage. J. Theor. Biol.
65, 79-99.

Hiramoto, Y.(1958) A quantitative description of protoplasmic movement
during cleavage in the sea urchin egg. Exp. Biol. 35, 407-424.

Hiramoto, Y.(1963) Mechanical properties of sea urchin eggs. II.Changes
in mechanical properties from fertilization to cleavage. Expl.
cell. Res. 32, 76-89.

Hiramoto, Y.(1968) The mechanics and mechanism of cleavage in the sea
urchin egg. Symp. Soc. Exp. Biol. 22, 311-327.

Hiramoto, Y.(1970) Rheological properties of sea urchin eggs. Biorheology 6, 201-234.

Hiramoto, Y.(1976a) Mechanical properties of starfish oocytes. Develop. Growth Differ. 18, 205-209.

Hiramoto, Y.(1976b) Mechanical properties of sea urchin eggs. III. Viscoelasticity of the cell surface. Develop. Growth Differ. 18, 377-386.

Mitchison, J.M. and Swann, M.M.(1955) The mechanical properties of the cell surface. III. The sea urchin egg from fertilization to cleavage. J.exp. Biol. 32, 734-750.

Nakamura, S. and Hiramoto, Y.(1978) Mechanical properties of the cell surface in starfish eggs. Develop. Growth Differ. 20, 317-327.

Nicolson, G.L., Poste, G. and Ji, T.H.(1977) The dynamics of cell membrane organization. In: Dynamic Aspects of Cell Surface Organization, pp. 1-73. (Edited by Nicolson, G.L. and Poste, G.) Elsevier, North Holland Biomedical Press.

Opas, J. and Soltynska, M.S.(1978) Reorganization of the cortical Layer during cytokinesis in mouse blastomeres. Exp. Cell Res. 113, 208-211.

Pujara, P.(1978) Analysis of Finite Deformations of Membranes. Unpublished Ph.D Thesis, University of Illinois at Urbana-Champaign, Urbana, Illinois (Directed by T.J. Lardner).

Pujara, P. and Lardner, T.J.(1979) A model for cell division. J. Biomechanics 12, 293-299.

Rappaport, R.(1971) Cytokinesis in animal cells. Int. Rev. Cytol. 31, 169-213.

Rappaport, R.(1974) Cleavage. In: Concepts of Development (Edited by J. Lash and J.R. Whittaker) pp. 76-98. Sinauer, Connecticut.

Rappaport, R.(1975) Establishment and organization of the cleavage mechanism. In: Molecules and Cell Movement (Edited by S.Inoue and R.E. Stephens) pp. 287-304. Raven Press, New York.

Robbins, J.C. and Nicolson, G.L.(1975) Surface of normal and transformed cells. In: Cancer, A Comprehensive Treatise. Vol. 4, pp. 3-54. (Edited by Becker, F.F.) Plenum Press, New York.

Schroeder, T.E.(1975) Dynamics of the contractile ring. In: Molecules and Cell Movement, pp. 305-334. (Edited by Inoue, S. and Stephens, R.E.) Raven Press, New York.

Singer, S.J. and Nicolson, G.L.(1972) The fluid mosaic model of the structure of cell membranes. Science, N.Y. 175,720.

Skalak, R., Tözeren, A., Zarda, R., and Chien, S.(1973) Strain energy function of red blood cell membranes. Biophys. J. 13, 245-264.

Taylor, D. L. and Condeelis, J.S.(1979) Cytoplasmic structure and contractility in amoeboid cells. Int. Rev. Cytol. 56, 57-144.

Wolosewick, J.L. And Porter, K.R.(1979) Microtrabecular lattice of the cytoplasmic ground substance-Artifact or reality. J. Cell Biology 82, 114-139.

Yoneda, M.(1973) Tension at the surface of sea urchin eggs on the basis of "liquid-drop" concept. Adv. Biophys. 4, 153-190.

INDEX